SPRINGER HANDBOOK OF AUDITORY RESEARCH

Series Editors: Richard R. Fay and Arthur N. Popper

SPRINGER HANDBOOK OF AUDITORY RESEARCH

Volume 1: The Mammalian Auditory Pathway: Neuroanatomy
Edited by Douglas B. Webster, Arthur N. Popper, and Richard R. Fay

Volume 2: The Mammalian Auditory Pathway: Neurophysiology
Edited by Arthur N. Popper and Richard R. Fay

Volume 3: Human Psychophysics
Edited by William Yost, Arthur N. Popper, and Richard R. Fay

Volume 4: Comparative Hearing: Mammals
Edited by Richard R. Fay and Arthur N. Popper

Volume 5: Hearing by Bats
Edited by Arthur N. Popper and Richard R. Fay

Volume 6: Auditory Computation
Edited by Harold L. Hawkins, Teresa A. McMullen, Arthur N. Popper, and Richard R. Fay

Volume 7: Clinical Aspects of Hearing
Edited by Thomas R. Van De Water, Arthur N. Popper, and Richard R. Fay

Volume 8: The Cochlea
Edited by Peter Dallos, Arthur N. Popper, and Richard R. Fay

Volume 9: Development of the Auditory System
Edited by Edwin W Rubel, Arthur N. Popper, and Richard R. Fay

Volume 10: Comparative Hearing: Insects
Edited by Ronald Hoy, Arthur N. Popper, and Richard R. Fay

Volume 11: Comparative Hearing: Fish and Amphibians
Edited by Richard R. Fay and Arthur N. Popper

Volume 12: Hearing by Whales and Dolphins
Edited by Whitlow W.*L. Au, Arthur N. Popper, and Richard R. Fay*

Volume 13: Comparative Hearing: Birds and Reptiles
Edited by Robert Dooling, Arthur N. Popper, and Richard R. Fay

Volume 14: Genetics and Auditory Disorders
Edited by Bronya J.B. Keats, Arthur N. Popper, and Richard R. Fay

Volume 15: Integrative Functions in the Mammalian Auditory Pathway
Edited by Donata Oertel, Richard R. Fay, and Arthur N. Popper

Volume 16: Acoustic Communication
Edited by Andrea Simmons, Arthur N. Popper, and Richard R. Fay

Volume 17: Compression: From Cochlea to Cochlear Implants
Edited by Sid P. Bacon, Richard R. Fay, and Arthur N. Popper

Volume 18: Speech Processing in the Auditory System
Edited by Steven Greenberg, William Ainsworth, Arthur N. Popper, and Richard R. Fay

Volume 19: The Vestibular System
Edited by Stephen M. Highstein, Richard R. Fay, and Arthur N. Popper

Volume 20: Cochlear Implants: Auditory Prostheses and Electric Hearing
Edited by Fan-Gang Zeng, Arthur N. Popper, and Richard R. Fay

Volume 21: Electroreception
Edited by Theodore H. Bullock, Carl D. Hopkins, Arthur N. Popper, and Richard R. Fay

Continued after index

Matthew W. Kelley
Doris K. Wu
Arthur N. Popper
Richard R. Fay
Editors

Development of the Inner Ear

With 29 illustrations and 4 color illustrations

Matthew W. Kelley
Section on Developmental Neuroscience
Porter Neuroscience Center
NIDCD, National Institutes of Health
Bethesda, MD 20892
USA
kelleymt@nidcd.nih.gov

Doris K. Wu
Section on Hair Cell Development and Regeneration
NIDCD, National Institutes of Health
Rockville, MD 20850
USA
wud@nidcd.nih.gov

Arthur N. Popper
Department of Biology
University of Maryland
College Park, MD 20742
USA
apopper@umd.edu

Richard R. Fay
Parmly Hearing Institute and Department of Psychology
Loyola University of Chicago
Chicago, IL 60626
USA
rfay@wpo.it.luc.edu

Series Editors:
Richard R. Fay
Parmly Hearing Institute and Department of Psychology
Loyola University of Chicago
Chicago, IL 60626
USA

Arthur N. Popper
Department of Biology
University of Maryland
College Park, MD 20742
USA

Cover illustration: Paint-fills of the developing otocyst in mouse (*top row*), chick (*second row*), frog (*third row*), and zebrafish (*bottom row*) (see Fig. 3.2A from Mansour and Schoenwolf). Paint-fill figures compiled courtesy of M. Bever and D. Fekete. Mouse paint-fills were originally published in Morsli et al. (1998, reprinted with permission of the Society for Neuroscience © 1998), chick paint-fills in Bissonnette and Fekete (1996, reprinted with permission of Wiley-Liss, Inc., a subsidiary of John Wiley & Sons © 1996), frog paint-fills in Bever et al. (2003, reprinted with permission of Wiley-Liss, Inc., a subsidiary of John Wiley & Sons © 2003), and zebrafish paint-fills in Bever and Fekete (2002, reprinted with permission of Wiley-Liss, Inc., a subsidiary of John Wiley & Sons © 2002).

Library of Congress Control Number: 2005925504

ISBN 10: 0-387-25068-9 Printed on acid-free paper.
ISBN 13: 978-0387-25068-7

Printed in the United States of America. (MP)

9 8 7 6 5 4 3 2 1

springeronline.com

Series Preface

The *Springer Handbook of Auditory Research* presents a series of comprehensive and synthetic reviews of the fundamental topics in modern auditory research. The volumes are aimed at all individuals with interests in hearing research including advanced graduate students, postdoctoral researchers, and clinical investigators. The volumes are intended to introduce new investigators to important aspects of hearing science and to help established investigators to better understand the fundamental theories and data in fields of hearing that they may not normally follow closely.

Each volume presents a particular topic comprehensively, and each serves as a synthetic overview and guide to the literature. As such, the chapters present neither exhaustive data reviews nor original research that has not yet appeared in peer-reviewed journals. The volumes focus on topics that have developed a solid data and conceptual foundation rather than on those for which a literature is only beginning to develop. New research areas will be covered on a timely basis in the series as they begin to mature.

Each volume in the series consists of a few substantial chapters on a particular topic. In some cases, the topics will be ones of traditional interest for which there is a substantial body of data and theory, such as auditory neuroanatomy (Vol. 1) and neurophysiology (Vol. 2). Other volumes in the series deal with topics that have begun to mature more recently, such as development, plasticity, and computational models of neural processing. In many cases, the series editors are joined by a co-editor having special expertise in the topic of the volume.

RICHARD R. FAY, Chicago, Illinois
ARTHUR N. POPPER, College Park, Maryland

Volume Preface

The last century ended with a renewed interest in the developmental biology of the mammalian inner ear. This arose as a result of the emergence of molecular biological techniques that allowed investigators to work with one of the smallest organs in the mammalian body. These new investigations, many of which are summarized in this volume, have resulted in a striking increase in the pace of discovery and remarkable progress in our understanding of the developmental biology of this organ. Indeed, as a result of the many new discoveries on inner ear biology, the development of the inner ear has been referred to as one of the most striking examples of cellular morphogenesis in any biological system.

This volume provides a detailed overview of the development of the inner ear, particularly as our understanding has increased in the last decade of the twentieth century and the first five years of the twenty-first. In the first chapter of this volume, Kelley and Wu provide an overview of these recent discoveries as well as an overview of the volume. They complete their chapter with suggestions for areas of future research and discovery.

In Chapter 2, Groves concisely describes the classic experiments of the past century and then provides a critical interpretation of these experiments in light of the emerging molecular data regarding the same developmental process. In Chapter 3, Mansour and Schoenwolf describe the ongoing assembly of a series of genetic cascades, both in surrounding tissues and in the otocyst itself, that play a role in these crucial developmental events. This chapter also highlights the power of mouse genetics as a tool for the study of early developmental events in ear formation.

One of the most striking events that occurs during the initial formation of the otocyst is the specification and delamination of a group of neuroblasts from its ventral region. These interactions, as well as stimulating new hypotheses regarding the specification of the initial neuroblast population and its relationship with other cell types in the ear, are described in Chapter 4 by Pauley, Matei, Beisel, and Fritzsch.

The *Notch* pathway is a nearly ubiquitously expressed signaling cascade that is used in multiple developing and mature systems to sort homogeneous progenitor cells into different cell fates (reviewed in Schweisguth 2004). In Chapter

5, Lanford and Kelley examine the role of *Notch* in the ear in light of expression and functional data and recent progress in our understanding of the different cofactors and signaling events that mediate this intriguing signaling pathway.

The final two chapters in this volume examine an exciting emerging field in inner ear development, the development of the stereociliary bundle located on all mechanosensory hair cells. In Chapter 6, Bryant, Forge, and Richardson describe the morphological process of hair cell differentiation, including the development of the stereociliary bundle, while in Chapter 7, Hertzano and Avraham review insights into the development of the inner ear that have been obtained through the identification of genetic mutations that underlie human nonsyndromic and syndromic deafness.

Development of the inner ear, particularly at the molecular level, has not heretofore been considered in this series. However, development of other aspects of the auditory system has been of considerable interest in earlier volumes and these complement the chapters in this volume. Indeed, many chapters in Volume 15 (*Development and Plasticity of the Central Auditory System*) consider development of the auditory portions of the central nervous system as well as plasticity during development. Similarly, Volume 9 (*Development of the Auditory System*) has chapters that consider the overall embryology and development of the cochlea and central nervous system, including behavioral and functional data.

MARTTHEW W. KELLEY, Bethesda, Maryland
DORIS K. WU, Rockville, Maryland
ARTHUR N. POPPER, College Park, Maryland
RICHARD R. FAY, Chicago, Illinois

Contents

Contributors

KAREN B. AVRAHAM
Department of Human Genetics and Molecular Medicine, Sackler School of Medicine, Tel Aviv University, Ramat Aviv, Tel Aviv 69978, Israel

KIRK W. BEISEL
Department of Biomedical Sciences, Creighton University, Omaha, NE 68178, USA

JANE E. BRYANT
School of Life Sciences, University of Sussex, Falmer, Brighton BN1 9QR, United Kingdom

ANDREW FORGE
UCL Center for Auditory Research, University College London 322, London WC1X 8EE, United Kingdom

BERND FRITZSCH
Department of Biomedical Sciences, Creighton University, Omaha, NE 68178, USA

ANDREW K. GROVES
House Ear Institute, Gonda Department of Cell and Molecular Biology, Los Angeles, CA 90057, USA

RONNA HERTZANO
Department of Human Genetics and Molecular Medicine, Sackler School of Medicine, Tel Aviv University, Ramat Aviv, Tel Aviv 69978, Israel

MATTHEW W. KELLEY
Section on Developmental Neuroscience, Porter Neuroscience Center, NIDCD, National Institutes of Health, Bethesda, MD 20892, USA

PAMELA J. LANFORD
Biological Sciences Program, University of Maryland, College Park, MD 20742, USA

SUZANNE L. MANSOUR
Department of Human Genetics, University of Utah, Salt Lake City, UT 84112, USA

VERONICA MATEI
Department of Biomedical Sciences, Creighton University, Omaha, NE 68178, USA

SARAH PAULEY
Department of Biomedical Sciences, Creighton University, Omaha, NE 68178, USA

GUY P. RICHARDSON
School of Life Sciences, University of Sussex, Falmer, Brighton BN1 9QR, United Kingdom

GARY C. SCHOENWOLF
Department of Neurobiology and Anaotmy, University of Utah School of Medicine, Salt Lake City, UT 84132, USA

DORIS K. WU
Section on Hair Cell Development and Regeneration, NIDCD, National Institutes of Health, Rockville, MD 20850, USA

1

Developmental Neurobiology of the Ear: Current Status and Future Directions

Matthew W. Kelley and Doris K. Wu

The close of the twentieth century marked the dawn of a renaissance in inner ear developmental biology. During the preceding 100 years, the number of scientists willing to accept the challenges of working with one of the smaller structures in the body had remained comparatively small. More recently, however, the emergence of molecular biological techniques, combined with a greater appreciation of the elegance and importance of the inner ear, has led to an increase in the number of scientists who actively study inner ear biology and, more importantly, to a striking increase in the pace of discovery. As a result, now five years into the new century, it seems appropriate to review the remarkable progress that has occurred and to discuss the challenges that still await researchers.

The development of the inner ear has been referred to as one of the most striking examples of cellular morphogenesis in any biological system (Barald and Kelley 2004). From a rather humble beginning as a patch of unremarkable ectodermal cells, the developing ear expands to give rise to a spectrum of specialized cell types and structures that encompass neural, epithelial, secretory, and mechanosensory phenotypes. Along the way, different regions and, ultimately, individual cells, become specified to develop as different parts of the ear through a series of inductive signaling events that require both autocrine and paracrine functions.

The first step in the formation of the ear is the specification of a region of ectoderm as the otic placode (reviewed in Riley and Phillips 2003). Classic embryological studies had examined the role of adjacent tissues in the formation of this placodes. More recently, however, these tissue interactions have been reexamined at a molecular level (Ladher et al. 2000; Mackereth et al. 2004). In Chapter 2 of this volume, Andrew Groves concisely describes the classic experiments of the past century and then provides a critical interpretation of these experiments in light of the emerging molecular data regarding the same developmental process. Following its formation, the otic placode invaginates and ultimately separates from the surface ectoderm to form the spherical otic vesicle (also called the otocyst) (reviewed in Torres and Giraldez 1998). Moreover, as soon as it forms, the otocyst is already regionalized into different developmental compartments or zones (reviewed in Fekete and Wu 2002). As was the case for the placode, classic embryological studies had demonstrated an important role for surrounding tissues, in particular the developing hindbrain and periotic mesenchyme, in formation and regionalization of the vesicle (reviewed by Fritzsch and Beisel 2001; Fekete and Wu 2002; Liu et al. 2003). In Chapter 3, Mansour and Schoenwolf describe the ongoing assembly of a series of genetic cascades, both in surrounding tissues as well as the otocyst itself, that play a role in these crucial developmental events. This chapter also highlights the power of mouse genetics as a tool for the study of early developmental events in ear formation. By generating animals that carry compound mutations in multiple genes, researchers have been able to demonstrate the existence of redundant genetic signaling that apparently exists to ensure the formation of a relatively normal ear even in the presence of disruptions in individual genes (Maroon et al. 2002; Wright and Mansour 2003). Finally, Chapter 3 also provides an intriguing comparison between the formation of the otocyst and the neural tube. Both structures undergo similar developmental events, including the transition from an initially flat sheet of cells (neural plate and otic placode) to a closed three-dimensional structure with a central lumen. As one might guess, there are commonalties and differences in the molecular pathways used to achieve similar morphogenetic goals.

One of the most striking events that occurs during the initial formation of the otocyst is the specification and delamination of a group of neuroblasts from its ventral region. Almost as soon as these cells leave, they begin to extend neurites back into the otocyst as if they are unwilling to separate fully from their old companions (Carney and Silver 1983; Hemond and Morest 1991a,b). These cells, and their progeny, will go on to give rise to the neurons of the acousticovestibular (VIIIth cranial) nerve that provide afferent innervation for all aspects of both the auditory and vestibular regions of the ear. Over the past few years, significant progress has been made in establishing the cellular events and molecular cascades that direct these cells from unspecified neuroblasts to mature neurons forming elaborate and precise connections with mechanosensory hair cells in the periphery and auditory and vestibular nuclei in the central nervous system (CNS) (Ma et al. 1998; Liu et al. 2000; Raft et al. 2004). These interactions, as well as stimulating new hypotheses regarding the specification of the initial neuroblast population and its relationship with other cell types in the ear, are described by Pauley, Matei, Beisel, and Fritzsch in Chapter 4.

As its morphogenesis continues, a subset of epithelial cells within the otocyst become specified to develop as the sensory patches that will actually perceive sound and movement (reviewed in Whitfield et al. 1997). Subsequently, individual cells within these patches become specialized to develop as either mechanosensory hair cells or as surrounding nonsensory cells, which are collectively referred to as supporting cells. Intriguingly, but perhaps not surprisingly considering the limited number of developmental signaling pathways, the same molecular pathway appears to regulate both types of decisions (Adam et al. 1998; Haddon et al. 1998; Lanford et al. 1999). The Notch pathway is a nearly ubiquitously expressed signaling cascade that is used in multiple developing and mature systems to sort homogeneous progenitor cells into different cell fates (reviewed in Schweisguth 2004). In Chapter 5, Lanford and Kelley examine the role of Notch in the ear in light of expression and functional data in the ear and recent progress in our understanding of the different cofactors and signaling events that mediate this intriguing signaling pathway.

The final two chapters in this volume examine an exciting emerging field in inner ear development, the development of the stereociliary bundle located on all mechanosensory hair cells. In Chapter 6, Bryant, Forge, and Richardson describe the morphological process of hair-cell differentiation, including the development of the stereociliary bundle, while in Chapter 7, Hertzano and Avraham review insights into the development of the inner ear that have been obtained through the identification of genetic mutations that underly human nonsyndromic and syndromic deafness. Surprisingly, a number of these mutations have direct effects on the development of stereociliary bundles. Although the stereociliary bundle is comprised of modified microvilli that are not dissimilar from the microvilli located on many developing and mature epithelial cells, the striking arrangement of these cells into a staircase pattern with a specific plane of polarization, and the exquisite sensitivity of this structure, suggests that it may be unique. In fact, as Hertzano and Avraham discuss, the recent explosion in the identification and understanding of the molecular factors that regulate the formation of these bundles has its roots in the field of human genetics, and more specifically in the study of mutations that lead to auditory and/or vestibular defects (A. Wang et al. 1998; Zheng et al. 2000; Naz et al. 2004). Many of these genes play a crucial role in the formation of the stereociliary bundle, and the ongoing studies of their molecular function has led to valuable insights into the cell biology of bundle development (Belyantseva et al. 2003; Sekerkova et al. 2004; Rzadzinska et al. 2004). These studies also demonstrate the power of genetics in developmental biology and highlight the opportunity to learn about unique cell types or structures through the identification of nonsyndromic genetic mutations in both humans and mice.

Although the chapters in this book strikingly describe the progress that has occurred in recent years, it is important to consider that many questions remain unanswered and that there is much work to be done. Perhaps the most glaring deficits exist in our understanding, or lack of understanding, of the factors that generate heterogeneity throughout the ear. For instance, while considerable efforts have been devoted to the examination of the factors that specify the sensory patches, we know relatively little about the level and degree of heterogeneity in the nonsensory regions of the ear. As an example, the endolymphatic duct and the semicircular canals are both located in the dorsal region of the inner ear. Fate mapping studies in chicken show that the endolymphatic duct is derived from the dorsal region of the otic cup, whereas cells in the three semicircular canals are derived mostly from the posterolateral region of the otic cup (Brigande et al. 2000). These two regions of the otic cup are molecularly distinct from each other, suggesting that their fates are restricted early in development

(W. Wang et al. 1998, 2001; Acampora et al. 1999; Depew et al. 1999), but the factors that specify either structure are unclear. Similarly, the specification of the two nonsensory structures in the mammalian cochlea, the stria vascularis and Reissner's membrane, are largely unknown.

Similar heterogeneities exist among various sensory patches. Ampullae differ from saccule or utricle and both clearly differ from auditory epithelia. Recent results have suggested that the Wnt signaling pathway, and more specifically b-catenin, may play a role in the determination of vestibular versus auditory sensory patches (Stevens et al. 2003), but this discovery serves as only a potential tip of the iceberg. Heterogeneities are even found within individual sensory patches. Vestibular epithelia contain type I and type II hair cells, while auditory epithelia such as the avian basilar papilla and the mammalian cochlea contain at least two types of hair cells (tall and short in birds, inner and outer in mammals). Similarly, at least four different types of supporting cells can be identified in the mammalian cochlea, and it seems likely that similar supporting cell heterogeneities exist in other sensory patches.

A second area of uncertainty is the developmental relationship between mechanosensory hair cells and the neurons that innervate them. Existing molecular data suggest that the progenitors for both populations of cells arise from the same anterior–ventral region of the otocyst (reviewed in Fritzsch and Beisel 2001; Fekete and Wu 2002; Fritzsch et al. 2002), but it is unclear whether any clonal relationship, such as has been observed for sensory cells and innervating neurons in invertebrates (Hartenstein and Posakony 1990; Ghysen and Dambly-Chaudiere 1993; Parks and Muskavitch 1993; Jan and Jan 1995; Zeng et al. 1998), exists between the two cell types. Lineage data in chicken generated using replication-incompetent retroviruses indicate that common precursors can give rise to both neurons and hair cells (Satoh and Fekete 2004), but the number of reported clones is small, and it is unclear whether the neurons and hair cells that derive from a common precursor actually communicate with one another, as would be expected by analogy with invertebrates.

This question also highlights a greater need for studies of lineage, fate mapping, and cell movement, especially in mammals, in which the relative inaccessibility of the inner ear has limited our ability to generate meaningful data about these important questions. The inner ear undergoes dynamic morphogenesis during development. Gene expression data alone are insufficient for the full comprehension of the developmental processes involved in the formation of this intricate organ. Two recent fate mapping studies in *Xenopus laevis* (Kil and Collazo 2001) and chicken (Brigande et al. 2000) have indicated that dynamic cell movements occur during inner ear development, suggesting that there is much to be learned from these approaches. Encouragingly, the first cell lineage studies in a mammalian (mouse) ear have recently been reported using ultrasound backscatter microscopy techniques (Brigande and Fekete, personal communication).

As the chapters in this book emphasize, the pace of discovery at the molecular level has increased dramatically, in particular in terms of our understanding of the earliest events in ear development. Ironically, however, the crucial roles for many of these genes in early ear development have also proven to be a major impediment to our understanding of the molecular factors that regulate later developmental events. All biologically developing systems utilize a combination of a relatively limited number of molecular signaling pathways, but unique contextually based responses to those pathways generate diverse heterogeneities at all levels from the determination of the three basic germ layers through organogenesis. Therefore, disruption of a single molecular signaling pathway may have multiple profound effects of different developmental events even within a single organ, but the first effect of this disruption may negate the analyses of later effects of this same pathway.

A good example of this is the role of Fgfr1. Complete deletion of *Fgfr1* leads to early embryonic lethality prior to inner ear formation (Deng et al. 1994; Yamaguchi et al. 1994), but a conditional deletion of *Fgfr1* that is limited to the inner ear and small number of other structures reveals a specific role for this gene in the development of the cochlea (Pirvola et al. 2002). The repeated use of conserved signaling pathways highlights the need for the generation of tissue-specific mutants and the examination of specific pathways at different developmental time points.

The importance of the development of these tools is emphasized by multiple examples of studies that attempted to generate mouse models for human diseases by simply disrupting genes that were known to cause human syndromic or nonsyndromic deafness, but instead resulted in novel and unexpected consequences. For instance, mutations in *EYA1*, *PAX2*, and *PENDRIN* have been associated with branchio-oto-renal, renal coloboma, and Pendred syndromes, respectively, all of which can cause syndromic forms of human deafness (Abdelhak et al. 1997; Everett et al. 1997; Li et al. 1998; Sanyanusin et al. 1995). The knockout mouse models for each of these genes, however, show more severe inner ear defects than observed in human patients (Torres et al. 1996; Xu et al. 1999; Everett et al. 2001; Burton et al. 2004). Perhaps the most striking example of this phenomenon is the observation that mutations in the gap junction protein *GJB2* lead to nonsyndromic deafness in humans while deletion of the mouse homolog, *Connexin 26*, results in lethality prior to implantation as a result of placental defects (Gabriel et al. 1998). All of these phenotypic differences could be attributable to species differences or to the fact that some of the mutations in the human genes result in hypomorphic versions of the genes rather than the functional nulls generated in the mouse models. These results demonstrate the crucial need for the ability to regulate gene deletion both spatially and temporally using mice that express Cre-recombinase under the control of ear specific promoters.

Finally, as more and more candidate molecules that are important for normal inner ear functions are identified, it will be crucial to gain an in-depth understanding of the cellular events that are mediated by these molecules. A number of studies, particularly in the area of hair cell biology and stereociliary bundle formation, have certainly advanced in this direction (Belyantseva et al. 2003;

Rzadzinska et al. 2004; Sekerkova et al. 2004). Therefore, despite the technical challenges and the requirement for the development of novel and unique methodologies, the recent advances in our current understanding of molecular basis for mechanotransduction clearly make these efforts worthwhile.

The potential benefits of an increased understanding of the cell biology of the inner ear are perhaps no more obvious than when one considers the potential application of this knowledge to the generation of therapies for both congenital and acquired deafness. Genetic analyses in both humans and mice have demonstrated that both the hair cells and the supporting cells are crucial for normal auditory function. Yet, our understanding of how these cells develop and function is still extremely limited. Considering the potential impact of regenerative therapies for auditory or vestibular dysfunction, a more comprehensive understanding of both hair cells and supporting cells is crucial.

References

Abdelhak S, Kalatzis V, Heilig R, Compain S, Samson D, Vincent C, Weil D, Cruaud C, Sahly I, Leibovici M, Bitner-Glindzicz M, Francis M, Lacombe D, Vigneron J, Charachon R, Boven K, Bedbeder P, Van Regemorter N, Weissenbach J, Petit C (1997) A human homologue of the *Drosophila* eyes absent gene underlies branchio-oto-renal (BOR) syndrome and identifies a novel gene family. Nat Genet 15: 157–164.

Acampora D, Merlo GR, Paleari L, Zerega B, Postiglione MP, Mantero S, Bober E, Barbieri O, Simeone A, Levi G (1999) Craniofacial, vestibular and bone defects in mice lacking the *Distal-less*-related gene *Dlx5*. Development 126:3795–3809.

Adam J, Myat A, Le Roux I, Eddison M, Henrique D, Ish-Horowicz D, Lewis J (1998) Cell fate choices and the expression of Notch, Delta and Serrate homologues in the chick inner ear: parallels with *Drosophila* sense-organ development. Development 125:4645–4654.

Barald KF, Kelley MW (2004) From placode to polarization: new tunes in inner ear development. Development 131:4119–4130.

Belyantseva IA, Boger ET, Friedman TB (2003) Myosin XVa localizes to the tips of inner ear sensory cell stereocilia and is essential for staircase formation of the hair bundle. Proc Natl Acad Sci USA 100:13958–13963.

Brigande JV, Iten LE, Fekete DM (2000) A fate map of chick otic cup closure reveals lineage boundaries in the dorsal otocyst. Dev Biol 227:256–270.

Burton Q, Cole LK, Mulheisen M, Chang W, Wu DK (2004) The role of *Pax2* in mouse inner ear development. Dev Biol 272:161–175.

Carney PR, Silver J (1983) Studies on cell migration and axon guidance in the developing distal auditory system of the mouse. J Comp Neurol 215:359–369.

Deng CX, Wynshaw-Boris A, Shen MM, Daugherty C, Ornitz DM, Leder P (1994) Murine FGFR-1 is required for early postimplantation growth and axial organization. Genes Dev 8:3045–3057.

Depew MJ, Liu JK, Long JE, Presley R, Meneses JJ, Pedersen RA, Rubenstein JL (1999) Dlx5 regulates regional development of the branchial arches and sensory capsules. Development 126:3831–3846.

Everett LA, Glaser B, Beck JC, Idol JR, Buchs A, Heyman M, Adawi F, Hazani E,

Nassir E, Baxevanis AD, Sheffield VC, Green ED (1997) Pendred syndrome is caused by mutations in a putative sulphate transporter gene (PDS). Nat Genet 17:411–422.
Everett LA, Belyantseva IA, Noben-Trauth K, Cantos R, Chen A, Thakkar SI, Hoogstraten-Miller SL, Kachar B, Wu DK, Green ED (2001) Targeted disruption of mouse Pds provides insight about the inner-ear defects encountered in Pendred syndrome. Hum Mol Genet 10:153–161.
Fekete DM, Wu DK (2002) Revisiting cell fate specification in the inner ear. Curr Opin Neurobiol 12:35–42.
Fritzsch B, Beisel KW (2001) Evolution and development of the vertebrate ear. Brain Res Bull 55:711–721.
Fritzsch B, Beisel KW, Jones K, Farinas I, Maklad A, Lee J, Reichardt LF (2002) Development and evolution of inner ear sensory epithelia and their innervation. J Neurobiol 53:143–156.
Gabriel HD, Jung D, Butzler C, Temme A, Traub O, Winterhager E, Willecke K (1998) Transplacental uptake of glucose is decreased in embryonic lethal connexin26-deficient mice. J Cell Biol 140:1453–1461.
Ghysen A, Dambly-Chaudiere C (1993) The specification of sensory neuron identity in *Drosophila*. Bioessays 15:293–298.
Haddon C, Jiang YJ, Smithers L, Lewis J (1998) Delta-Notch signalling and the patterning of sensory cell differentiation in the zebrafish ear: evidence from the mind bomb mutant. Development 125:4637–4644.
Hartenstein V, Posakony JW (1990) A dual function of the Notch gene in *Drosophila* sensillum development. Dev Biol 142:13–30.
Hemond SG, Morest DK (1991a) Formation of the cochlea in the chicken embryo: sequence of innervation and localization of basal lamina-associated molecules. Brain Res Dev Brain Res 61:87–96.
Hemond SG, Morest DK (1991b) Ganglion formation from the otic placode and the otic crest in the chick embryo: mitosis, migration, and the basal lamina. Anat Embryol (Berl) 184:1–13.
Jan YN, Jan LY (1995) Maggot's hair and bug's eye: role of cell interactions and intrinsic factors in cell fate specification. Neuron 14:1–5.
Kil SH, Collazo A (2001) Origins of inner ear sensory organs revealed by fate map and time-lapse analyses. Dev Biol 233:365–379.
Ladher RK, Anakwe KU, Gurney AL, Schoenwolf GC, Francis-West PH (2000) Identification of synergistic signals initiating inner ear development. Science 290:1965–1967.
Lanford PJ, Lan Y, Jiang R, Lindsell C, Weinmaster G, Gridley T, Kelley MW (1999) Notch signalling pathway mediates hair cell development in mammalian cochlea. Nat Genet 21:289–292.
Li XC, Everett LA, Lalwani AK, Desmukh D, Friedman TB, Green ED, Wilcox ER (1998) A mutation in PDS causes non-syndromic recessive deafness. Nat Genet 18: 215–217.
Liu M, Pereira FA, Price SD, Chu MJ, Shope C, Himes D, Eatock RA, Brownell WE, Lysakowski A, Tsai MJ (2000) Essential role of BETA2/NeuroD1 in development of the vestibular and auditory systems. Genes Dev 14:2839–2854.
Liu D, Chu H, Maves L, Yan YL, Morcos PA, Postlethwait JH, Westerfield M (2003) Fgf3 and Fgf8 dependent and independent transcription factors are required for otic placode specification. Development 130:2213–2224.
Ma Q, Chen Z, del Barco Barrantes I, de la Pompa JL, Anderson DJ (1998) *neurogenin1*

is essential for the determination of neuronal precursors for proximal cranial sensory ganglia. Neuron 20:469–482.

Mackereth MD, Kwak SJ, Fritz A, Riley BB (2004) Zebrafish *pax8* is required for otic placode induction and plays a redundant role with *Pax2* genes in the maintenance of the otic placode. Development 132:371–382.

Maroon H, Walshe J, Mahmood R, Kiefer P, Dickson C, Mason I (2002) Fgf3 and Fgf8 are required together for formation of the otic placode and vesicle. Development 129: 2099–2108.

Naz S, Griffith AJ, Riazuddin S, Hampton LL, Battey JF, Jr., Khan SN, Wilcox ER, Friedman TB (2004) Mutations of ESPN cause autosomal recessive deafness and vestibular dysfunction. J Med Genet 41:591–595.

Parks AL, Muskavitch MA (1993) Delta function is required for bristle organ determination and morphogenesis in *Drosophila*. Dev Biol 157:484–496.

Pirvola U, Ylikoski J, Trokovic R, Hebert JM, McConnell SK, Partanen J (2002) FGFR1 is required for the development of the auditory sensory epithelium. Neuron 35:671–680.

Raft S, Nowotschin S, Liao J, Morrow BE (2004) Suppression of neural fate and control of inner ear morphogenesis by *Tbx1*. Development 131:1801–1812.

Riley BB, Phillips BT (2003) Ringing in the new ear: resolution of cell interactions in otic development. Dev Biol 261:289–312.

Rzadzinska AK, Schneider ME, Davies C, Riordan GP, Kachar B (2004) An actin molecular treadmill and myosins maintain stereocilia functional architecture and self-renewal. J Cell Biol 164:887–897.

Sanyanusin P, Schimmenti LA, McNoe LA, Ward TA, Pierpont ME, Sullivan MJ, Dobyns WB, Eccles MR (1995) Mutation of the *PAX2* gene in a family with optic nerve colobomas, renal anomalies and vesicoureteral reflux. Nat Genet 9:358–364.

Schweisguth F (2004) Notch signaling activity. Curr Biol 14:R129–138.

Sekerkova G, Zheng L, Loomis PA, Changyaleket B, Whitlon DS, Mugnaini E, Bartles JR (2004) Espins are multifunctional actin cytoskeletal regulatory proteins in the microvilli of chemosensory and mechanosensory cells. J Neurosci 24:5445–5456.

Stevens CB, Davies AL, Battista S, Lewis JH, Fekete DM (2003) Forced activation of Wnt signaling alters morphogenesis and sensory organ identity in the chicken inner ear. Dev Biol 261:149–164.

Torres M, Giraldez F (1998) The development of the vertebrate inner ear. Mech Dev 71:5–21.

Torres M, Gomez-Pardo E, Gruss P (1996) *Pax2* contributes to inner ear patterning and optic nerve trajectory. Development 122:3381–3391.

Wang A, Liang Y, Fridell RA, Probst FJ, Wilcox ER, Touchman JW, Morton CC, Morell RJ, Noben-Trauth K, Camper SA, Friedman TB (1998) Association of unconventional myosin *MYO15* mutations with human nonsyndromic deafness *DFNB3*. Science 280: 1447–1451.

Wang W, Van De Water T, Lufkin T (1998) Inner ear and maternal reproductive defects in mice lacking the *Hmx3* homeobox gene. Development 125:621–634.

Wang W, Chan EK, Baron S, Van de Water T, Lufkin T (2001) *Hmx2* homeobox gene control of murine vestibular morphogenesis. Development 128:5017–5029.

Whitfield T, Haddon C, Lewis J (1997) Intercellular signals and cell-fate choices in the developing inner ear: origins of global and of fine-grained pattern. Semin Cell Dev Biol 8:239–247.

Wright TJ, Mansour SL (2003) Fgf3 and Fgf10 are required for mouse otic placode induction. Development 130:3379–3390.

Xu PX, Adams J, Peters H, Brown MC, Heaney S, Maas R (1999) *Eya1*-deficient mice lack ears and kidneys and show abnormal apoptosis of organ primordia. Nat Genet 23:113–117.

Yamaguchi TP, Harpal K, Henkemeyer M, Rossant J (1994) *fgfr-1* is required for embryonic growth and mesodermal patterning during mouse gastrulation. Genes Dev 8: 3032–3044.

Zeng C, Younger-Shepherd S, Jan LY, Jan YN (1998) Delta and Serrate are redundant Notch ligands required for asymmetric cell divisions within the *Drosophila* sensory organ lineage. Genes Dev 12:1086–1091.

Zheng L, Sekerkova G, Vranich K, Tilney LG, Mugnaini E, Bartles JR (2000) The deaf jerker mouse has a mutation in the gene encoding the espin actin-bundling proteins of hair cell stereocilia and lacks espins. Cell 102:377–385.

2

The Induction of the Otic Placode

Andrew K. Groves

1. Introduction

The development of the inner ear has been studied actively for more than 100 years. On a purely practical level, the anlagen of the inner ear—the otic placode—is readily visible from an early age in most vertebrate embryos, making it an attractive tissue for developmental biologists to study in the late nineteenth and early twentieth centuries. However, part of the historical motivation to study inner ear development also undoubtedly arose from the fascination in seeing a highly complicated sensory organ produced from a simple patch of ectoderm. It is this transformation—from a very simple tissue to a Darwinian "organ of extreme perfection"—that modern researchers seek to understand. This chapter focuses on the very first stage of this transformation, in which cranial ectoderm is induced to form the otic placode. Later aspects of inner ear development, such as morphogenesis and cell type determination, are covered elsewhere in this volume. In addition, several other excellent reviews of early ear development have appeared recently (Fritzsch et al. 1997, 2002; Torres and Giraldez 1998; Baker and Bronner-Fraser 2001; Fritzsch and Beisel 2001; Kiernan et al. 2002; Whitfield et al. 2002; Brown et al. 2003; Riley and Phillips 2003).

2. Morphological and Molecular Events in Otic Placode Induction

The otic placode arises as a patch of thickened ectoderm adjacent to the hindbrain. As differentiation proceeds, the placodal ectoderm begins to invaginate to form a pit, which then deepens into a cup, finally closing over to form a vesicle (Alvarez et al. 1989). In some teleosts such as zebrafish (*Danio rerio*), the placode does not invaginate, but rather forms a thickened ball that becomes hollow by cavitation (Haddon and Lewis 1996). By this time, differentiation of specific cell types is already underway with the delamination of vestibulo-acoustic neurons from the otic epithelium. For the purposes of this chapter, the induction of the otic placode is considered complete when invagination to form an otic vesicle has begun.

Are the cells destined to give rise to the otic placode a physically discrete population, or do they intermingle with cells destined to form other tissues? Streit's studies in chick embryos have provided convincing evidence for the latter hypothesis (Streit 2002). Labeling small numbers of cells with vital dyes reveals that cells fated to give rise to the otic placode come from a wide region of the embryonic epiblast, and appear to converge toward the future placode as the embryo matures. During gastrulation, otic placode precursors appear to be mixed with cells destined to give rise to the central nervous system, neural crest, epibranchial placodes, and epidermis (Streit 2002). Similar results have also been obtained in zebrafish (Kozlowski et al. 1997). Interestingly, there is evidence to suggest that some otic placode cells derive from the neural folds at relatively late times after gastrulation (Mayordomo et al. 1998; Streit 2002). At present, it is not clear at what precise point the wandering otic placode precursors begin to receive signals that direct them toward their final fate, nor whether such signals cause the future placode cells to migrate toward their final location. This is discussed further in Section 3.

In the last 10 years, a variety of molecular markers have been identified that label the otic placode prior to invagination (Baker and Bronner-Fraser 2001). A comprehensive list of these markers in different species is given in Table 2.1, and a few main markers are discussed here. The earliest specific markers of the otic placode appear to be the transcription factors *Pax8* (Pfeffer et al. 1998; Heller and Brandli 1999), its close relative *Pax2* (Nornes et al. 1990; Krauss et al. 1991; Groves and Bronner-Fraser 2000), *Foxi1* (Nissen et al. 2003; Solomon et al. 2003a,b), and *Sox9* (Wright et al. 1995; Liu et al. 2003; Saint-Germain et al. 2004). On the basis of fate-mapping experiments, Streit (2002) has suggested that not all *Pax2*-expressing cells will contribute solely to the inner ear, a result confirmed recently by genetic lineage tracing of *Pax2*-expressing cells in mice. Few other markers have been examined across a range of different vertebrate species, but some transcription factors such as *Eya1*, *Gata3*, *Nkx5.1/Hmx3*, *Gbx2*, *Sox3*, and members of the *Dlx* family appear to be expressed in the otic placode prior to invagination. Signaling molecules such as *Bmp4* and *7* are also expressed during otic placode induction, while the transmembrane receptor *Notch* tends to be expressed shortly before the placode starts to invaginate (see Table 2.1).

Molecular markers of the otic placode are also providing support for theories about the evolutionary origins of the otic placode and inner ear. A widely held view has been that neural crest cells and sensory placodes are exclusively vertebrate structures not found in other chordates (Gans and Northcutt 1983). Cephalochordates such as *Amphioxus* do not appear to have any structures resembling the inner ear, and the *Amphioxus* homolog of *Pax2* and *Pax8*—*AmphiPax 2/5/8*—is not expressed in any otic placodelike structures (Kozmik et al. 1999). However, an ascidian *Pax* homolog—*HrPax2/5/8*—is expressed in the larval atrial primordia, which ultimately contain mechanosensory receptors

TABLE 2.1. List of molecular markers of the otic placode in the approximate order of appearance.

Gene	Description	Species examined	References
Pax-8	Transcription factor	Fish, frog, mouse	Pfeffer et al. 1998; Heller and Brändli 1999
Foxi1	Transcription factor	Fish	Solomon et al. 2003a,b
Sox9	Transcription factor	Fish, mouse	Wright et al. 1995; Liu et al. 2003; Saint-Germain et al. 2004
Pax2	Transcription factor	Fish, frog, chick, mouse	Krauss et al. 1991; Nornes et al. 1990; Groves and Bronner-Fraser 2000
Eya1	Transcription cofactor	Fish, mouse	Sahly et al. 1999; Xu et al. 1997
Cldna	Membrane protein	Fish	Kollmar et al. 2001
Scyba	CXC chemokine	Fish	Long et al. 2000
Erm	ETS-transcription factor	Fish	Munchberg et al. 1999; Raible and Brand 2001
Pea3	ETS-transcription factor	Fish	Munchberg et al. 1999; Raible and Brand 2001
Sprouty2	FGF antagonist	Fish	Chambers and Mason 2000
Sprouty4	FGF antagonist	Chick	Furthauer et al. 2001
Gata3	Transcription factor	Chick, mouse	George et al. 1994; Sheng and Stern 1999
Nkx5.1/Hmx3	Transcription factor	Fish, chick, mouse	Adamska et al. 2000; Rinkwitz-Brandt et al. 1995
Tbx2	Transcription factor	Fish, frog, chick	Logan et al. 1998; Ruvinsky et al. 2000; Takabatake et al. 2000.
Dlx3	Transcription factor	Fish, frog, chick, mouse	Akimenko et al. 1994; Ellies et al. 1997; Papalopulu and Kintner 1993; Pera and Kessel 1999; Robinson and Mahon 1994.
BMP7	Growth factor	Chick, mouse	Solloway and Robertson 1999; Groves and Bronner-Fraser 2000
Sox2	Transcription factor	Frog, chick, mouse	Mizuseki et al. 1998; Wood and Episkopou 1999; Groves, unpublished
Sox3	Transcription factor	Frog, chick, mouse	Groves and Bronner-Fraser 2000; Penzel et al. 1997; Wood and Episkopou 1999; Ishii et al. 2001; Abu-Emagd et al. 2001
Gbx2	Transcription factor	Frog, chick, mouse	Liu and Joyner 2001; Sanchez-Calderon et al. 2002; Shamim and Mason 1998; von Bubnoff et al. 1996.
Lmx1	Transcription factor	Chick	Giraldez 1998
Frz1	Wnt receptor	Chick	Stark et al. 2000
Frzb1	Wnt antagonist	Chick	Baranski et al. 2000; Duprez et al. 1999.
Fgf3	Growth factor	Chick, mouse	McKay et al. 1996; Mahmood et al. 1995.
c-*kit*	Growth factor receptor	Mouse	Orr-Urtreger et al. 1990
Groucho-related 4 and *5*	Transcription factor	Frog	Molenaar et al. 2000; Roose et al. 1998
Wnt5a	Growth factor	Chick	Baranski et al. 2000
Notch	Membrane receptor	Fish, chick, mouse	Groves and Bronner-Fraser 2000; Haddon et al. 1998; Lewis et al. 1998

with cupulae (Bone and Ryan 1978; Katz 1983; Baker and Bronner-Fraser 1997; Wada et al. 1998), and which have been suggested as the evolutionary precursor of the inner ear (Wada et al. 1998; Shimeld and Holland 2000; Holland and Holland 2001). Examination of more otic placode markers in nonvertebrate chordates will help resolve whether the otic placode is a fundamental chordate feature or not.

3. A Simple Embryological Profile of Otic Placode Induction

Embryonic induction has been defined as "an interaction between an inducing and a responding tissue that alters the path of differentiation of the responding tissue" (Gurdon 1987; Jacobson and Sater 1988). Historically, investigators relied on morphological landmarks as indicators of placode induction, which can be seen only some time after the first molecular markers of the placode are expressed. The use of molecular markers in addition to morphological landmarks has greatly improved the accuracy and resolution of monitoring events in otic placode induction. Any investigation into otic placode induction requires that we first answer three simple questions about the induction process—whether the ectoderm that forms the placode is in some way unique, when the induction starts, and when it is complete. Only when these questions have been addressed can meaningful experiments be performed to understand the mechanism of otic placode induction—for example, it serves little purpose to experimentally manipulate candidate otic placode-inducing factors after the induction of the placode is complete. Below, we describe the sorts of simple embryological experiments that have been performed to determine the timing of these events in otic placode induction. Illustrative data from the chick (Groves and Bronner-Fraser 2000) is shown in Figure 2.1.

3.1 Is the Responding Tissue Unique? The Concept of Competence

There are two distinct formal possibilities concerning the properties of the ectoderm that gives rise to the otic placode. On one hand, this ectoderm could be uniquely able to give rise to the otic placode from an early age. Alternatively, many regions of embryonic ectoderm could, in principle, give rise the otic placode if they received appropriate inducing signals. These two possibilities can be tested experimentally by transplanting different populations of ectoderm to the site where the otic placode normally forms, and testing whether such foreign populations can form an otic placode in this new location. Populations of foreign tissue that can respond in this way are said to be *competent* to give rise to the otic placode. A less rigorous variation on this experiment is to surgically

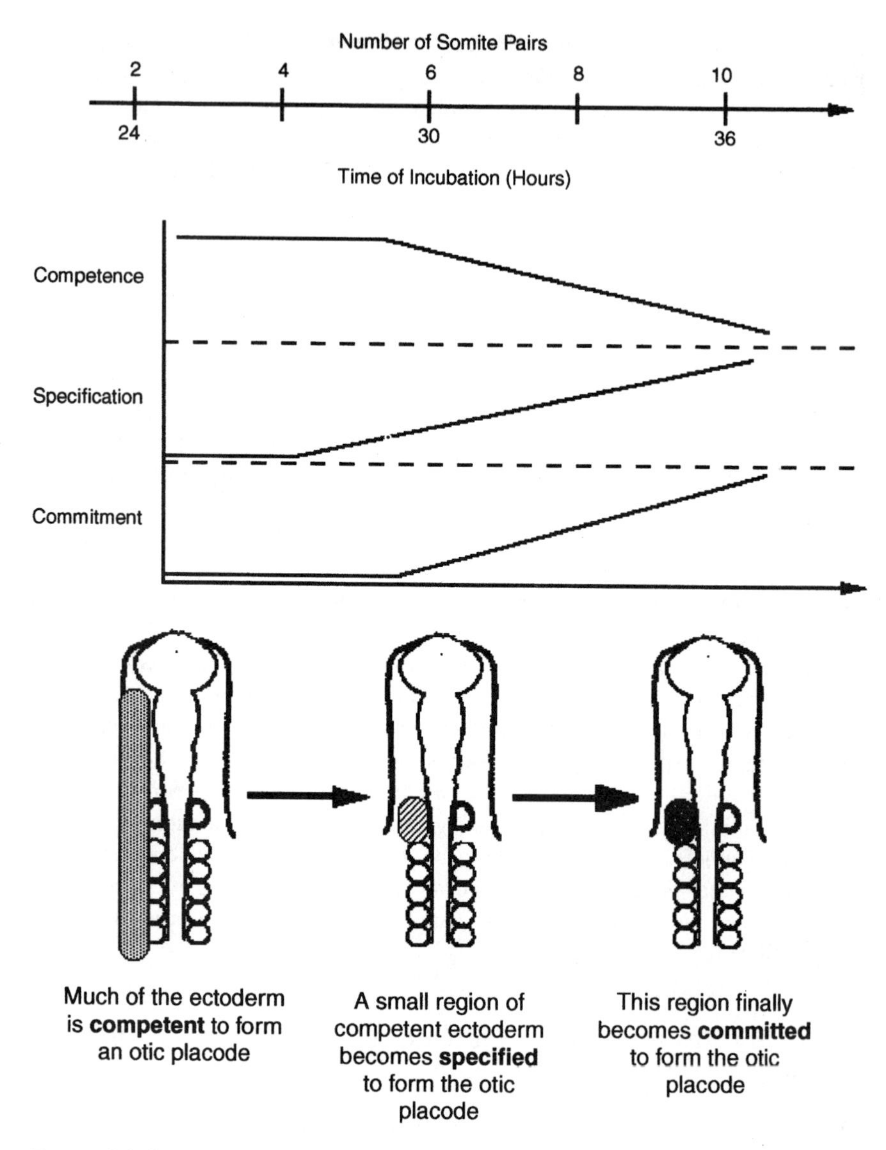

FIGURE 2.1. Schematic diagram indicating the time course and parameters of otic placode induction in the chick embryo, as described in Groves and Bronner-Fraser (2000). The figure shows how much of the embryonic ectoderm is initially competent to form the otic placode, with this competence decreasing over time. Local inductive signals specify the otic placode, which gradually becomes committed to a placode fate.

ablate the otic placode at different ages, and to determine whether the surrounding tissue is competent to regenerate the ablated placode.

These sorts of experiments have been carried out in amphibians and chick embryos for the past 75 years (Kaan 1926; Yntema 1933; Jacobson 1963a,b; Gallagher et al. 1996; Groves and Bronner-Fraser 2000). Although the specific details of these experiments are not relevant here, a common theme is strikingly clear. Much of the early embryonic ectoderm is competent to form the otic placode, provided both the host and donor embryos are sufficiently young. However, these diverse populations of ectoderm lose competence to form the otic placode when taken from progressively older embryos, or when grafted into much older hosts (Groves and Bronner-Fraser 2000). This illustrates a general principle of embryonic development—that the initially plastic and multipotent cells of the young embryo become progressively restricted in their fates as development proceeds. At present, the molecular basis of competence is not well understood, nor do we have a molecular picture of how competence is lost from cell populations with time. Nevertheless, these experiments suggest that it is the properties of the environment around the presumptive otic placode that direct ectoderm to a placodal fate, rather than the presumptive placodal ectoderm possessing some unique propensity for ear formation.

3.2 When Does the Induction Begin? The Concept of Specification

To evaluate candidate inducing tissues or inducing molecules for a particular induction, it is crucial to know approximately when the induction starts. Historically, researchers had to rely on morphological landmarks to guess when otic placode induction was occurring, such as the thickening of the placodal ectoderm or its invagination to form a pit or vesicle. More recently, the advent of molecular markers of the otic placode has made it clear that the placode starts to be induced well before it is morphologically distinct (see Section 2). Indeed, it is likely that presumptive otic ectoderm is exposed to inducing signals even before the first molecular markers of differentiation appear.

To determine when ectoderm begins to receive inducing signals, a standard assay is used in which pieces of presumptive otic ectoderm of different ages are explanted and maintained in culture in the absence of any inducing signals such as growth factors or serum. If the ectoderm has not yet received inducing signals, it is unlikely to express otic-specific markers in such a neutral culture environment. If, however, the explanted ectoderm has already begun to respond to inducing signals, it is possible that it will express otic-specific markers during the culture period. Such tissue is said to be *specified.* Unspecified tissue is suitable in principle for use in induction assays.

Historically, these experiments have been performed quite rarely, but some investigators have carried out specification experiments, such as those in the newt *Taricha* (Jacobson 1963a), *Xenopus* (Gallagher et al. 1996), and chick (Groves and Bronner-Fraser 2000). In this last study, the use of molecular markers revealed that specification of the otic placode as assayed by *Pax2* expression occurred just before the time *Pax2* normally begins to be expressed in vivo, at the four- or five-somite stage. It is important to note that specification must be described only with respect to the molecular markers used in the specification assay, and that any inference of the starting point of induction must by necessity be a provisional one. For example, in the case of the chick otic placode described above, it is formally possible that other, as yet undiscovered markers of the otic placode may push back the time of specification to earlier ages.

3.3 When Is the Induction Complete? The Concept of Commitment

At some point during the induction of a tissue, sufficient molecular changes accumulate in the responding tissue to make the process of induction irreversible. At present, we have no real understanding of the nature of such molecular changes, but their effects can be demonstrated experimentally by challenging the induced tissue with a variety of signals with the potential to divert the tissue from its fate. This is typically done by transplanting the tissue to be assayed to a variety of new locations in the embryo. If the transplanted tissue continues to recapitulate its original fate regardless of its new environment, it is said to be *committed* to that fate. It should be noted that commitment is difficult to demonstrate definitively, as one can never be sure that the tissue in question has been challenged with all possible alternative environments. Nevertheless, commitment assays represent a reasonable operational indication of when the induction of a particular tissue has proceeded beyond the point of no return.

Commitment of the otic placode has been studied in a number of species, including amphibians (Yntema 1933, 1939; Zwilling 1941; Ginsburg 1995) and chick (Waddington 1937; Vogel and Davies 1993; Herbrand et al. 1998; Groves and Bronner-Fraser 2000). Once again, several common features may be gleaned from these studies. First, the proportion of ectoderm transplants that successfully produce an ear after grafting (i.e., that demonstrate commitment) increases with the age of the ectoderm. This age can vary among species, as strikingly demonstrated by Ginsburg (1995) in her studies of different amphibian species. Second, presumptive otic tissue of a given age can give different results in a commitment assay depending on the region of the embryo to which it is transplanted—for example, experiments in the axolotl give three different estimates for the time of commitment of the otic placode depending on whether tissue was transplanted to the limb, the ventral body wall, or the preotic head region (Yntema 1933, 1939; Ginsburg 1946, 1995). This underlines the provisional and operational nature of commitment, and the need to test multiple environments in commitment assays.

As with specification, above, it is necessary to describe commitment with respect to a particular developmental end point. For example, if ectoderm of a particular age is shown experimentally to be committed to forming an otic vesicle, one cannot conclude that this tissue will also form more mature cell types such as sensory hair cells in the same assay. This reflects the fact that the induction of a tissue may occur in a series of steps, each of which may in principle be regulated independently from one another by different signals (Groves and Anderson 1996; Groves and Bronner-Fraser 2000). Indeed, several of the studies examining commitment of the otic placode showed that tissue taken from a particular age may be committed to one point in ear development (e.g., the induction of early placode markers), but not to later stages (e.g., the appearance of sensory patches (Waterman and Evans 1940; Evans 1943; Swanson et al. 1990; Ginsburg 1995; Herbrand et al. 1998).

It should be emphasized once more that the terms described above—competence, specification and commitment—are strictly *operational* definitions, and a molecular understanding of these three terms remains elusive. For example, do specification and commitment reflect different degrees of the same molecular process, or do they represent the culmination of two completely different molecular pathways? Nevertheless, despite the provisional and operational nature of these definitions, they provide the investigator with a set of tools to determine the most appropriate time to investigate otic placode induction in a particular species.

4. Experimental Investigation of Otic Placode Induction

Before we embark on a discussion of the candidate inducing tissues and molecules implicated in otic placode induction, it is worth spending some time discussing the experimental approaches used to investigate the inductive process. In particular, the following topics are useful when considering how to critically interpret studies on otic placode induction. This is especially important when evaluating claims made in early papers on the induction of the otic placode.

4.1 The Importance of Early Markers of Otic Placode Induction

It is important to be able to distinguish early events in inner ear development—such as induction of the otic placode—from later ones, such as the induction of neurons or sensory cells. Historical studies often score the presence of an otic vesicle as evidence of otic placode induction, but in other studies it is often less clear exactly what the criteria for assaying otic placode induction actually are. In some cases, the presence of differentiated sensory patches or semicircular canals are scored as representing otic placode induction, even though these structures appear long after the initial induction of the placode (e.g., Yntema 1950; see Section 5.1), and were likely induced by a very different set of signals. These problems were summarized succinctly by Gurdon: "I believe that the greatest obstacle to the molecular analysis of induction over many decades may have been the imprecision and late appearance of the assays used, which often depend on morphological assessment many days after the inductive response has started" (Gurdon 1987). Any attempt to understand the events leading to inner ear induction must therefore make use of early and specific markers of the otic placode, rather than later cell-type specific or morphological events.

4.2 Identification of Host and Donor Tissues

Some of the most frequently used approaches in experimental embryology are to transplant tissue from one location to another, or to combine tissues together in culture. In such experiments, it is critical to be able to distinguish host and donor tissues. Historically, this has been achieved by using two species whose cells can be distinguished by histology (Lewis 1907; Richardson 1933; Le Douarin and Kalcheim 1999) or species-specific antibodies (Le Douarin and Kalcheim 1999), by pigmented and unpigmented hosts and donors (Ginsburg 1995), or by labeling one population of tissue with vital dyes or tracers (Yntema 1939; Jacobson 1963a; Gallagher et al. 1996; Woo and Fraser 1997, 1998). Without using host–donor labels, it is impossible to distinguish whether an otic placode has been induced in the host by donor tissue, whether the donor tissue has simply formed an otic placode due to contaminating otic tissue (Stone 1931; Waddington 1937; Kuratani and Eichele 1993), or whether migration of otic placode precursors from a grafted hindbrain have formed the ear (Mayordomo et al. 1998; Streit 2002).

4.3 The Identity of Tissues Used in Induction Experiments

It is common for claims to be made concerning the inducing abilities of a particular tissue following transplantation. In many cases, however, it is hard to know exactly what tissues have been isolated and transplanted, and whether any contaminating tissue was included in the graft. Historically, the purity of such transplants were assessed by eye at the time of grafting, or by looking for characteristic tissue types (such as nervous tissue) at the end of the experiment (e.g., Zwilling 1940). More recently, the advent of molecular markers to specific tissue types has made it much easier to assess the purity of tissues used in transplants and tissue recombinations.

4.4 Necessity and Sufficiency for Otic Placode Induction

Experimental embryology frequently uses the formal logical concepts of necessity and sufficiency. A factor is *necessary* for a biological process if that process does not happen when one removes the factor in question. A factor is *sufficient* for a biological process if that process happens when one introduces the factor into a system that would otherwise remain constant. A "factor" in these definitions could be any one of a number of components in cellular and molecular signaling. It could be a piece of tissue, a secreted growth factor, a growth factor receptor, an enzyme, or a transcription factor. Examples of *necessity* experiments include knocking out or mutating a gene, surgically removing a piece of tissue, adding an inhibitor to a signaling pathway, or expressing a dominant-negative growth factor receptor. Examples of *sufficiency* experiments include adding a growth factor to tissue in a neutral culture environment or over-expressing a transcription factor in a piece of cultured tissue.

One advantage of thinking about induction using these concepts is that it directs the investigator away from certain experiments that may give ambiguous answers. For example, many historical studies of otic placode induction have involved transplanting a piece of tissue to a new location in the embryo and assaying whether the transplanted tissue induces an ectopic ear. The problem with such experiments is that one can never be confident that the transplanted tissue is acting directly and alone to induce the ear, or whether it is acting indirectly (e.g., by transforming another piece of adjacent tissue into an ear inducer), or in cooperation with other factors produced by the surrounding tissue. A better way to perform such experiments is to combine the candidate inducing tissue with a piece of unspecified responding ectoderm in a neutral culture environment. Expression of otic placode markers in the responding ectoderm would suggest that the candidate tissue is sufficient to induce the otic markers. A more recent example of such ambiguity is a study by Vendrell and colleagues (Vendrell et al. 2000), in which the *Fgf3* growth factor was over-expressed in the heads of chick embryos, leading to the formation of small ectopic otic vesicles. It is impossible to conclude from this study whether *Fgf3* was acting by itself (i.e., it was sufficient for the induction) or in cooperation with other factors, and whether it was acting directly on the host ectoderm, or indirectly by first affecting surrounding tissues.

With these experimental caveats in hand, which tissues may have a role in inducing the otic placode are now examined.

5. What Tissues Induce the Otic Placode?

The last 70 years have seen many studies that have attempted to address which tissues induce the otic placode. The two main candidates are the hindbrain, which lies adjacent to the otic placode, and the cranial paraxial mesoderm, which comes to lie under the otic placode. A few studies have also suggested roles for axial mesoderm and endoderm in the induction process, although in many species these two tissues never directly contact presumptive otic ectoderm. The relative contributions of these different tissues to otic placode induction are discussed. Rather than simply providing an undigested list of historical studies for the reader, the discussion makes clear which studies exhibit flaws or ambiguities in the experimental design, and which studies may shed light on the tissue interactions leading to otic placode induction.

5.1 The Contributions of Hindbrain and Mesoderm to Otic Placode Induction—Historical Studies

In this section, studies from the early part of the century through to the early 1980s are discussed. None of these studies were able to take advantage of modern molecular markers of the early otic placode, instead relying on morphological or histological identification of ear tissue. For this reason, many historical studies which refer to "induction" of the otic placode actually examined a combination of induction and later events such as formation of sensory patches or semicircular canals, which likely have their own set of inductive signals and events. In general the historical studies fall into two groups. First, a variety of studies ablated one or more candidate inducing tissues, and then examined the development of the otic placode in the absence of these tissues. A second variety of studies transplanted one or more tissues to ectopic sites in the embryo and assayed whether an otic placode formed in the new location. This second sort of studies are more difficult to interpret, as many of the studies did not use host–donor markers to determine whether ectopic ears were truly induced or merely carried over in the transplanted tissue. Moreover, as discussed in Section 4, it is difficult to conclude from such experiments that the transplanted tissue is acting alone to induce ectopic otic vesicles, or whether it is instead cooperating with host tissues at the transplant site.

A variety of early studies attempted to demonstrate that the hindbrain can induce the otic placode when grafted to ectopic locations, but in the absence of clear markers to distinguish host from donor tissue, these studies should be interpreted with caution (Stone 1931; Harrison 1936; Albaum and Nestler 1937; Waddington 1937; Kohan 1944; Harrison 1945). Other experiments attempted to examine the role of the hindbrain in placode induction by replacing the hindbrain with other neural tissue (Detwiler 1948; Detwiler and van Dyke 1950). They concluded that the hindbrain was necessary for induction of the inner ear, although these experiments were actually performed after the otic ectoderm became committed to an ear fate (see Ginsburg, 1995).

A role for mesoderm in the induction of the otic placode was first suggested in the early 1930s (Dalcq 1933; Holtfreter 1933) and confirmed by others (Harrison 1936, 1938, 1945; Albaum and Nestler 1937; Kohan 1944), but once again, a lack of host and donor markers in these experiments should be noted. Several studies in different species of *Rana*, chick, and *Discoglossus* conclude that the ear is committed before the hindbrain is morphologically visible, and interpret this as evidence for an exclusive action of mesoderm, rather than the hindbrain, in the induction of the otic placode (Szepsenwol 1933; Pasteels 1939; Zwilling 1941; Ginsburg 1995). In the absence of molecular markers of the nervous system, however, it is again hard to rule out signaling from the presumptive hindbrain in these experiments.

Jacobson performed explant cultures of the presumptive otic placode with various combinations of hindbrain, mesoderm, and endoderm. He concluded that the mesoderm was capable of inducing the otic placode in a small number of cases, but that much better induction was seen if the hindbrain or neural plate was included in the explants (Jacobson 1963a, 1966). Since these experiments did not distinguish between inducing and responding tissues, however, it is possible that the ears observed were derived from precursors present in the neural plate (Mayordomo et al. 1998; Streit 2002). Finally, two ablation studies in chick suggested that paraxial mesoderm may induce the otic placode, as removal of paraxial mesoderm precursors at an early stage blocked otic placode induction, even in the presence of the hindbrain (Orts-Llorca and Jimenez-Collado 1971; Cuevas 1977).

The difficulty in separating the relative contributions of hindbrain and mesoderm has lent support to the idea that both tissues may be involved in either redundant or sequential functions. In particular, the idea that otic placode induction occurs by a sequential series of inductive influences emanating first from cranial mesoderm and then the hindbrain has become embedded in the literature. It is worth spending some time examining the evidence for these claims. The two papers cited to promote this idea are by Yntema (Yntema 1950) and Jacobson (Jacobson 1963a; reviewed in Jacobson 1966). In the Yntema paper, grafts of gill ectoderm replaced the host's presumptive otic ectoderm, and were then cultured for 2 to 3 weeks. In the Jacobson paper (1963a), explants containing unspecified otic ectoderm and various other tissues were cultured for 11 to 21 days. In both cases, specimens were stained histologically and examined for signs of differentiation. Yntema devised an elaborate scoring system in which the size of the ear, the presence of sensory areas, semicircular canals, cartilage, and the endolymphatic sac all counted toward the final score in various proportions, together with the degree to which these structures were correctly positioned with respect to each other (Yntema 1950). Thus, Yntema's scoring system, although admirable in its thoroughness, did not actually score placode induction per se, but was rather a measure of induction together with much later stages of inner ear differentiation such as sensory patch formation and semicircular canal morphogenesis. A number of studies have made it clear that induction of early placode markers, morphogenesis, and sensory patch formation can be uncoupled from each other experimentally (Ginsburg 1995; Groves and Bronner-Fraser 2000), and as such, it is likely that different sets of signals mediate each aspect of development. When viewed in this light it is understandable why Yntema concluded that both mesoderm and hindbrain signals were sequentially necessary for the "induction" of the otic placode as he used the term in his paper. Since early placode markers were unavailable to him at the time, Yntema was unable to distinguish between signals that simply induced early otic placode markers prior to invagination of the otic placode and those that promoted later aspects of ear differentiation such as sensory patch formation or semicircular canal development.

Jacobson's study of lens, nose, and ear induction did not discuss the specific criteria for scoring his specimens as having otic vesicles versus olfactory vesicles. Although his study clearly showed differences in the ability of different tissue combinations to induce otic vesicles, it is not clear how he concluded that these different tissues act at different times in vivo, as opposed to acting at the same time or possessing redundant inducing activities.

We believe that many authors have been insufficiently critical in interpreting these studies, and that as a result, the suggestion that mesoderm and hindbrain tissue act separately and sequentially to induce the otic placode is still an open question, which needs to be addressed using early molecular markers for the otic placode.

5.2 The Contributions of Hindbrain and Mesoderm to Otic Placode Induction—Recent Studies

The availability of molecular markers for the otic placode and the emergence of new genetic organisms such as zebrafish have prompted new studies of otic placode induction. Several of these studies have tended to emphasize the role of cranial paraxial mesoderm in otic placode induction. For example, Mendonsa and Riley (Mendonsa and Riley 1999) examined a series of zebrafish mutants that affected development of cranial mesoderm or the hindbrain. They found that *cyclops* and *one-eyed pinhead* mutants, which both have deficiencies in cranial mesoderm, had delayed formation of the otic placode. Total disruption of maternal and zygotic *one-eyed pinhead* mRNA using morpholino knockdown lead to embryos with little or no Pax8 expression, and significantly smaller otic vesicles (Phillips et al. 2001). In contrast, mutants affecting differentiation of axial mesoderm (such as *no tail* or *floating head*) did not affect otic placode induction. Moreover, the *valentino* mutation (a mutant of the *kreisler/MafB* gene; Moens et al. 1998), which disrupts formation of rhombomeres 5 and 6, develops an otic placode on schedule even though subsequent differentiation of the ear is highly abnormal (Mendonsa and Riley 1999). Fish in which the entire hindbrain adopts a rhombomere 1 identity also have small otic vesicles (Waskiewicz et al. 2002), although it is not yet clear whether the "pan-r1" hindbrain in these studies has any residual positional information left. Similar results have been obtained in a very different experimental system, the vitamin A-deficient quail. Such embryos lack rhombomeres 5 to 7 (Maden et al. 1996; Gale et al. 1999; Dupe and Lumsden 2001), yet the otic placode continues to form in approximately the correct position (Dupe and Lumsden 2001; A Groves, unpublished results). Lastly, ablation of the paraxial mesoderm that will come to lie beneath the otic placode either delays or abolishes induction of early otic placode markers if performed prior to specification of the placode (A. Streit, unpublished observations), and transplantation of this mesoderm to more rostral locations in the head can induce some otic placode markers (A. Streit, unpublished observations).

These results suggest a role for mesoderm in otic placode induction. However, other recent studies suggest a role for the hindbrain as well. Woo and Fraser transplanted the germ ring to different locations in zebrafish, and observed that germ ring transplants that came to lie close to the forebrain were able to transform forebrain tissue to a hindbrain fate. By marking the donor tissue, they demonstrated that host tissue next to the ectopic hindbrain was induced to form otic vesicles (Woo and Fraser 1997). Interestingly, grafts of hindbrain to the forebrain region (in the absence of germ ring) were not able to form otic vesicles, suggesting that the germ ring itself (which will form mesendodermal derivatives) may be inducing the otic vesicles, or cooperating with the ectopic hindbrain to induce otic tissue. In a second series of experiments, prospective hindbrain progenitors were grafted to the future ventral side of zebrafish embryos. Once again, ectopic otic vesicles were induced in the host ventral ectoderm by the hindbrain tissue (Woo and Fraser 1998). Significantly, little or no mesoderm lay next to the transplanted hindbrain at the end of the experiment, suggesting that the hindbrain may have induced ectopic otic vesicles by itself. However, the hindbrain grafts were placed close to host mesoderm at the start of the experiment, so it is hard to completely rule out an influence of mesoderm in these grafts. Studies in zebrafish in which both *Fgf3* and *Fgf8* are disrupted (Phillips et al. 2001; Leger and Brand 2002; Maroon et al. 2002) led to an abnormal hindbrain lacking rhombomeres 5 and 6 (Maves et al. 2002; Walshe et al. 2002) and virtually no evidence of an otic placode. As discussed in Section 6.1, it is not clear if the absence of an otic placode is due to a loss of *Fgf3/8* activity from the hindbrain, or to a more general loss of r5 and r6. Other studies have also demonstrated ectopic otic vesicles next to transplanted hindbrain (Kuratani and Eichele 1993; Sechrist et al. 1994), although since neither study used markers to distinguish host from donor tissue, these results should be interpreted with caution. Finally, a recent paper examining the effect of the hindbrain on the otic placode marker *Lmx1* concluded that the hindbrain was necessary at early stages for the appearance of this gene in the placode (Giraldez 1998). However, in the absence of data as to when the expression of *Lmx1* is specified, it is unclear whether the hindbrain was required for the induction or maintenance of *Lmx1* in these experiments.

5.3 Conclusions: The Relative Contributions of Mesoderm and Hindbrain

Both gain- and loss-of-function experiments described above suggest a role for cranial paraxial mesoderm in the induction of the otic placode. The role, if any, of neural tissue in the induction is harder to establish at present. Part of the reason for this is the demonstration that at least some parts of the otic placode may actually be derived from the early neural plate (Mayordomo et al. 1998; Streit 2002), making it hard to tell the difference between induction of placode cells and migration of placode cells from the hindbrain in the absence of good markers of the inducing and responding tissues. It may also be the case that some inducing factors expressed in paraxial mesoderm are also expressed later in the hindbrain (e.g., *Fgf19*; and *Fgf3*; Ladher et al., 2000; A. Groves, unpublished observations). We believe that one way to resolve this question is to perform simple tissue recombination experiments (similar to those of Jacobson 1963a) with inducing and responding tissues clearly labeled to distinguish one from the other, and assaying for early markers of otic placode induction. It is also possible that otic placode-inducing molecules are expressed in different tissues in different species. We discuss this possibility in the next section.

6. What Molecules Induce the Otic Placode?

In this section, what is known of the molecular basis of otic placode induction is described. Most attention has been focused on soluble signaling molecules in the process, and this topic is addressed first. Some of the transcriptional regulators that may be required to convert ectoderm to an otic placode fate are then briefly considered.

6.1 Signaling Molecules Implicated in Otic Placode Induction

The last 10 years have seen a number of candidate molecules proposed to induce the otic placode. In particular, attention focused on a member of the fibroblast growth factor family, *Fgf3* (formerly known as *int-2*), owing to its expression in rhombomeres 4 to 6 of the caudal hindbrain, adjacent to where the otic placode will form. The first study to propose that *Fgf3* was necessary for the induction of the otic placode was based on experiments with antisense oligonucleotides and neutralizing antibodies (Represa et al. 1991). This study is now considered flawed on a number of grounds. First, the antisense oligonucleotides were designed against human *Fgf3* (the chick gene had not been cloned at the time of the experiments), and no controls were performed to demonstrate that the antisense treatment actually reduced FGF3 protein levels. Subsequent investigations failed to repeat these results with the same reagents (Mahmood et al. 1995), and it is now clear that several mismatches exist between the human antisense oligonucleotides used by Represa and colleagues and the actual chick *Fgf3* gene. Finally, the experiments were performed using stage 10 chicken embryos—an age at which the otic placode is already committed (Groves and Bronner-Fraser 2000) and therefore the study does not actually address the process of placode induction at all. Other studies have since examined the effect of *Fgf3* in vivo, either by implanting beads soaked in Fgf3 into the head region of *Xenopus* embryos (Lombardo et al. 1998), or by overexpressing *Fgf3* in the heads of chick embryos with an *Fgf3*-expressing herpes virus vector (Vendrell et al. 2000). In both these cases, *Fgf3* expression led to the induction of otic placode markers and small epithelial vesicles. As discussed above, however, these studies do not rule out the possibility that Fgf3 is cooperating with other factors in the implanted or infected embryos, or is inducing ectopic otic tissue indirectly by first activating an adjacent tissue at the site of implant or infection.

To date no one has tested the sufficiency of *Fgf3* for otic placode induction by adding it to isolated unspecified ectoderm in a neutral culture medium. Similar criticisms can be leveled at a more recent paper in which implanting Fgf2 beads caudal to the otic placode caused tiny ectopic patches of placode tissue to form next to the normal ear in a small number of cases (Adamska et al. 2001).

The necessity of *Fgf3* for otic placode induction has been tested by inactivating the gene in mice (Mansour et al. 1988, 1993). Otic placode induction appears to proceed normally in these mice, although subsequent development of the ear appears to be severely affected, with homozygous mutant mice displaying an undeveloped cystic ear phenotype with variable penetrance and expressivity. These results suggest that *Fgf3* is not necessary for otic placode induction, but do not rule out the possibility that another member of the *Fgf* family is compensating for the loss of *Fgf3* in these mutant embryos. One possible candidate is *Fgf10,* which activates the same set of FGF receptors as *Fgf3*, and is expressed at an appropriate time and location to participate in placode induction. *Fgf10* mutant mice have small otic vesicles (Ohuchi et al. 2000; Pauley et al. 2003), and mouse mutants for *FgfrIIIb*, through which both *Fgf3* and *Fgf10* signal, have even more severe ear phenotypes than either the mutants of *Fgf3* or *Fgf10* alone (De Moerlooze et al. 2000). Recently, Wright and Mansour have examined mice homozygous for both *Fgf3* and *Fgf10* mutations, and the otic vesicles of such embryos are either missing completely or greatly reduced. Otic markers such as *Pax2* are generally absent, although ventral patches of thickened ectoderm expressing both *Dlx5* and *Gbx2* remain, which may contribute to the epibranchial placodes. Importantly, the hindbrain of the *Fgf3/10* double homozygous mice appears to be normal, suggesting that *Fgf3* and *Fgf10* may act directly on presumptive otic epithelium to specify the otic placode (Wright and Mansour 2003). Interestingly, whereas *Fgf3* is expressed in the mouse hindbrain, *Fgf10* is expressed in the paraxial mesoderm underlying the future ear (Wright and Mansour 2003; Wright et al. 2003). This raises the possibility that redundant signals from redundant tissues may induce the otic placode in mice. There a perplexing lack of conservation of *Fgf* family member expression between different species. For example, *Fgf3* and *Fgf8* are expressed in the hindbrain of zebrafish (Phillips et al. 2001; Leger and Brand 2002; Maroon et al. 2002; Maves et al. 2002; Walshe et al. 2002). In contrast, chick and quail *Fgf3* is expressed in both the hindbrain and cranial mesoderm (Mahmood et al. 1995), whereas *Fgf8* is restricted to the endoderm (Karabagli et al. 2002a,b). Moreover, *Fgf19* and *Fgf4* are expressed in the cranial mesoderm underlying the caudal hindbrain in birds, but not fish or mice (Shamim and Mason 1999; Ladher et al. 2000; Karabagli et al. 2002a,b). These observations may well explain why different tissues have been implicated in otic placode induction in different species.

A similar potential redundancy between *Fgf3* and *Fgf8* has been proposed in zebrafish. Three studies depleted *Fgf3* and *Fgf8* mRNA with either morpholino oligonucleotides to these two genes or by using *Fgf3* morpholinos in *ace* (*acerebellar*) fish that carry null mutations for *Fgf8* (Phillips et al. 2001; Leger and Brand 2002; Maroon et al. 2002). In all cases, depletion of both *Fgf3* and *Fgf8* gave otic vesicles that were either extremely small or absent altogether. Significant numbers of treated embryos displayed little or no expression of the placode markers *Pax2*, *Pax8*, and *Dlx3b*. These results clearly suggest a role for these two FGF family members in zebrafish otic placode development, although the location and time of their action in this process is less clear. Both genes are expressed in a variety of tissues implicated in otic placode induction (such as the germ ring, prechordal plate, cranial paraxial mesendoderm, and rhombomere 4 of the hindbrain, and could thus induce the otic placode at a variety of developmental stages. Moreover, since *Fgf3* and *Fgf8* activities in rhombomere 4 are required for an organizing activity that patterns rhombomere 5 and 6 (Maves et al. 2002; Walshe et al. 2002), it is possible that the ear phenotypes observed in *Fgf3/8* fish mutants are caused by an indirect effect of *Fgf3* and *Fgf8* on hindbrain patterning, rather than by a direct effect on placodal ectoderm.

An alternative approach to examining the necessity of FGF signaling in the induction of the otic placode is to block all signaling through FGF receptors. This has been attempted with the pharmacological inhibitor SU5402, which blocks kinase activity in all FGF receptors (Mohammadi et al. 1997). Incubating zebrafish embryos in this compound for 5 hr was able to block subsequent expression of *Pax2*, although *Pax8* expression appeared unaffected (Maroon et al. 2002). However, a second study, using lower concentrations of the SU5402 inhibitor at comparable times successfully blocked *Pax8* expression (Leger and Brand 2002). At the present time, it is not clear why these two zebrafish studies obtained different results using the same inhibitor, although a number of possibilities are considered in more detail in a recent review by Brown and colleagues (Brown et al. 2003). It is formally possible that different aspects of otic placode induction are controlled by different inducing factors, and further dissection of the necessity and sufficiency for FGF signaling in this process may prove fruitful.

Another member of the Fgf family implicated in otic placode induction is *Fgf19* (in chick and humans; the homolog in mice appears to be *Fgf15*). *Fgf19* is expressed in mesoderm under the prospective neural plate during chick gastrulation, and in paraxial mesoderm and the hindbrain after otic placode specification has started (Ladher et al. 2000; Karabagli et al. 2002a, 2002b). *Fgf19*-expressing mesoderm is able to induce some otic markers when combined with ectoderm and neural tissue. The observed requirement for neural tissue in these experiments led the authors to propose that *Wnt8c* (which is expressed in rhombomere 4 (Hume and Dodd 1993) may cooperate with *Fgf19* to induce otic markers. *Fgf19* and *Wnt8c* together induce a series of otic markers when applied to unspecified presumptive otic ectoderm; it is not clear whether similar levels of induction are seen with other competent ectoderm populations (Ladher et al. 2000). Since *Fgf19* can also induce *Wnt8c* in neural tissue, these authors propose that *Fgf19* may activate *Wnt8c* expression in the hindbrain, and that these two factors synergize to activate otic placode markers in the adjacent ectoderm.

Interestingly, treatment of *Xenopus* embryos with lithium (which inhibits glycogen synthase kinase 3, a negatively acting component of the canonical Wnt signaling pathway; [Klein and Melton 1996; Stambolic et al. 1996]) causes enlarged or multiple anastomosing ear vesicles (Gutknecht and Fritzsch 1990). It is not clear, however, whether this phenotype is due to induction of extraplacodal tissue, or of proliferation of the placode.

It is not yet clear whether *Fgf19* and *Wnt8c* are acting with other factors in the induction process, or whether they are acting in a truly synergistic or sequential manner. Several observations suggest that some other factors may be involved. Fgf19 signals exclusively through the *Fgfr4* receptor (Xie et al. 1999), but mice in which *Fgfr4* has been inactivated have apparently normal ears (Weinstein et al. 1998). Inactivation of the mouse homolog of *Fgf19*, *Fgf15*, also appears to give normal ears (S. Mansour, unpublished observations). Studies in zebrafish suggest that blocking canonical Wnt signaling does not disrupt placode induction (Phillips et al. 2004). Furthermore, transgenic reporter mice and fish in which Lef/TCF binding sites provide a readout of canonical Wnt pathway signaling do not show reporter gene expression in presumptive otic tissue (A. Groves, unpublished observations; Phillips et al. 2004), suggesting that canonical Wnt signaling is not acting on presumptive otic tissue. Lastly, in avian embryos in which retinoic acid signaling is blocked, *Wnt8c* is no longer expressed adjacent to where the otic placode forms, although *Fgf3*, *Fgf19*, and *Fgf4* are still expressed in the underlying cranial paraxial mesoderm (Dupe and Lumsden 2001; A. Groves, unpublished observations). These observations do not rule out that other Fgf family members or noncanonical Wnt signaling may function in otic placode induction, and a clearer understanding of this signaling scheme must wait for better loss-of-function experiments.

6.2 Transcriptional Regulators of Otic Placode Induction

In the preceding section, we have considered the tissues that may induce the otic placode and the signaling molecules that may be produced by these tissues in the course of the induction process. Relatively little is known concerning the transcriptional cascades that are activated in ectoderm cells as they become specified to a placodal fate. Many of the earliest markers of the otic placode are transcription factors. Table 2.2 shows a list of mutants of early otic placode marker genes and the phenotypes they display. Gain of function experiments in the fish medaka suggest that both *Six3* and *Sox3* may be able to induce ectopic otic vesicles in a small percentage of cases (Loosli et al. 1998; Koster et al. 2000).

Zebrafish mutants have also helped shed light on this problem. In the course of a γ-ray mutagenesis screen, a large 6-cM deletion called b380 was isolated, in which the otic (and olfactory) placodes are almost completely absent (Fritz and Westerfield 1996; Solomon and Fritz 2002; Liu et al. 2003). This region encodes a series of genes, including *Dlx3b*, *Dlx4b*, and *Sox9*. Antisense morpholino oligonucleotides to *Dlx3b* give a moderate otic placode phenotype,

TABLE 2.2. Mouse mutants of genes that mark the otic placode.

Gene	Knockout ear phenotype	Reference
Pax8	No ear phenotype reported	Mansouri et al. 1998
Pax2	Agenesis of cochlea and vestibulo-acoustic ganglion	Torres et al. 1996
Sox9	None reported	Bi et al. 1999, 2001; Akiyama et al. 2002; MoriAkiyama et al. 2003
Eya1	Ear fails to develop past otic vesicle stage	Xu et al. 1999
Gata3	Ear fails to develop past otic vesicle stage	Karis et al. 2001
Nkx5.1/HMX3	Semicircular canal defects	Hadrys et al. 1998; Wang et al. 1998
Dlx3	Mice die at otic vesicle stage; ear phenotype not reported	Morasso et al. 1999
Dlx5	Vestibular defects	Acampora et al. 1999; Depew et al. 1999
Dlx5 and *Dlx6*	Patterning defects	Robledo et al. 2002
BMP7	None reported	Dudley et al. 1995
Fgf10	Ear fails to develop past otic vesicle stage	Ohuchi et al. 2000; Pauley et al. 2003
Fgf3	Ear fails to develop past otic vesicle stage	Mansour et al. 1993
Fgf3 and *Fgf10*	Almost no evidence of otic placode; variable phenotype	Wright et al. 2003
Six1	Ear fails to develop past otic vesicle stage	Ozaki et al. 2004; Zheng et al. 2003
Six4	None	Ozaki et al. 2001
Notch	Mice die at late otic vesicle stage. Heterozygotes have hair cell patterning defects	Swiatek et al. 1994; Zhang et al. 2000

which is made more severe by morpholinos to *Dlx4b* (Solomon and Fritz 2002). *Dlx3b* and *4b* are expressed in a semicircle around the border between the neural plate and ectoderm (a region from which all craniofacial placodes derive; see Section 7) and *Sox9* is one of the earliest genes to be expressed specifically in the otic placode. In mice and birds, *Dlx5* and *Dlx6* are expressed in a similar broad semicircle around the neural plate, but their expression is not seen in the otic placode proper until relatively late stages (Quint et al. 2000). This may explain why mouse mutants of both *Dlx5* and *Dlx6* have mild ear phenotypes compared to zebrafish (Robledo et al. 2002). Morpholinos to *Dlx3b*, *Dlx4b*, and *Sox9* together appear to phenocopy the b380 mutation (Liu et al. 2003). Disruption of *Sox9* in *Xenopus* also leads to a severe otic placode phenotype (Saint-Germain et al. 2004).

Another transcription factor expressed very early in zebrafish otic placode development is the Forkhead family member *Foxi1* (Lee et al. 2003; Nissen et al. 2003; Solomon et al. 2003a, b). *Foxi1* expression precedes *Pax8* expression in zebrafish, making it the earliest specific marker of the otic region in this species. *Foxi1* mutants do not express *Pax8* or *Dlx3b* and frequently have no otic vesicles at all, although expression of later otic markers such as *Dlx4b* and *Dlx5a* is seen, resulting in small otic vesicles (Solomon et al. 2003a). This result offers more support for the concept that otic placode induction is the result of a series of inductive signals and regulatory pathways, and that some aspects of the pathway can be experimentally uncoupled from others (Groves and Bronner-Fraser 2000; Maroon et al. 2002). Amphibian and mouse *Foxi1* homologs do not seem to be such early markers of the otic placode (Hulander et al. 1998; Pohl et al. 2002) and mouse mutants of *Foxi1* form an otic placode, although the mice later develop ear abnormalities as a result of malformation of the endolymphatic duct (Hulander et al. 2003).

7. Toward a Model for Otic Placode Induction

In the last few years, several lines of evidence have converged to support the idea that induction of the otic placode may actually be a two-stage process. In this model, a "pan-placodal" or "pre-placodal" domain is induced at the border of the neural plate and future epidermis, extending down the length of the head. All craniofacial placodes are proposed to arise from this domain. Individual placodes are then induced by local specific signals. The interesting feature of this model is implication that induction of a "pan-placodal" or "pre-placodal" cell state is a prerequisite for subsequent induction of each placode. Below, we review the main lines of evidence for and against this two-step model of otic placode induction. For further discussion on this topic, the reader is referred to Baker and Bronner-Fraser (2001), Graham and Begbie (2000), and Begbie and Graham (2001).

7.1 Morphological Evidence for a Common Placodal Domain?

The physical nature of placodes is reflected in the etymology of their name—they occur as platelike thickenings on the side of the embryonic head. A number of studies have pointed out that multiple placodes can derive from a single thickening on the head (Knouff 1935; Braun 1996; Schlosser and Northcutt 2000). Indeed, some studies have suggested that all placodes can be traced to a continuous sheet of thickened ectoderm—for example, in the cod embryo (Miyake et al. 1997), *Necturus* (Platt 1896), mice (van Oostrom and Verwoerd 1972; Verwoerd and van Oostrom 1979), and humans (O'Rahilly and Müller 1985). On the other hand, evidence from other vertebrates such as zebrafish, *Xenopus*, and chicks does not suggest a common morphological thickening.

7.2 Evolutionary Evidence for a Common Placodal Domain?

It has been proposed that the neurogenic placodes evolved together, and that they represent a characteristic feature of vertebrates (Gans and Northcutt 1983). However, there is some evidence from urochordates and cephalochordates that some placodelike derivatives may have evolved at different times. For example, as described previously, some ascidians possess an "atrial primordium" that may be related to the otic placode on the basis of the presence of ciliated sensory cells (Bone and Ryan 1978; Katz 1983; Baker and Bronner-Fraser 1997) and the expression of *HrPax-258* (Wada et al. 1998). Other ascidians and the cephalochordate *Amphioxus* also have some sensory areas that resemble olfactory placodes (Lacalli and Hou 1999; Manni et al. 1999). Since other placodes, such as the epibranchial placodes, seem to be truly specific to vertebrates, it may be the case that different placodes evolved at different times (Shimeld and Holland 2000; Manni et al. 2001).

7.3 Molecular Evidence for a Common Placodal Domain?

Recently a number of genes have been identified that appear to mark the border region between the neural plate and epidermis. In zebrafish these include *Dlx3b* and *4b*, *Six4.1*, *Eya1*; in *Xenopus*, *Six1*, and in chick, *Six4*, *Six1*, *ERN1*, *Dlx5*, *Dlx6*, *BMP4* and *Msx1* (Akimenko et al. 1994; Streit and Stern 1999; Streit et al. 2000; Streit 2002). These markers are expressed prior to the appearance of specific markers for individual placodes, suggesting that they may mark a preplacodal domain. It should be noted that in chick at least, the expression patterns of these "border" genes are not in complete registration, as the rostrocaudal extent of the different expression domains varies (Streit 2002). Nevertheless, the expression of so many different genes in a region that corresponds quite precisely to the placodal fate map is intriguing. To date however, the only loss-of-function data that suggest a role for these "border" genes in the formation of multiple placodes come from the zebrafish b380 mutation, in which *Dlx3b*, *Dlx4b*, and several other genes are deleted. As described above, this leads to an almost complete loss of the otic and olfactory placodes, but not of any other placodes. The mouse knockout of the locus spanning *Dlx5* and *Dlx6* develops an abnormal ear, although the placode itself appears to form normally (Robledo et al. 2002). The *Six4* knockout also appears normal (Ozaki et al. 2001), but mutants of *Six1* develop a placode, but exhibit later defects in the development of the otocyst (Zheng et al. 2003; Ozaki et al. 2004). It will be of great interest to determine whether these genes are individually or collectively necessary or sufficient for the induction of craniofacial placodes.

7.4 Experimental Evidence for a Common Placodal Domain?

Fate mapping experiments demonstrate that all craniofacial placodes derive from the neural plate–epidermis border, and some fate maps have shown that the precursors for particular placodes may be mixed with nonplacodal cells, or with precursors of other placodes (Kozlowski et al. 1997; Streit 2002). Significantly, grafting experiments suggest that head ectoderm is generally more competent to generate placodes than ectoderm from more caudal regions of the embryo (reviewed in Baker and Bronner-Fraser, 2001; Groves and Bronner-Fraser, 2000). Overexpression of *Fgf3* also suggested that ectoderm from this border region was more competent to generate an otic placode than ectoderm from the trunk (Vendrell et al. 2000). Interestingly, however, anterior epiblast from gastrulating chick embryos—which does not express any of the "border" genes mentioned above—is readily competent to form both trigeminal and otic placodes when grafted in place of the host placodes (Baker et al. 1999; Groves and Bronner-Fraser 2000). It is possible that the grafted epiblast first up-regulates "border" genes and only then responds to local placode-specific inducing signals in these transplants. In this respect, it is particularly interesting that anterior epiblast cannot express otic placode markers when cultured alone in the presence of Fgf2, but that unspecified trigeminal ectoderm does express otic placode markers under these same culture conditions (K. Martin and A. Groves, unpublished observations). In other words, two populations of ectoderm that are equally competent to express ear markers in vivo are not equally competent to express ear markers in the presence of *Fgf2* in vitro. Since the epiblast in these experiments does not express the "border" genes, but the trigeminal ectoderm does, this may be the first tentative evidence that cells must first adopt a "panplacodal" state prior to differentiating into a particular placode.

8. Summary: A Model for Otic Placode Induction

It is clear from a number of studies that different populations of ectoderm have different capacities to respond to otic placode-inducing signals. We find the idea of a common preplacodal domain an attractive one to interpret these experiments. Under such a model, interactions between the neural plate and epidermis establish a border region in the anterior part of the embryo marked by the border genes described above (Streit and Stern 1999). The exact mechanism for this process has yet to be determined. Once this border identity has been established, localized signals along the rostro-caudal axis induce specific placodes, including the otic placode (Fig. 2.2).

This two-step model makes a number of predictions concerning ectoderm that is competent to form the otic placode, but that does not normally express border genes, such as anterior epiblast in chick embryos (Groves and Bronner-Fraser 2000). We predict that such ectoderm will up-regulate border genes prior to

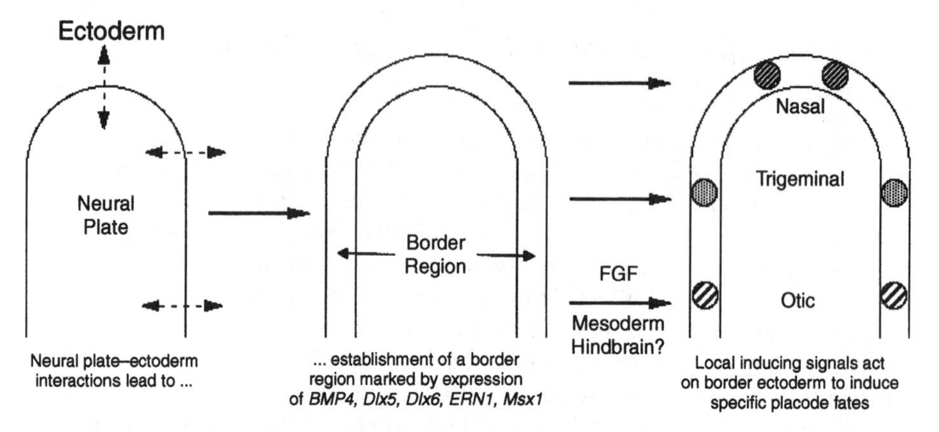

FIGURE 2.2. Simple two-step model for the induction of the otic placode. Neural plate–ectoderm interactions lead to the establishment of a border region marked by genes such as *Six4*, *Dlx5*, *Dlx6*, *ERN1*, *BMP4*, and *Msx1*. Within this region, local signals from the mesoderm and hindbrain, which may include Fgf family members, act to specify the otic placode.

up-regulating otic placode-specific genes when grafted in place of the host's otic placode. Moreover, we predict that such ectoderm will not respond to otic placode-inducing signals in culture, unless it is able to up-regulate border genes first. A corollary of this prediction is that overexpression of some border genes in cultured anterior epiblast may be sufficient to allow the epiblast to respond to otic placode inducing signals.

The identity of the tissues and molecules that specifically induce the otic placode is still uncertain. However, at least some of the signals likely are produced by cranial paraxial mesoderm, and may include members of the Fgf family of growth factors. It is likely that the hindbrain also produces placode-inducing signals including Fgfs in some species, and that the hindbrain and mesoderm may have different contributions to otic placode induction in different species. The bewildering variety of Fgf family members implicated in ear induction, together with their expression in different tissues in different species, suggests that a quest for a detailed model of otic placode induction common to all vertebrate species is likely to prove fruitless. Instead, by accepting the central role of Fgf family members in this process, further investigations should focus on the necessity and sufficiency of Fgf signaling in otic placode induction, and should determine the targets of Fgf signaling in this process.

The coming years will see these questions resolved using a combination of classic embryological techniques, molecular biology, and genetics. In particular, the use of molecular markers, tissue explant cultures, and the possibility of gain- and loss-of-function experiments in mice and fish will aid our understanding of how a simple patch of ectoderm is transformed into arguably the most complicated sensory structure in the body.

References

Abu-Elmagd M, Ishii Y, Cheung M, Rex M, Le Rouedec D, Scotting PJ (2001) *cSox3* expression and neurogenesis in the epibranchial placodes. Dev Biol 237:258–269.

Acampora D. Merlo GR, Palearil, Zerega B, Postiglione MP, Mantero S, Bober E, Barbieri O, Simeone A, Levi G (1999) Craniofacial, vestibular and bone defects in mice lacking the Distalless related gene *Dlx5*. Development 126:3795–3809.

Adamska M, Léger S, Brand M, Hadrys T, Braun T, Bober E (2000) Inner ear and lateral line expression of a zebrafish *Nkx5-1* gene and its downregulation in the ears of FGF8 mutant, *ace*. Mech Dev 97:161–165.

Adamska M, Herbrand H, Adamski M, Kruger M, Braun T, Bober E (2001) FGFs control the patterning of the inner ear but are not able to induce the full ear program. Mech Dev 109:303–313.

Akimenko MA, Ekker M, Wegner J, Lin W, Westerfield M (1994) Combinatorial expression of three zebrafish genes related to *distal-less:* part of a homeobox gene code for the head. J Neurosci 14:3475–3486.

Akiyama H, Chaboissier MC, Martin JF, Schedl A, de Crombrugghe B (2002) The transcription factor *Sox9* has essential roles in successive steps of the chondrocyte differentiation pathway and is required for expression of *Sox5* and *Sox6*. Genes Dev 16: 2813–2828.

Albaum HG, Nestler HA (1937) Xenoplastic ear induction between *Rana pipiens* and *Amblystoma punctatum*. J Exp Zool 75:1–9.

Alvarez IS, Martin-Partido G, Rodriguez-Gallardo L, Gonzalez-Ramos C, Navascues J (1989) Cell proliferation during early development of the chick embryo otic anlage: quantitative comparison of migratory and nonmigratory regions of the otic epithelium. J Comp Neurol 290:278–288.

Baker CV, Bronner-Fraser M (1997) The origins of the neural crest. Part II: an evolutionary perspective. Mech Dev 69:13–29.

Baker CVH, Bronner-Fraser M (2001) Vertebrate cranial placodes I. Embryonic induction. Dev Biol 232:1–61.

Baker CVH, Stark MR, Marcelle C, Bronner-Fraser M (1999) Competence, specification and induction of *Pax-3* in the trigeminal placode. Development 126:147–156.

Baranski M, Berdougo E, Sandler JS, Darnell DK, Burrus LW (2000) The dynamic expression pattern of *frzb-1* suggests multiple roles in chick development. Dev Biol 217:25–41.

Begbie J, Graham A (2001) The ectodermal placodes: a dysfunctional family. Philos Trans R Soc Lond B Biol Sci 356:1655–1660.

Bi W, Deng JM, Zhang Z, Behringer RR, de Crombrugghe B (1999) *Sox9* is required for cartilage formation. Nat Genet 22:85–89.

Bi W, Huang W, Whitworth DJ, Deng JM, Zhang Z, Behringer RR, de Crombrugghe B (2001) Haploinsufficiency of *Sox9* results in defective cartilage primordia and premature skeletal mineralization. Proc Natl Acad Sci USA 98:6698–6703.

Bone Q, Ryan KP (1978) Cupular sense organs in *Ciona* (Tunicata: Ascidiacea). Zool 186:417–429.

Braun CB (1996) The sensory biology of the living jawless fishes: a phylogenetic assessment. Brain Behav Evol 48:262–276.

Brown ST, Martin K, Groves AK (2003) Molecular basis of inner ear induction. Curr Top Dev Biol 57:115–149.

Chambers D, Mason I (2000) Expression of *sprouty2* during early development of the chick embryo is coincident with known sites of FGF signalling. Mech Dev 91:361–364.

Cuevas P (1977) [Presence of a mesodermical inductor in chick's otic placode (author's transl)]. Experientia 33:660–661.

Dalcq A (1933) La détermination de la vésicule auditive chez le discoglosse. Arch Anat Microsc 29:389–420.

De Moerlooze L, Spencer-Dene B, Revest J, Hajihosseini M, Rosewell I, Dickson C (2000) An important role for the IIIb isoform of fibroblast growth factor receptor 2 (FGFR2) in mesenchymal-epithelial signalling during mouse organogenesis. Development 127:483–492.

Depew MJ, Liu JK, Long JE, Presley R, Meneses JJ, Pedersen RA, Rubenstein JL (1999) *Dlx5* regulates regional development of the branchial arches and sensory capsules. Development 126:3831–3846.

Detwiler SR (1948) Further quantitative studies on locomotor capacity of larval *Amblystoma* following surgical procedures upon embryonic brain. J Exp Zool 108:45–74.

Detwiler SR, van Dyke RH (1950) The role of the medulla in the differentiation of the otic vesicle. J Exp Zool 113:179–199.

Dudley AT, Lyons KM, Robertson EJ (1995) A requirement for bone morphogenetic protein-7 during development of the mammalian kidney and eye. Genes Dev 9:2795–2807.

Dupe V, Lumsden A (2001) Hindbrain patterning involves graded responses to retinoic acid signalling. Development 128:2199–2208.

Duprez D, Leyns L, Bonnin MA, Lapointe F, Etchevers H, De Robertis EM, Le Douarin N (1999) Expression of *Frzb-1* during chick development. Mech Dev 89:179–183.

Ellies DL, Stock DW, Hatch G, Giroux G, Weiss KM, Ekker M (1997) Relationship between the genomic organization and the overlapping embryonic expression patterns of the zebrafish *dlx* genes. Genomics 45:580–590.

Evans HJ (1943) The independent differentiation of the sensory areas of the avian inner ear. Biol Bull 84:252–262.

Fritz A, Westerfield M (1996) Analysis of two mutants affecting neuroectodermal patterning in zebrafish. In: Zebrafish. Cold Spring Harbor, NY: Cold Spring Harbor Laboratory Press, p. 216.

Fritzsch B, Beisel KW (2001) Evolution and development of the vertebrate ear. Brain Res Bull 55:711–721.

Fritzsch B, Barald KF, Lomax MI (1997) Early embryology of the vertebrate ear. In: Rubel EW, Popper AN, Fay RR (eds), Development of the Auditory System. New York: Springer-Verlag, pp. 80–145.

Fritzsch B, Beisel KW, Jones K, Farinas I, Maklad A, Lee J, Reichardt LF (2002) Development and evolution of inner ear sensory epithelia and their innervation. J Neurobiol 53:143–156.

Furthauer M, Reifers F, Brand M, Thisse B, Thisse C (2001) *sprouty4* acts in vivo as a feedback-induced antagonist of FGF signaling in zebrafish. Development 128:2175–2186.

Gale E, Zile M, Maden M (1999) Hindbrain respecification in the retinoid-deficient quail. Mech Dev 89:43–54.

Gallagher BC, Henry JJ, Grainger RM (1996) Inductive processes leading to inner ear formation during *Xenopus* development. Dev Biol 175:95–107.

Gans C, Northcutt RG (1983) Neural crest and the origin of vertebrates: a new head. Science 220:268–274.

George KM, Leonard MW, Roth ME, Lieuw KH, Kioussis D, Grosveld F, Engel JD (1994) Embryonic expression and cloning of the murine *GATA-3* gene. Development 120:2673–2686.

Ginsburg AS (1946) Specific differences in the determination of the internal ear and other ectodermal organs in certain *Urodela*. C R (Dokl) Acad Sci USSR 54:557–560.

Ginsburg AS (1995) Determination of the labyrinth in different amphibian species and its correlation with determination of the other ectoderm derivatives. Roux Arch Dev Biol 204:351–358.

Giraldez F (1998) Regionalized organizing activity of the neural tube revealed by the regulation of *lmx1* in the otic vesicle. Dev Biol 203:189–200.

Graham A, Begbie J (2000) Neurogenic placodes: a common front. Trends Neurosci 23: 313–316.

Groves AK, Anderson DJ (1996) Role of environmental signals and transcriptional regulators in neural crest development. Dev Genet 18:64–72.

Groves AK, Bronner-Fraser M (2000) Competence, specification and commitment in otic placode induction. Development 127:3489–3499.

Gurdon JB (1987) Embryonic induction—molecular prospects. Development 99:285–306.

Gutknecht D, Fritzsch B (1990) Lithium can transform ear placodes of *Xenopus* into multiple otic vesicles connected by tubes. Naturwissenschaften 77:235–237.

Haddon C, Lewis J (1996) Early ear development in the embryo of the zebrafish, *Danio rerio*. J Comp Neurol 365:113–128.

Haddon C, Jiang YJ, Smithers L, Lewis J (1998) Delta-Notch signalling and the patterning of sensory cell differentiation in the zebrafish ear: evidence from the mind bomb mutant. Development 125:4637–4644.

Hadrys T, Braun T, Rinkwitz-Brandt S, Arnold HH, Bober E (1998) Nkx5-1 controls semicircular canal formation in the mouse inner ear. Development 125:33–39.

Harrison RG (1936) Relations of symmetry in the developing ear of *Amblystoma punctatum*. Proc Natl Acad Wash 22:238–247.

Harrison RG (1938) Further investigations of the factors concerned in teh development of the ear. Anat Rec (Suppl) 70:35.

Harrison RG (1945) Relations of symmetry in the developing embryo. Trans Conn Acad Arts Sci 36:277–330.

Heller N, Brändli AW (1999) *Xenopus Pax-2/5/8* orthologues: novel insights into *Pax* gene evolution and identification of *Pax-8* as the earliest marker for otic and pronephric cell lineages. Dev Genet 24:208–219.

Herbrand H, Guthrie S, Hadrys T, Hoffman S, Arnold HH, Rinkwitz-Brandt S, Bober E (1998) Two regulatory genes, *cNkx5-1* and *cPax2*, show different responses to local signals during otic placode and vesicle formation in the chick embryo. Development 125:645–654.

Holland LZ, Holland ND (2001) Evolution of neural crest and placodes: amphioxus as a model for the ancestral vertebrate? J Anat 199:85–98.

Holtfreter J (1933) Der Einfluss von Wirtsalter und verschiedenen Organbezirken auf die Differenzierung von angelagertem Gastrulaektoderm. Roux Arch Entw-mech 127: 619–775.

Hulander M, Wurst W, Carlsson P, Enerback S (1998) The winged helix transcription factor Fkh10 is required for normal development of the inner ear. Nat Genet 20:374–376.

Hulander M, Kiernan AE, Blomavist SR, Carlsson P, Samuelsson EJ, Johansson BR, Steel KP, Enerback S (2003) Lack of pendrin expression leads to deafness and expansion of the endolymphatic compartment in inner ears of *Foxi1* null mutant mice. Development 130:2013–2025.

Hume CR, Dodd J (1993) *Cwnt-8C*: a novel *Wnt* gene with a potential role in primitive streak formation and hindbrain organization. Development 119:1147–1160.

Ishii Y, Abu-Elmagd M, Scotting PJ (2001) *Sox3* expression defines a common primordium for the epibranchial placodes in chick. Dev Biol 236:344–353.

Jacobson AG (1963a) The determination and positioning of the nose, lens and ear. I. Interactions within the ectoderm, and between the ectoderm and underlying tissues. J Exp Zool 154:273–283.

Jacobson AG (1963b) The determination and positioning of the nose, lens and ear. III. Effects of reversing the antero–posterior axis of epidermis, neural plate and neural fold. J Exp Zool 154:293–303.

Jacobson AG (1966) Inductive processes in embryonic development. Science 152:25–34.

Jacobson AG, Sater AK (1988) Features of embryonic induction. Development 104:341–359.

Kaan HW (1926) Experiments on the development of the ear of *Amblystoma punctatum*. J Exp Zool 46:13–61.

Karabagli H, Karabagli P, Ladher RK, Schoenwolf GC (2002a) Comparison of the expression patterns of several fibroblast growth factors during chick gastrulation and neurulation. Anat Embryol (Berl) 205:365–370.

Karabagli H, Karabagli P, Ladher RK, Schoenwolf GC (2002b) Survey of fibroblast growth factor expression during chick organogenesis. Anat Rec 268:1–6.

Karis A, Pata I, van Doorninck JH, Grosveld F, de Zeeuw CI, de Caprona D, Fritzsch B (2001) Transcription factor *GATA-3* alters pathway selection of olivocochlear neurons and affects morphogenesis of the ear. J Comp Neurol 429:615–630.

Katz MJ (1983) Comparative anatomy of the tunicate tadpole *Ciona intestinalis*. Biol Bull 164:1–27.

Kiernan AE, Steel KP, Fekete DM (2002) Development of the mouse inner ear. In: Rossant J, Tam PL (eds), Mouse Development: Patterning, Morphogenesis and Organogenesis. San Diego: Academic Press, pp. 539–566.

Klein PS, Melton DA (1996) A molecular mechanism for the effect of lithium on development. Proc Natl Acad Sci USA 93:8455–8459.

Knouff RA (1935) The developmental pattern of ectodermal placodes in *Rana pipiens*. J Comp Neurol 62:17–71.

Kohan R (1944) The chordomesoderm as an inductor of the ear vesicle. C R (Dokl) Acad Sci USSR 45:39–41.

Kollmar R, Nakamura SK, Kappler JA, Hudspeth AJ (2001) Expression and phylogeny of claudins in vertebrate primordia. Proc Natl Acad Sci USA 98:10196–10201.

Koster RW, Kuhnlein RP, Wittbrodt J (2000) Ectopic *Sox3* activity elicits sensory placode formation. Mech Dev 95:175–187.

Kozlowski DJ, Murakami T, Ho RK, Weinberg ES (1997) Regional cell movement and tissue patterning in the zebrafish embryo revealed by fate mapping with caged fluorescein. Biochem Cell Biol 75:551–562.

Kozmik Z, Holland ND, Kalousova A, Paces J, Schubert M, Holland LZ (1999) Characterization of an amphioxus paired box gene, *AmphiPax2/5/8:* developmental expression patterns in optic support cells, nephridium, thyroid-like structures and pharyngeal gill slits, but not in the midbrain-hindbrain boundary region. Development 126:1295–1304.

Krauss S, Johansen T, Korzh V, Fjose A (1991) Expression of the zebrafish paired box gene *pax[zf–b]* during early neurogenesis. Development 113:1193–1206.

Kuratani SC, Eichele G (1993) Rhombomere transplantation repatterns the segmental organization of cranial nerves and reveals cell-autonomous expression of a homeodomain protein. Development 117:105–117.

Lacalli TC, Hou S (1999) A reexamination of the epithelial sensory cells of amphioxus (*Branchiostoma*). Acta Zool (Stockh) 80:125–134.

Ladher RK, Anakwe KU, Gurney AL, Schoenwolf GC, Francis-West PH (2000) Identification of synergistic signals initiating inner ear development. Science 290:1965–1967.

Le Douarin NM, Kalcheim C (1999) The Neural Crest, 2nd ed. Cambridge: Cambridge University Press.

Lee SA, Shen EL, Fiser A, Sali A, Guo S (2003) The zebrafish forkhead transcription factor *Foxi1* specifies epibranchial placode-derived sensory neurons. Development 130:2669–2679.

Leger S, Brand M (2002) *Fgf8* and *Fgf3* are required for zebrafish ear placode induction, maintenance and inner ear patterning. Mech Dev 119:91–108.

Lewis AK, Frantz GD, Carpenter DA, de Sauvage FJ, Gao WQ (1998) Distinct expression patterns of notch family receptors and ligands during development of the mammalian inner ear. Mech Dev 78:159–163.

Lewis WH (1907) On the origin and differentiation of the otic vesicle in amphibian embryos. Anat Rec 1:141–145.

Liu A, Joyner AL (2001) EN and GBX2 play essential roles downstream of FGF8 in patterning the mouse mid/hindbrain region. Development 128:181–191.

Liu D, Chu H, Maves L, Yan YL, Morcos PA, Postlethwait JH, Westerfield M (2003) *Fgf3* and *Fgf8* dependent and independent transcription factors are required for otic placode specification. Development 130:2213–2224.

Logan M, Simon HG, Tabin C (1998) Differential regulation of T-box and homeobox transcription factors suggests roles in controlling chick limb-type identity. Development 125:2825–2835.

Lombardo A, Isaacs HV, Slack JM (1998) Expression and functions of FGF-3 in *Xenopus* development. Int J Dev Biol 42:1101–1107.

Long Q, Quint E, Lin S, Ekker M (2000) The zebrafish *scyba* gene encodes a novel CXC-type chemokine with distinctive expression patterns in the vestibulo-acoustic system during embryogenesis. Mech Dev 97:183–186.

Loosli F, Koster RW, Carl M, Krone A, Wittbrodt J (1998) *Six3*, a medaka homologue of the *Drosophila* homeobox gene *sine oculis* is expressed in the anterior embryonic shield and the developing eye. Mech Dev 74:159–164.

Maden M, Gale E, Kostetskii I, Zile M (1996) Vitamin A-deficient quail embryos have half a hindbrain and other neural defects. Curr Biol 6:417–426.

Mahmood R, Kiefer P, Guthrie S, Dickson C, Mason I (1995) Multiple roles for FGF-3 during cranial neural development in the chicken. Development 121:1399–1410.

Manni L, Lane NJ, Sorrentino M, Zaniolo G, Burighel P (1999) Mechanism of neurogenesis during the embryonic development of a tunicate. J Comp Neurol 412:527–541.

Manni L, Lane NJ, Burighel P, Zaniolo G (2001) Are neural crest and placodes exclusive to vertebrates? Evol Dev 3:297–298.

Mansour SL, Thomas KR, Capecchi MR (1988) Disruption of the proto-oncogene *int-2* in mouse embryo-derived stem cells: a general strategy for targeting mutations to non-selectable genes. Nature 336:348–352.

Mansour SL, Goddard JM, Capecchi MR (1993) Mice homozygous for a targeted disruption of the proto-oncogene *int-2* have developmental defects in the tail and inner ear. Development 117:13–28.

Mansouri A, Chowdhury K, Gruss P (1998) Follicular cells of the thyroid gland require *Pax8* gene function. Nat Genet 19:87–90.

Maroon H, Walshe J, Mahmood R, Kiefer P, Dickson C, Mason I (2002) *Fgf3* and *Fgf8* are required together for formation of the otic placode and vesicle. Development 129: 2099–2108.

Maves L, Jackman W, Kimmel CB (2002) FGF3 and FGF8 mediate a rhombomere 4 signaling activity in the zebrafish hindbrain. Development 129:3825–3837.

Mayordomo R, Rodriguez-Gallardo L, Alvarez IS (1998) Morphological and quantitative studies in the otic region of the neural tube in chick embryos suggest a neuroectodermal origin for the otic placode. J Anat 193:35–48.

McKay IJ, Lewis J, Lumsden A (1996) The role of FGF-3 in early inner ear development: an analysis in normal and *kreisler* mutant mice. Dev Biol 174:370–378.

Mendonsa ES, Riley BB (1999) Genetic analysis of tissue interactions required for otic placode induction in the zebrafish. Dev Biol 206:100–112.

Miyake T, von Herbing IH, Hall BK (1997) Neural ectoderm, neural crest, and placodes: contribution of the otic placode to the ectodermal lining of the embryonic opercular cavity in Atlantic cod (Teleostei). J Morphol 231:231–252.

Mizuseki K, Kishi M, Matsui M, Nakanishi S, Sasai Y (1998) *Xenopus Zic-related-1* and *Sox-2*, two factors induced by chordin, have distinct activities in the initiation of neural induction. Development 125:579–587.

Moens CB, Cordes SP, Giorgianni MW, Barsh GS, Kimmel CB (1998) Equivalence in the genetic control of hindbrain segmentation in fish and mouse. Development 125: 381–391.

Mohammadi M, McMahon G, Sun L, Tang C, Hirth P, Yeh BK, Hubbard SR, Schlessinger J (1997) Structures of the tyrosine kinase domain of fibroblast growth factor receptor in complex with inhibitors. Science 276:955–960.

Molenaar M, Brian E, Roose J, Clevers H, Destree O (2000) Differential expression of the *Groucho*-related genes 4 and 5 during early development of *Xenopus laevis*. Mech Dev 91:311–315.

Morasso MI, Grinberg A, Robinson G, Sargent TD, Mahon KA (1999) Placental failure in mice lacking the homeobox gene *Dlx3*. Proc Natl Acad Sci USA 96:162–167.

Mori-Akiyama Y, Akiyama H, Rowitch DH, de Crombrugghe B (2003) Sox9 is required for determination of the chondrogenic cell lineage in the cranial neural crest. Proc Natl Acad Sci USA 100:9360–9365.

Munchberg SR, Ober EA, Steinbeisser H (1999) Expression of the Ets transcription factors erm and *pea3* in early zebrafish development. Mech Dev 88:233–236.

Nissen RM, Yan J, Amsterdam A, Hopkins N, Burgess SM (2003) Zebrafish foxi one modulates cellular responses to Fgf signaling required for the integrity of ear and jaw patterning. Development 130:2543–2554.

Nornes HO, Dressler GR, Knapik EW, Deutsch U, Gruss P (1990) Spatially and temporally restricted expression of *Pax2* during murine neurogenesis. Development 109: 797–809.

Ohuchi H, Kimura S, Watamoto M, Itoh N (2000) Involvement of fibroblast growth factor

(FGF)18-FGF8 signaling in specification of left-right asymmetry and brain and limb development of the chick embryo. Mech Dev 95:55–66.

O'Rahilly R, Müller F (1985) The origin of the ectodermal ring in staged human embryos of the first 5 weeks. Acta Anat 122:145–157.

Orr-Urtreger A, Avivi A, Zimmer Y, Givol D, Yarden Y, Lonai P (1990) Developmental expression of c-*kit*, a proto-oncogene encoded by the *W* locus. Development 109:911–923.

Orts-Llorca F, Jimenez-Collado J (1971) Regulation of the embryo after the extirpation of Hensen's node. Consequences on the differentiation of the otic placode. Arch Anat Histol Embryol 54:1–11.

Ozaki H, Watanabe Y, Takahashi K, Kitamura A, Tanaka A, Urase K, Momoi T, Sudo K, Sakagami J, Asano M, Iwakura Y, Kawakami K (2001) *Six4*, a putative myogenin gene regulator, is not essential for mouse embryonal development. Mol Cell Biol 21: 3343–3350.

Ozaki H, Nakamura K, Funahashi J, Ikeda K, Yamada G, Tokano H, Okamura HO, Kitamura K, Muto S, Kotaki H, Sudo K, Horai R, Iwakura Y, Kawakami K (2004) Six1 controls patterning of the mouse otic vesicle. Development 131:551–562.

Papalopulu N, Kintner C (1993) *Xenopus* Distal-less related homeobox genes are expressed in the developing forebrain and are induced by planar signals. Development 117:961–975.

Pasteels J (1939) Les effets de la centrufigation axiale de l'oeuf fécondé et insegmenté chez les amphibiens anoures. Bull Acad Belg Cl Sci 25:334–345.

Pauley S, Wright TJ, Pirvola U, Ornitz D, Beisel K, Fritzch B (2003) Expression and function of FGF10 in mammalian inner ear development. Dev Dyn 227:203–215.

Penzel R, Oschwald R, Chen Y, Tacke L, Grunz H (1997) Characterization and early embryonic expression of a neural specific transcription factor *xSOX3* in *Xenopus laevis*. Int J Dev Biol 41:667–677.

Pera E, Kessel M (1999) Expression of *DLX3* in chick embryos. Mech Dev 89:189–193.

Pfeffer PL, Gerster T, Lun K, Brand M, Busslinger M (1998) Characterization of three novel members of the zebrafish *Pax2/5/8* family: dependency of *Pax5* and *Pax8* expression on the *Pax2.1* (*noi*) function. Development 125:3063–3074.

Phillips BT, Bolding K, Riley BB (2001) Zebrafish *fgf3* and *fgf8* encode redundant functions required for otic placode induction. Dev Biol 235:351–365.

Phillips BT, Storch EM, Lekven AC, Riley BB (2004) A direct role for *Fgf* but not *Wnt* in otic placode induction. Development 131:923–931.

Platt JB (1896) Ontogenetic differentiation of the ectoderm in *Necturus*. Study II. On the development of the peripheral nervous system. Q J Micr Sci 38:485–547.

Pohl BS, Knochel S, Dillinger K, Knochel W (2002) Sequence and expression of FoxB2 (XFD-5) and FoxI1c (XFD-10) in *Xenopus* embryogenesis. Mech Dev 117: 283–287.

Quint E, Zerucha T, Ekker M (2000) Differential expression of orthologous *Dlx* genes in zebrafish and mice: implications for the evolution of the *Dlx* homeobox gene family. J Exp Zool 288:235–241.

Raible F, Brand M (2001) Tight transcriptional control of the ETS domain factors *Erm* and *Pea3* by Fgf signaling during early zebrafish development. Mech Dev 107:105–117.

Represa J, León Y, Miner C, Giraldez F (1991) The *int-2* proto-oncogene is responsible for induction of the inner ear. Nature 353:561–563.

Richardson D (1933) Some effects of heteroplastic transplantation of the ear vesicle in *Amblystoma.* J Exp Zool 63:413–445.

Riley BB, Phillips BT (2003) Ringing in the new ear: resolution of cell interactions in otic development. Dev Biol 261:289–312.

Rinkwitz-Brandt S, Justus M, Oldenettel I, Arnold HH, Bober E (1995) Distinct temporal expression of mouse *Nkx-5.1* and *Nkx-5.2* homeobox genes during brain and ear development. Mech Dev 52:371–381.

Robinson GW, Mahon KA (1994) Differential and overlapping expression domains of *Dlx-2* and *Dlx-3* suggest distinct roles for *Distal-less* homeobox genes in craniofacial development. Mech Dev 48:199–215.

Robledo RF, Rajan L, Li X, Lufkin T (2002) The *Dlx5* and *Dlx6* homeobox genes are essential for craniofacial, axial, and appendicular skeletal development. Genes Dev 16:1089–1101.

Roose J, Molenaar M, Peterson J, Hurenkamp J, Brantjes H, Moerer P, van de Wetering M, Destree O, Clevers H (1998) The *Xenopus Wnt* effector *XTcf-3* interacts with *Groucho*-related transcriptional repressors. Nature 395:608–612.

Ruvinsky I, Oates AC, Silver LM, Ho RK (2000) The evolution of paired appendages in vertebrates: T-box genes in the zebrafish. Dev Genes Evol 210:82–91.

Sahly I, Andermann P, Petit C (1999) The zebrafish *eya1* gene and its expression pattern during embryogenesis. Dev Genes Evol 209:399–410.

Saint-Germain N, Lee YH, Zhang Y, Sargent TD, Saint-Jeannet JP (2004) Specification of the otic placode depends on *Sox9* function in *Xenopus*. Development 131:1755–1763.

Sanchez-Calderon H, Martin-Partido G, Hidalgo-Sanchez M (2002) Differential expression of *Otx2, Gbx2, Pax2,* and *Fgf8* in the developing vestibular and auditory sensory organs. Brain Res Bull 57:321–323.

Schlosser G, Northcutt RG (2000) Development of neurogenic placodes in *Xenopus laevis*. J Comp Neurol 418:121–146.

Sechrist J, Scherson T, Bronner-Fraser M (1994) Rhombomere rotation reveals that multiple mechanisms contribute to the segmental pattern of hindbrain neural crest migration. Development 120:1777–1790.

Shamim H, Mason I (1998) Expression of *Gbx-2* during early development of the chick embryo. Mech Dev 76:157–159.

Shamim H, Mason I (1999) Expression of *Fgf4* during early development of the chick embryo. Mech Dev 85:189–192.

Sheng G, Stern CD (1999) *Gata2* and *Gata3*: novel markers for early embryonic polarity and for non- neural ectoderm in the chick embryo. Mech Dev 87:213–216.

Shimeld SM, Holland PW (2000) Vertebrate innovations. Proc Natl Acad Sci USA 97: 4449–4452.

Solloway MJ, Robertson EJ (1999) Early embryonic lethality in *Bmp5*; *Bmp7* double mutant mice suggests functional redundancy within the 60A subgroup. Development 126:1753–1768.

Solomon KS, Fritz A (2002) Concerted action of two *dlx* paralogs in sensory placode formation. Development 129:3127–3136.

Solomon KS, Kudoh T, Dawid IB, Fritz A (2003a) Zebrafish *foxi1* mediates otic placode formation and jaw development. Development 130:929–940.

Solomon KS, Logsdon JM, Jr., Fritz A (2003b) Expression and phylogenetic analyses of three zebrafish *FoxI* class genes. Dev Dyn 228:301–307.

Stambolic V, Ruel L, Woodgett JR (1996) Lithium inhibits glycogen synthase kinase-3 activity and mimics wingless signalling in intact cells. Curr Biol 6:1664–1668.

Stark MR, Biggs JJ, Schoenwolf GC, Rao MS (2000) Characterization of avian *frizzled* genes in cranial placode development. Mech Dev 93:195–200.
Stone LS (1931) Induction of the ear by the medulla and its relation to experiments on the lateralis system in Amphibia. Science 74:577.
Streit A (2002) Extensive cell movements accompany formation of the otic placode. Dev Biol 249:237–254.
Streit A, Stern CD (1999) Establishment and maintenance of the border of the neural plate in the chick: involvement of FGF and BMP activity. Mech Dev 82:51–66.
Streit A, Berliner AJ, Papanayotou C, Sirulnik A, Stern CD (2000) Initiation of neural induction by FGF signalling before gastrulation. Nature 406:74–78.
Swanson GJ, Howard M, Lewis J (1990) Epithelial autonomy in the development of the inner ear of a bird embryo. Dev Biol 137:243–257.
Swiatek PJ, Lindsell CE, del Amo FF, Weinmaster G, Gridley T (1994) *Notch1* is essential for postimplantation development in mice. Genes Dev 8:707–719.
Szepsenwol J (1933) recherches sur les centres organisatueyrs des vésicules auditives chez des embryons de poulets omphalocéphales obtenus expérimentalement. Arch Anat Microsc 29:5–94.
Takabatake Y, Takabatake T, Takeshima K (2000) Conserved and divergent expression of T-box genes *Tbx2-Tbx5* in *Xenopus*. Mech Dev 91:433–437.
Torres M, Giraldez F (1998) The development of the vertebrate inner ear. Mech Dev 71:5–21.
Torres M, Gómez-Pardo E, Gruss P (1996) *Pax2* contributes to inner ear patterning and optic nerve trajectory. Development 122:3381–3391.
van Oostrom CG, Verwoerd CDA (1972) The origin of the olfactory placode. Acta Morphol Neerl Scand 9:160.
Vendrell V, Carnicero E, Giraldez F, Alonso MT, Schimmang T (2000) Induction of inner ear fate by FGF3. Development 127:2011–2019.
Verwoerd CDA, van Oostrom CG (1979) Cephalic neural crest and placodes. Adv Anat Embryol Cell Biol 58:1–75.
Vogel KS, Davies AM (1993) Heterotopic transplantation of presumptive placodal ectoderm changes the fate of sensory neuron precursors. Development 119:263–276.
von Bubnoff A, Schmidt JE, Kimelman D (1996) The *Xenopus laevis* homeobox gene *Xgbx-2* is an early marker of anteroposterior patterning in the ectoderm. Mech Dev 54:149–160.
Wada H, Saiga H, Satoh N, Holland PWH (1998) Tripartite organization of the ancestral chordate brain and the antiquity of placodes: insights from ascidian *Pax-2/5/8, Hox* and *Otx* genes. Development 125:1113–1122.
Waddington CH (1937) The determination of the auditory placode in the chick. J Exp Biol 14:232–239.
Walshe J, Maroon H, McGonnell IM, Dickson C, Mason I (2002) Establishment of hindbrain segmental identity requires signaling by FGF3 and FGF8. Curr Biol 12: 1117–1123.
Wang W, Van De Water T, Lufkin T (1998) Inner ear and maternal reproductive defects in mice lacking the *Hmx3* homeobox gene. Development 125:621–634.
Waskiewicz AJ, Rikhof HA, Moens CB (2002) Eliminating zebrafish pbx proteins reveals a hindbrain ground state. Dev Cell 3:723–733.
Waterman AJ, Evans HJ (1940) Morphogenesis of the avian ear rudiment in chorioallantoic grafts. J Exp Zool 84:53–71.
Weinstein M, Xu X, Ohyama K, Deng CX (1998) FGFR-3 and FGFR-4 function cooperatively to direct alveogenesis in the murine lung. Development 125:3615–3623.

Whitfield TT, Riley BB, Chiang MY, Phillips B (2002) Development of the zebrafish inner ear. Dev Dyn 223:427–458.

Woo K, Fraser SE (1997) Specification of the zebrafish nervous system by nonaxial signals. Science 277:254–257.

Woo K, Fraser SE (1998) Specification of the hindbrain fate in the zebrafish. Dev Biol 197:283–296.

Wood HB, Episkopou V (1999) Comparative expression of the mouse *Sox1, Sox2* and *Sox3* genes from pre-gastrulation to early somite stages. Mech Dev 86:197–201.

Wright E, Hargrave MR, Christiansen J, Cooper L, Kun J, Evans J, Gangadharan U, Greenfield A, Koopman P (1995) The *Sry*-related gene *Sox9* is expressed during chondrogenesis in mouse embryos. Nat Genet 9:15–20.

Wright TJ, Mansour SL (2003) *Fgf3* and *Fgf10* are required for mouse otic placode induction. Development 130:3379–3390.

Wright TJ, Hatch EP, Karabagli H, Karabagli P, Schoenwolf GC, Mansour SL (2003) Expression of mouse fibroblast growth factor and fibroblast growth factor receptor genes during early inner ear development. Dev Dyn 228:267–272.

Xie MH, Holcomb I, Deuel B, Dowd P, Huang A, Vagts A, Foster J, Liang J, Brush J, Gu Q, Hillan K, Goddard A, Gurney AL (1999) FGF-19, a novel fibroblast growth factor with unique specificity for FGFR4. Cytokine 11:729–735.

Xu PX, Cheng J, Epstein JA, Maas RL (1997) Mouse *Eya* genes are expressed during limb tendon development and encode a transcriptional activation function. Proc Natl Acad Sci USA 94:11974–11979.

Xu PX, Adams J, Peters H, Brown MC, Heaney S, Maas R (1999) *Eya1*-deficient mice lack ears and kidneys and show abnormal apoptosis of organ primordia. Nat Genet 23:113–117.

Yntema CL (1933) Experiments on the determination of the ear ectoderm in the embryo of *Amblystoma punctatum*. J Exp Zool 65:317–357.

Yntema CL (1939) Self-differentiation of heterotopic ear ectoderm in the embryo of *Amblystoma punctatum*. J Exp Zool 80:1–17.

Yntema CL (1950) An analysis of induction of the ear from foreign ectoderm in the salamander embryo. J Exp Zool 113:211–244.

Zhang N, Martin GV, Kelley MW, Gridley T (2000) A mutation in the *Lunatic fringe* gene suppresses the effects of a *Jagged2* mutation on inner hair cell development in the cochlea. Curr Biol 10:659–662.

Zheng W, Huang L, Wei ZB, Silvius D, Tang B, Xu PX (2003) The role of *Six1* in mammalian auditory system development. Development 130:3989–4000.

Zwilling E (1940) An experimental analysis of the development of the anuran olfactory organ. J Exp Zool 84:291–324.

Zwilling E (1941) The determination of the otic vesicle in *Rana pipiens*. J Exp Zool 86:333–342.

3

Morphogenesis of the Inner Ear

Suzanne L. Mansour and Gary C. Schoenwolf

1. Introduction

The early development of the inner ear—the process of otogenesis—occurs in three phases. The first phase, formation of the otic placode, which is the earliest rudiment of the inner ear, occurs as a result of inductive interactions with surrounding tissues (see Chapter 2 by Groves for a discussion of this process). From a morphogenetic standpoint, several interesting events underlie formation of the placode, and these are discussed briefly below. The second phase of early development of the inner ear consists of the morphogenesis of the otic placode to form the otocyst, a spherical vesicle that gives rise to both the auditory and vestibular components of the inner ear. Transformation of the otic placode into the otocyst is the focus of the first part of this chapter. Based on our knowledge of a similar morphogenetic event—neurulation—a model for this process is proposed. This part of the chapter also focuses on the chick embryo, because most of the studies providing insight into mechanisms of morphogenesis of the otic epithelium were done in chick embryos, from which the otic epithelium can be readily manipulated both in ovo and in vitro. The final phase of early development of the inner ear involves regional patterning of the otocyst, which consists of formation of the anteroposterior, dorsoventral, and mediolateral axes and localized morphogenesis along these axes, resulting in the complex three-dimensional morphology underlying the specialized functions of the mature membranous labyrinth. Regional morphogenesis is the focus of the second part of this chapter. Several excellent reviews consider one or more of these phases of otogenesis (Noden and Van de Water 1992; Fekete 1996, 1999; Whitfield et al. 1997, 2002; Fritzsch et al. 1998; Torres and Giráldez 1998; Brigande et al. 2000b; Chang et al. 2002; Kiernan et al. 2002).

2. From Placode to Vesicle—Origin of the Otic Placode

The origin of the otic placode and its transformation into the otocyst occur similarly in amphibians, birds, and mammals (shown for the chick, *Gallus*

gallus, in Fig. 3.1). These processes resemble the major events of a phase of neurulation called primary neurulation. They involve the formation of the otic placode (Fig. 3.1A–C); changes in its overall shape (i.e., shaping of the placode); bending of the placode to form an otic pit or cup, which is encircled by a rimlike lip (Fig. 3.1D–G); and apposition and fusion of the rim cells (Fig. 3.1H), closing the pit to establish the otocyst (Fig. 3.1I). Formation of the otocyst in fish involves comparable events, with the notable exception that the otic placode becomes a multilayered mass of cells that subsequently cavitates (rather than invaginates) to form the otocyst (Haddon and Lewis 1996; Bever and Fekete 2002; Whitfield et al. 2002). In this respect, formation of the fish otocyst resembles a second phase of neurulation called secondary neurulation, a process that occurs in birds and mammals to establish the caudal portion of the neural tube. During secondary neurulation, a solid cord of cells derived from the tail bud undergoes cavitation to form a hollow tube.

Fate mapping studies in avian embryos have revealed the approximate positions of the prospective otic placode at only essentially two stages of early development: stage 3d/4 (Garcia-Martinez et al. 1993) and stage 8−/9− (D'Amico-Martel and Noden 1983; Couly and Le Douarin 1990). At the earlier stage, during gastrulation, a linear primitive streak has formed, marking both the future anteroposterior and mediolateral axes of the early embryo. The dorsoventral axis of the embryo at this stage is identified by the outer position of the ectoderm (marking dorsal) and the inner position of the endoderm (marking ventral). Although the overall axes of the embryo have been established, the axes of its constituent rudiments, including those of the prospective otic placode, remain plastic. At the later stage, during neurulation, the prospective otic placodes flank the elevating and converging neural folds at the future hindbrain region. It is at this time that the anteroposterior axis of the otic rudiment begins to become fixed. Other axes of the otic rudiment become fixed at progressively later stages of early organogenesis (see Section 3).

2.1 An Overview of the Process of Neurulation: A Model for Understanding the Morphogenesis of the Otocyst

To provide a better understanding of possible events underlying the formation and morphogenesis of the otic epithelium or rudiment (i.e., the placode, pit or cup, and vesicle), the chapter first provides an overview of the events of neurulation, a morphogenetic process that has been studied far more extensively than has otogenesis. The underlying thesis is that neurulation can serve as an excellent model system for understanding the early phases of otogenesis, providing insight into the morphogenetic mechanisms forming the rudiment of the inner ear. The major events of neurulation and how similar events might occur in otogenesis are then discussed in more detail. Because the focus of this chapter is on development of the inner ear and not neurulation, rather than providing references to the primary literature on neurulation only a few relevant reviews are cited (Schoenwolf and Smith 1990; Smith and Schoenwolf 1997; Colas and Schoenwolf 2001).

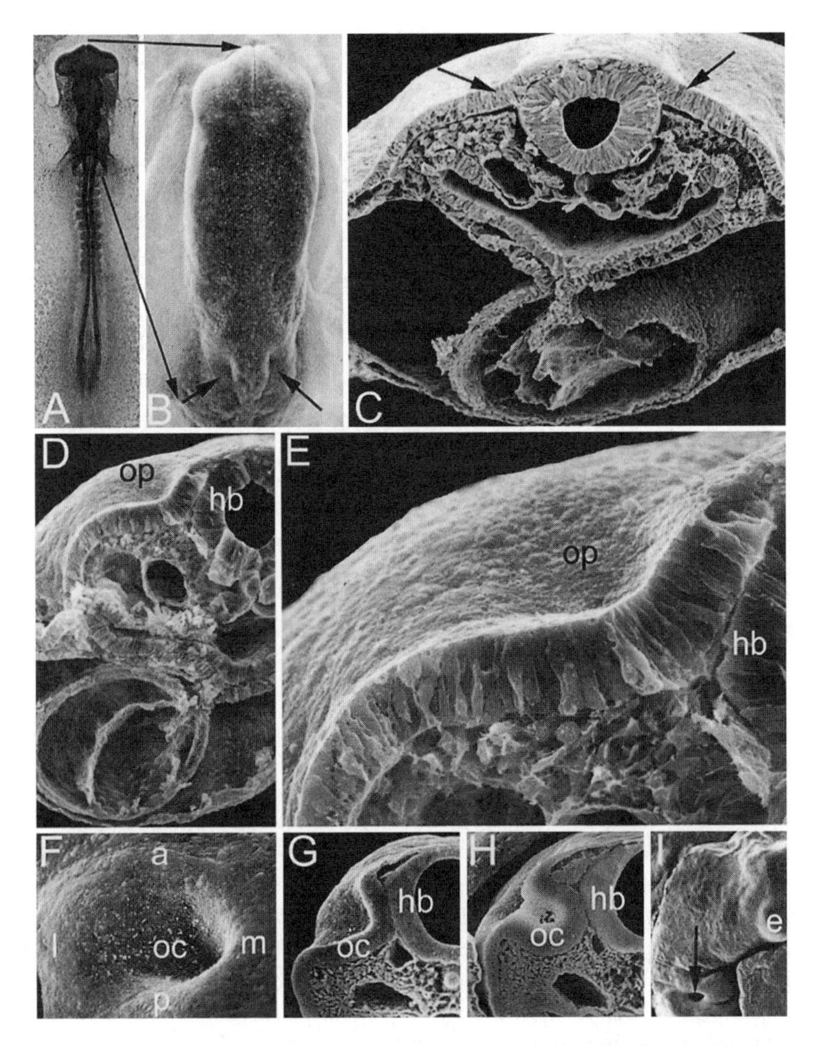

FIGURE 3.1. Micrographs (**A**, light; **B–I**, scanning electron) showing major events in formation of the otocyst in the chick (*Gallus gallus*) embryo. (**A**) Whole mount at the time that the otic placodes are first forming. (**B**) Enlargement of the head showing the otic placodes (*arrows*). (**C**) Cross section through the level of the otic placodes (*arrows*). (**D**) Cross section of the left side showing early invagination of the placode. (**E**) Enlargement of (**D**) showing the invaginating otic placode and the attachment of its dorsomedial side to the wall of the dorsal hindbrain. (**F**) View of the otic cup from its dorsal side. (**G**) Cross section through the fully invaginated otic cup on the left side of the embryo. Note the attachment of the dorsomedial half of the otic epithelium to the hindbrain. (**H**) Cross section through the closing otic cup on the left side of the hindbrain. A portion of the otic epithelium remains attached to the wall of the hindbrain. (**I**) View of the right side of the head of an embryo at a stage when the otic cup (*arrow*) is closing. a, anterior; e, eye; hb, hindbrain; l, lateral; m, medial; oc, otic cup; op, otic placode; p, posterior.

Neurulation results in the formation of the neural tube, the rudiment of the entire adult central nervous system (CNS). In birds and mammals this process occurs in two phases, termed primary and secondary neurulation. Primary neurulation results in formation of the entire brain and most of the length of the spinal cord. It occurs in four characteristic, but temporally and spatially overlapping, stages: (1) formation, (2) shaping, and (3) bending of the neural plate, with formation of the neural groove, flanked by the bilateral neural folds, and (4) closure of the neural groove through the apposition and fusion of the neural folds. Secondary neurulation involves a very different sequence of events. Near the end of gastrulation, persisting cells of the primitive streak cluster together to form the mesenchymal tail bud. Secondary neurulation begins as the most dorsal cells of the tail bud undergo a mesenchymal to epithelial transformation to form a primitive neuroepithelial structure (e.g., the medullary cord of birds), continuous cranially with the caudal end of the primary neural tube. This is followed by the cavitation of the cord to form (depending on the organism) either a single lumen or multiple lumina, which rapidly coalesce, eventually establishing a hollow secondary neural tube, identical morphologically to the primary neural tube. Thus, secondary neurulation gives rise to the caudal end of the spinal cord, beginning in birds at about the lumbosacral level.

2.1.1 Formation of the Neural Plate—Parallels with Formation of the Otic Placode

Formation of the neural plate involves cell pallisading, in which cells become apicobasically elongated and organized into a single-layered (in higher vertebrates), pseudostratified, columnar epithelium. Cells of the placode are mitotically active and their nuclei undergo interkinetic nuclear migration as the cell cycle is traversed, with the M phase of the cell cycle being restricted at the apical (future luminal) side of the epithelium, and the S phase occurring toward its basal side. A basal lamina forms at the basal side of the neural plate, separating this epithelial layer from underlying mesenchymal cells. Because of interkinetic nuclear migration, the neural plate has a multilayered appearance, with multiple (three or four) layers of cell nuclei stacked along its apicobasal extent. However, each interphase cell actually spans the entire thickness of the neural plate, and depending on the exact position of its nucleus, each cell can have one of three general shapes. Spindle-shaped cells are the most frequent ones in the neural plate; each spindle-shaped cell has both an elongated apical and basal cell process, with a centrally located nucleus. Cells with a basally located nucleus (i.e., cells lacking an elongated basal process) are said to be wedge shaped. A long process extends from the base of these cells to connect to the apex of the neural plate. Inverted wedge-shaped cells have an apically located nucleus and an elongated process that extends basally toward the basal lamina. Finally, cells in the M phase of the cell cycle have rounded up toward the apex of the epithelium; such cells lack both apically and basally elongated processes.

All cells of the neural plate are joined along their lateral surfaces to neighboring cells by occasional, small intercellular junctions, consisting principally of gap junctions. Apically, gap junctions are much more extensive, both in size and number. In addition, cells of the neural plate express cell adhesion molecules along their surfaces, including both calcium-dependent (e.g., N-cadherin) and calcium-independent (e.g., neural cell adhesion molecule [NCAM]) varieties.

Cells of the nascent neural plate are characterized by the presence of circumferential apical arrays of microfilaments and numerous apicobasally elongated microtubules (so-called paraxial microtubules). The chief morphogenetic event underlying formation of the neural plate is the apicobasal thickening of the ectoderm, which in turn is mediated solely by the apicobasal elongation of its cells (i.e., thickening of the epithelium occurs via cell elongation rather than by cell stratification). The process of cell elongation can occur in the absence of microfilaments, but it is blocked in the absence of microtubules. Other factors likely involved in formation of the neural plate include increases in cell packing, owing to sustained mitosis within the confines of the neural plate, and increased cell-to-cell adhesion.

Morphologically, formation of the otic placode seems to occur in a manner highly similar to that of the neural plate (Fig. 3.1C–E). In birds and mammals, the nascent otic placode consists of a single layer of apicobasally elongated ectodermal cells, arranged like the neural plate, into a pseudostratified, columnar epithelium. Like cells of the neural plate, cells of the otic placode undergo interkinetic nuclear migration, with cells undergoing mitosis adjacent to its apical surface (Alvarez et al. 1989). Based on morphometric analysis of the chick otic placode, a model of placode formation was proposed (Meier 1978a). Such analysis showed that while the height of the placode (i.e., the apicobasal extent of placodal cells) increased during placode formation and invagination, its surface area remained constant. Moreover, the population doubling time of placodal cells (8.5 hr) and their mitotic index (2.5%) remained similar to that of adjacent nonplacodal ectodermal cells, revealing that the formation of the otic placode does not involve localized accelerated cell division in the future placode. This led to the still untested model proposing that placodal cells are restricted from spreading laterally as they undergo division, thereby resulting in their accumulation within the limits of the placode and, consequently, to pseudostratification of the placodal epithelium (Meier 1978a).

Electron microscopic observations of forming placodal cells revealed that, as in the neural plate, apical bands of microfilaments and large apical intercellular junctions characterize formation of the otic placode (Meier 1978b). Furthermore, paraxial microtubules and a well-defined basal lamina are also present. The basal lamina expresses both fibronectin and cytotactin (tenascin-C), and both NCAM and LCAM (E-cadherin) are expressed throughout the forming placode (Richardson et al. 1987). Unfortunately, the potential roles of none of these factors in formation of the otic placode have been determined experimentally. Each of the four genes has been mutated in mice, but the resulting animals have either no or only subtle abnormalities (tenascin-C and NCAM [Cremer et al. 1994; Forsberg et al. 1996]), or they die prior to initiation of ear development (E-cadherin and fibronectin [Riethmacher et al. 1995; Romberger 1997]). Based on an understanding of the role of paraxial microtubules in apicobasal cell elongation in the neural plate, it is predicted that functionally intact paraxial microtubules are also necessary for formation of the otic placode.

2.1.2 Shaping of the Neural Plate—Parallels with Shaping of the Otic Placode

As the neural tube is forming, the second stage of neurulation begins: shaping of the neural plate. The early neural plate has a spade-shield-like shape when viewed from its apical (dorsal) surface. During shaping, the neural plate narrows mediolaterally and extends rostrocaudally. Such a coordinated and orientated change in the overall shape of an organ rudiment is referred to as a convergent extension movement. In the amphibian neural plate, convergent extension is driven chiefly by mediolateral cell to cell intercalation. In birds and mammals, oriented cell division also plays a role in convergent extension.

A process similar to shaping of the neural plate has been described in the avian otic placode (Fig. 1 in Alvarez and Navascues 1990). In contrast to the directionality of the convergent extension movement occurring during shaping of the neural plate, shaping of the otic rudiment involves a shortening of its anteroposterior axis and a concomitant lengthening of its mediolateral axis. The mechanisms underlying shaping of the otic placode have not been analyzed in detail. However, based on the presence of long apical cytokinetic bridges spanning several cell diameters (Meier 1978a; Alvarez and Navascues 1990), it seems likely that daughter cells undergo cell to cell intercalation within the plane of the epithelium, leading to cell rearrangements. Similar structures (often called beaded threads) are found in chick neural plate, where cell tracking studies have demonstrated that extensive cell rearrangements occur (Hilfer et al. 1990). Whether or not oriented cell divisions play roles in shaping of the otic placode, as they do in convergent extension of the chick and mouse neural plate, has not been analyzed. As an entrée into understanding the role of convergent extension in early otic morphogenesis, it might be profitable to examine those genes involved in convergent extension of the neural plate (Keller 2002; Wallingford et al. 2002; Copp et al. 2003), one of which (*Ltap/Vangl2*) has already been shown to be required for the elongation of the cochlear duct (Montcouquiol et al. 2003).

2.1.3 Bending of the Neural Plate—Parallels with Invagination of the Otic Placode and Formation of the Otic Pit

Bending of the neural plate is a process that is far more complicated than originally believed. As a result of improvements in electron microscopy in the late 1960s and early 1970s, bands of microfilaments were identified in the apices of a number of different types of embryonic epithelial sheets undergoing bending movements. This led to the hypothesis that bending of epithelial sheets was generated solely by intrinsic forces (i.e., forces originating solely within the epithelial sheet) through microfilament-mediated cell wedging. Thus it was believed that bending of the neural plate resulted from a change in cell shape from column-like (in the flat neural plate) to wedge-like (in the bending neural plate). The identification of circumferential apical bands of microfilaments led to the pursestring hypothesis, in which apical constriction caused cell wedging, thereby driving bending. The use of the drug cytochalasin B, which disrupts microfilaments, supported this hypothesis, as following treatment bending of epithelial sheets was severely disrupted.

Further studies revealed that the pursestring hypothesis was far too simplistic. First, as described above, cells of ectodermal rudiments such as the early neural plate and otic placode are not strictly column shaped; rather, they have a variety of shapes. Second, as bending of such ectodermal rudiments occurs, a wholesale change in cell shape to wedge-like does not occur across the bending rudiment; rather, at least in the chick neural plate where this has been analyzed in detail, roughly the same proportion of spindle-shaped, wedge-shaped, inverted wedge-shaped, and mitotic (i.e., spherical cells rounded up at the apex of the epithelium) are retained. Moreover, in both the chick neural plate and forming otic pit, loci of bending are present, called hinge points. In the neural plate, three hinge points are formed: a median hinge point, overlying the notochord; and paired dorsolateral hinge points, just proximal to the neural folds (the latter hinge points form mainly at future brain levels and are absent throughout most of the length of the spinal cord, except at the most caudal end of the neural groove). Most of the cells within these hinge points become wedge-shaped as the hinge points are forming, but dramatic reductions in their apical size do not occur. Instead, changes in cell shape seem to be far more affected by the position of the nucleus within the cells—a process linked to its phase in the cell cycle owing to interkinetic nuclear migration—rather than to apical constriction per se. This has led to the revised hypothesis that furrowing of epithelial sheets, that is, localized bending within hinge points, is generated by cell wedging, and that this cell wedging occurs as a result of two main events: microfilament-mediated apical constriction/stabilization and basal expansion, owing to an increase in the percentage of the cell cycle (as well as the absolute length of time) that the nucleus is retained at the base of the cell.

Additional studies on chick neurulation have further challenged the pursestring hypothesis. Microsurgical isolation of the neural plate from surrounding tissues has revealed that intrinsic forces generated by changes in cell shapes within the neural plate, although sufficient for furrowing of the neural plate are insufficient for its subsequent folding, that is, elevation of the neural folds—the rotation of the neural plate around the median hinge point, and convergence of the neural folds—the rotation of the neural folds around the dorsolateral hinge points. Both elevation and convergence of the neural folds require extrinsic forces, that is, forces generated by changes in the surrounding tissue. Further experiments have revealed that the epidermal ectoderm immediately lateral to the neural folds undergoes lateral-to-medial expansion during elevation and convergence, and that such expansion is required for folding.

Does invagination of the otic placode also involve such a complicated series of events? The existing evidence is quite convincing that it does indeed. Based on experimental studies of the chick otic epithelium, mainly by Hilfer and colleagues (Meier 1978a,b; Sinning and Olson 1988; Hilfer et al. 1989, 1990; Alvarez and Navascues 1990; Hilfer and Randolph 1993; Gerchman et al. 1995; Brown et al. 1998; Brigande et al. 2000a; Moro-Balbas et al. 2000; Visconti and Hilfer 2002), a model has been developed in which both intrinsic and extrinsic factors drive invagination of the otic placode. This model is consistent with the morphological changes observed in the otic rudiment during its bending (Fig. 3.1E). Bending of the otic placode begins with the formation of a longitudinal hinge point (Meier 1978a; Richardson et al. 1987; Hilfer et al. 1989; Alvarez and Navascues 1990; Hilfer and Randolph 1993; Moro-Balbas et al. 2000). Placodal cells within the hinge point display basally located nuclei (suggesting that they are wedge shaped), and the basal lamina underlying the hinge point is attached to the adjacent hindbrain through fibrous connections of extracellular matrix. Similarly, the basal lamina underlying the entire dorsomedial half of the otic placode becomes attached via extracellular matrix to the more dorsal hindbrain at the onset of invagination, but cells within this portion of the placode seem to be principally spindle shaped (i.e., centrally located nuclei) rather than wedge shaped (i.e., basally located nuclei). Although the differences in cell shapes likely play a role in determining whether the placodal epithelium is flattened (as it is dorsomedial and ventrolateral to the longitudinal hinge point) or furrowed (as it is within the longitudinal hinge point), intrinsic forces generated by cytoskeletal mediated cell wedging are not required for invagination of the otic placode, as experimental studies have shown that invagination is both calcium and ATP independent (Hilfer et al. 1989). Moreover, experiments in which the otic placode was detached from the adjacent hindbrain, using antibodies to laminin and integrins, showed that subsequent invagination was inhibited, providing further evidence that extrinsic forces are necessary for invagination (Visconti and Hilfer 2002). Furthermore, perturbation of the extracellular matrix surrounding the invaginating otic placode and/or the matrix participating in the formation of the otic folds at the rim of the invaginating cup by using either enzymes degrading hyaluronate or chondroitin sulfate, or β-xyloside to inhibit the synthesis of chondroitin sulfate proteoglycan, also inhibits invagination, suggesting that the extracellular matrix plays a critical role in invagination (Gerchman et al. 1995). It may be possible to address the role of chondroitin-sulfated proteoglycans genetically, as a mouse gene-trap mutation in *Chondroitin-4-sulfotransferase* (*Chst11/C4ST*), which is expressed by the otic vesicle, has been isolated recently (Klüppel et al. 2002). A similar role for hyaluronate has been described in neurulation at rostral levels of the chick neural tube (Schoenwolf and Fisher 1983). Finally, microinjection of heparinase III, degrading heparan sulfate proteoglycan present in the basal lamina of the otic placode (Gould et al. 1995; Moro-Balbas et al. 2000), blocks invagination (Moro-Balbas et al. 2000). Thus, several factors seem to interact to transform the flat otic placode into the closed otocyst.

As invagination of the otic placode occurs, a copious surface coat appears on its apical side (Sinning and Olson 1988). The role of this coat in invagination and subsequent closure of the otocyst is unknown, as is its molecular composition. In addition, during invagination the otic epithelium expresses NCAM. Disruption of NCAM with blocking antibodies reveals two distinct effects: separation of the otic epithelium from the adjacent hindbrain and failure of formation of the otic folds (Brown et al. 1998). Finally, as invagination is underway, additional folds appear in the otic epithelium, converting the otic cup into a boxlike shape. Especially prominent are rostral and caudal folds (Hilfer et al. 1989; Alvarez and Navascues 1990). With the formation of these folds the closing "otic pore" transforms from circular to oval, being narrow in the anteroposterior axis and broad in the mediolateral axis. The dynamics of otic sup invagination and closure (see next section) can be viewed at http://sdb.bio.purdue.edu/temp/otic_cup_closure.html (Brigande et al. 2000a)

2.1.4 Closure of the Neural Groove and Formation of the Neural Tube—Parallels with Closure of the Otic Pit and Formation of the Otocyst

During bending of the neural plate the neural folds are eventually brought into contact in the dorsal midline, where they adhere to one another and undergo fusion. Almost nothing is known about these processes, which result in the formation of a hollow tube, separated from the overlying epidermal ectoderm and a relatively small but highly active population of cells, the neural crest. Studies utilizing electron microscopy combined with fixation protocols that preserve cell-surface coats have revealed that the apical surfaces of the neural folds are coated with cell surface materials, but their molecular composition remains unknown. In addition, in some species, the apical surfaces of the ectodermal cells comprising the neural folds display numerous membrane blebs, ruffles, and filopodia, but their significance in fusion and cell rearrangements during formation of distinct tissue layers (i.e., epidermal ectoderm, roof plate of the neural tube, and neural crest) are unknown. Finally, the layer of the neural folds that come into first contact to close the neural groove vary at different rostrocaudal levels. Thus, in the area of the anterior neuropore, the epidermal ectodermal layer of each neural folds come into first contact, whereas at other levels of the developing neural axis, first contact occurs either between the neuroectodermal layers or at the interface between the two layers. Again, the significance of these differences is not obvious.

Similarly, little is known about closure of the otic cup to form the otocyst (Fig. 3.1H, I). The surfaces of the otic folds, like that of the entire apical surface of the otic epithelium, are covered by a dense cell coat (Sinning and Olson 1988), but like their potential role in invagination, their role in closure remains unknown. Analysis of cell movements during closure of the otic cup has revealed that cells at the rim of the cup change positions during closure. These studies have revealed that the entire dorsal rim of the otic cup becomes the endolymphatic duct, whereas the posteroventral rim becomes the lateral otocyst wall. Compartment boundaries are established through lineage restriction at the dorsal pole of the closing cup, and it has been hypothesized that signaling across compartment boundaries may play a critical role in specification of the endolymphatic duct during closure (Brigande et al. 2000a).

2.1.5 Cavitation of the Medullary Cord—Parallels with Formation of the Fish Otocyst

Above, it was argued that formation of the neural tube during primary neurulation can serve as a model system, providing insight into the tissue, cellular, and molecular mechanisms underlying formation and early morphogenesis of the otocyst in amphibians, birds, and mammalians. Here, this argument is extended by suggesting that secondary neurulation might serve as a model system for gaining insight into mechanisms underlying formation of the otocyst in fish.

Secondary neurulation occurs in both birds and mammals. The process is described in chick, where it begins with the formation of the tail bud, a mesenchymal mass of cells formed at the caudal end of the embryo from persisting remnants of the primitive streak. Next, the most dorsal cells of the tail bud condense into an epithelial cord called the medullary cord. A basal lamina forms around the medullary cord, separating it from the surrounding mesenchymal cells. The cord then undergoes cavitation to form multiple lumina, which ultimately coalesce into a single central lumen.

Formation of the zebrafish otocyst from the otic placode seems to occur in a similar manner, with the otocyst forming not by invagination but rather through cavitation of a solid mass of cells (Haddon and Lewis 1996). During formation of the otocyst, the otic placode sinks below the epidermal ectoderm and cavitates, with cells rapidly showing apicobasal polarization as revealed by the presence of apically localized actin-containing microfilaments (Whitfield et al. 2002). Similarly, the entire length of the zebrafish neural tube forms by cavitation of a solid epithelial rod, the neural keel. Future identification of zebrafish mutants that affect cavitation of the neural keel and subsequent identification and characterization of the relevant mutant genes should provide insights into potential molecular players controlling morphogenesis of the zebrafish otocyst.

2.2 *Neural Tube Defects—Parallels with Otocyst Closure Defects?*

As might be expected with the complexity described above, formation of the primary neural tube often goes awry, resulting in neural tube defects—a failure of the neural folds to come into contact and fuse to establish a closed CNS covered by epidermis. In fact, neural tube defects are among the most common birth defects occurring in humans (i.e., approximately 1:1000 live births). Owing to a similar complexity in the formation of the otocyst as suggested here, it might be expected that otocyst closure defects would also frequently occur, yet only one such defect has been recognized in the literature. Mice lacking expression of both cell death genes *Apaf1* and *Bcl2l* fail to complete otic vesicle closure and lack endolymphatic ducts, which ordinarily form at the site of vesicle closure. This defect has not yet been analyzed in embryos at the time of vesicle closure, but the result suggests an important role for cell death in this process (Cecconi et al. 2004).

Two possible explanations for the relative dearth of otic vesicle closure defects are proposed. First, based on patterns of gene expression during early embryogenesis, it is clear that virtually all genes expressed in the otic rudiment during its early morphogenesis are also expressed in other major organ rudiments, such as the neural tube, heart, and so forth. Thus, it might be expected that critical genes controlling formation and morphogenesis of the otocyst are essential for survival of the embryo and when mutated might be embryonic lethal. Second, also based on patterns of gene expression, it is clear that many genes (and many members of the same families of genes) are expressed in the early otocyst. This suggests that formation and early morphogenesis of the otocyst are likely controlled by redundant mechanisms, or that compensatory changes might occur in the absence of a single gene regulating these processes, masking potential otic phenotypes. Examples of these scenarios were given in the section above on the roles of cell-adhesion molecules in otic placode formation and for the vesicle closure defect seen in *Apaf1/Bcl2l* double mutants, Nevertheless, because otic defects are not always readily recognized, or even considered, it might be profitable to search for otic closure phenotypes in animals that carry mutations in genes that affect neural tube closure and that are also expressed in tissues relevant to otic development.

3. Otic Axis Formation

As noted in the Introduction, the ear is an asymmetric structure, with distinct morphological characteristics defining each of its three axes. When and how are the developmental axes of the ear determined? The timing of otic anteroposterior (AP) versus dorsoventral (DV) axis formation in salamanders was addressed in two classic experiments by Harrison (1936). In both cases he cut out squares of donor otic ectoderm at different stages, transplanted them into the otic region of same-stage embryos, and observed the subsequent development of the transplanted otic tissue. In the first type of experiment, by grafting "upright" (nonrotated) ectoderm from one side of the donor to the opposite side of the host embryo, only the AP orientation of the transplanted tissue was reversed. In the second type of experiment, by grafting "inverted" (180° rotated) ectoderm from one side of the donor to the opposite side of the host, only the DV orientation of the transplanted tissue was reversed. As a control, "upright" tissue from one side of the donor was transplanted to the same side of the host, which maintained all axial relationships between donor and host.

Harrison found that when either type of experimental transplantation was carried out at the medullary plate stage, when the embryo had just begun to elevate the neural folds and did not have any morphological signs of a placode, the transplanted tissue developed into a normally oriented ear. This suggests that the preplacodal ectoderm has not acquired any irreversible positional information and is still subject to axial cues from the host embryo. If the transplantations were performed a little later, when the neural folds were closing and the otic ectoderm had thickened, the results depended on the type of transplantation. Otic tissue grafted such that the AP orientation was reversed (experiments of the first type) developed with the donor's original AP asymmetry (i.e., "backwards" relative to the host). In contrast, when the DV orientation was inverted at this stage (experiments of the second type), the grafted ears acquired a DV axis that was aligned with that of the host and developed relatively normally. DV inversions done at later stages, when the otocyst was almost closed, never gave rise to ears with the asymmetries expected of the host embryo (Harrison 1936). These results suggest that the AP axis of the amphibian ear is established by the placode stage, and that this occurs before establishment of the DV axis.

More recently, Wu et al. (1998) carried out similar experiments in the chick. They came to the same conclusions as Harrison with respect to the order of axis formation, namely, that the AP axis is fixed before the DV axis. They also showed, however, that the time course of otic axis formation in avians may differ from that in amphibians. It seems that the AP axis in chick may be fixed during the otic cup stage and that DV fixation may occur after endolymphatic duct formation. Axis fixation in chicks thus occurs later than in amphibians. This issue has not been addressed yet in zebrafish or mice.

Whereas the timing of AP and DV otic axis formation in two species is relatively clear, the molecular basis of otic axis formation remains relatively unexplored. If otic axes are established by differential responses to gradients of signaling molecules, as is thought to be the case for the embryonic axes (i.e., bone morphogenetic proteins [BMPs] and Sonic hedgehog for the DV axis of the neural tube [Altmann and Brivanlou 2001]), then one might expect that inversion of the appropriate gradient at the appropriate time would effect an inversion of the otic axis and that removal of the graded signal would most strongly disrupt development of the otic structures closest to the source of the signal. Indeed, Sonic hedgehog (Shh) signals, which are produced by ventral midline tissues (notochord and floorplate), appear to be required for establishment of the mouse DV otic axis. Ventral otic structures (the cochlea and otic ganglion) fail to develop in *Shh*$^{msl-}$ mutants. The endolymphatic duct (dorsomedial) and lateral semicircular canal are also missing in the *Shh* mutants, but these morphogenetic defects seem to arise subsequent to normal patterning of the mediolateral axis and initial outgrowth of these structures (Riccomagno et al. 2002). Strikingly, when the embryonic chick ventral neural tube or notochord (also sources of Shh) is surgically removed, the ear develops without ventral structures. In addition, surgical inversion of the chick neural tube relative to the otocyst abolishes expression of dorsal genes and induces expression of ventral genes in the dorsal otocyst (Bok et al. 2005). It is not yet clear whether this manipulation truly inverts the otic DV axis, but taken together, the data suggest that at least in mouse and chick, Shh signaling may be one determinant of the otic DV axis.

The role of Hedgehog (Hh) signaling in otic axis formation in zebrafish is somewhat different from that in mouse and chick. Hh signals appear to be required to establish the AP axis of zebrafish otic vesicles. Strong inhibition of the Hh pathway leads to ears in which posterior structures and marker genes are missing and anterior structures and marker genes are present as partial mirror image duplications, whereas activation of Hh signal throughout the embryo causes a loss of anterior structures and partial mirror image duplication of posterior structures (Hammond et al. 2003). This result is rather surprising considering that Hedgehog signals are present along the entire AP axis of the developing neural tube floor plate and notochord and would seem to be better candidates for influencing the development of the DV (or mediolateral) otic axis. Asymmetric posterior concentration of the Hh receptor gene, *patched1*, in the otic vesicle may explain the differential effects of Hh along the AP otic axis (Hammond et al. 2003). Alternatively, it is possible that ventral or medial otic vesicle cells move posteriorly subsequent to receiving the Hh signal; the existing zebrafish fate map was not produced late enough to assess this possibility (Kozlowski et al. 1997).

It is interesting to note that there may be two independent pathways that regulate otic DV axis determination in the mouse. Mice that lack *Six1*, which encodes a transcription factor expressed most prominently in the ventral otic cup, have an otic phenotype similar to that of *Shh* mutants. Marker gene analysis of the mutant vesicles suggests that the DV axis is perturbed. Interestingly, *Shh* does not regulate *Six1* expression and vice versa (Ozaki et al. 2004). It will be interesting to learn whether the *Six1* homologs also play roles in otic axial determination in other species.

Fgf3, which is expressed from the hindbrain in all of the experimental species under consideration, is also an excellent candidate for a DV otic axis-inducing signal (Wright and Mansour 2003), but whether it actually forms a concentration gradient is unknown, and the manipulations required to reverse the putative gradient have not yet been attempted. In addition, mouse *Fgf16,* which is expressed in the posterior otic cup, could potentially play a role in AP axis formation (Wright et al. 2003).

4. Normal Morphogenesis of the Closed Otic Vesicle to the Mature Membranous Labyrinth

The normal developmental progression of the closed otic vesicle to the mature membranous labyrinth in mice (*Mus musculus*), chicks (*Gallus gallus*), frogs (*Xenopus laevis*), and zebrafish (*Danio rerio*) is illustrated with paint-filled specimens in Figure 3.2. Figure 3.2A shows all ears at the same scale, so that the substantial growth of the epithelium during morphogenesis can be appreciated.

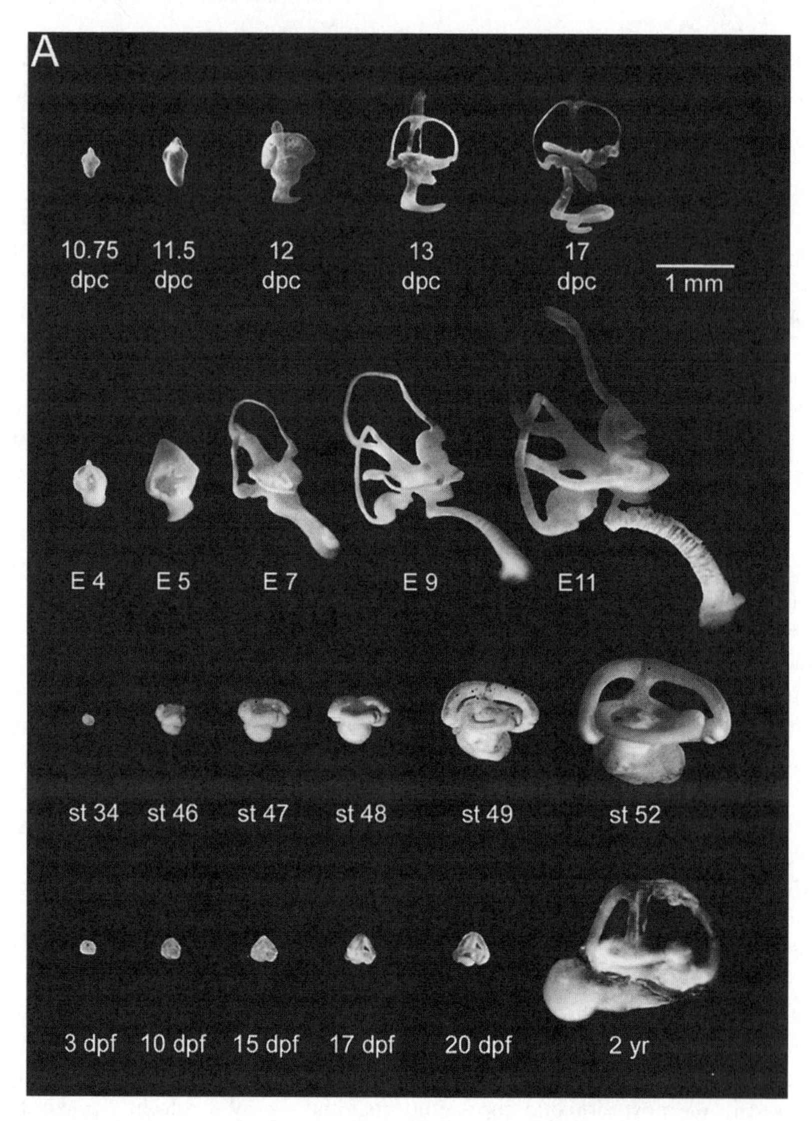

FIGURE 3.2. (**A**) Paint-fills of the developing otocyst in mouse (*Mus musculus, top row*; 10.75 to 17 days postcoitum [dpc]), chick (*Gallus gallus, second row*; 4 to 11 days of embryogenesis [E]), frog (*Xenopus laevis, third row*; stages [st] 34 to 52), and zebrafish (*Danio rerio, bottom row*; 3 days postfertilization [dpf] through 2 years [yr] of age). Otocysts are shown to scale (1 mm bar) to indicate the overall patterns of growth during development. (**B**) Paint-fills of selected developing otocysts shown in (**A**) from mouse (*top row*), chick (*second row*), frog (*third row*), and zebrafish (*bottom row*). Regardless of stage and species, all otocyts are shown at the same size to facilitate examination of their morphologies. A, anterior; aa, anterior ampulla; acp, anterior canal pouch; asc, anterior semicircular canal; bp, basilar papilla; cc, common crus; cd, cochlear duct; D, dorsal; dcd, distal cochlear duct; dpc, days post-coitus; dpf, days post-fertilization; E, embryonic day; ed, endolymphatic duct; es, endolymphatic sac; ha, horizontal ampulla; hcp, horizontal canal; pouch; hsc, horizontal semicircular canal; l, lagena; la, lateral

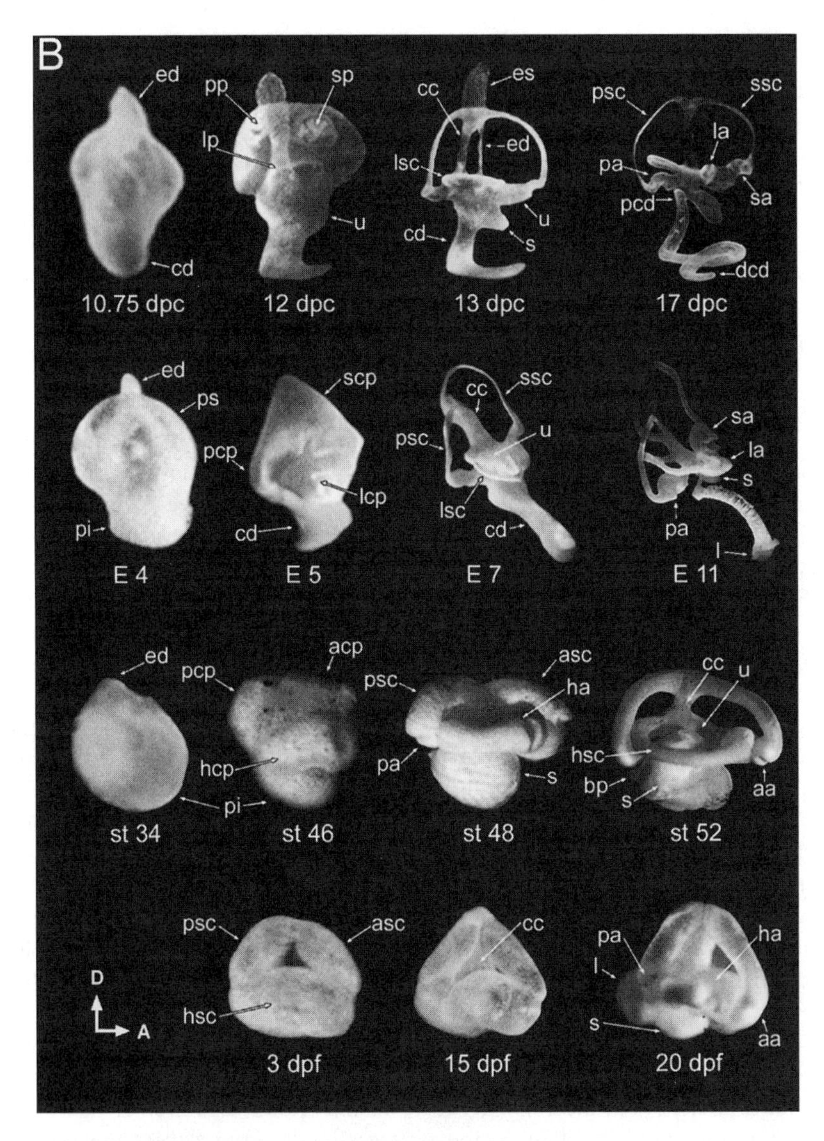

FIGURE 3.2. (*Continued*) ampulla; lcp, lateral canal pouch; lp, lateral pouch; lsc, lateral semicircular canal; pa, posterior ampulla; pcd, proximal cochlear duct; pi, pars inferior; pp, posterior pouch; pcp, posterior canal pouch; ps, pars superior; psc, posterior semicircular canal; s, saccule; sa, superior ampulla; scp, superior canal pouch; sp, superior pouch; ssc, superior semicircular canal; st, stage; u, utricle. Note that canal structures labeled as "superior" are more properly referred to as "anterior." Paint-fill figures compiled courtesy of M. Bever and D. Fekete. Mouse paint-fills were originally published in Morsli et al. (1998, reprinted with permission of the Society for Neuroscience © 1998), chick paint-fills in Bissonnette and Fekete (1996, reprinted with permission of Wiley-Liss, Inc., a subsidiary of John Wiley & Sons © 1996), frog paint-fills in Bever et al. (2003, reprinted with permission of Wiley-Liss, Inc., a subsidiary of John Wiley & Sons © 2003), and zebrafish paint-fills in Bever and Fekete (2002, reprinted with permission of Wiley-Liss, Inc., a subsidiary of John Wiley & Sons © 2002).

Figure 3.2B shows the same specimens, except that all ears are the same size, so that the morphological details can be appreciated. The details of otic vesicle morphogenesis in the four species have been described previously. Excellent descriptions of mouse otic morphogenesis can be found in Sher (1971), Sulik and Cotanche (1995), Morsli et al. (1998), and Kaufman and Bard (1999). These observations are nicely summarized in Kiernan et al. (2002). Chick otic morphogenesis is described in Knowlton (1967) and Bissonnette and Fekete (1996). *Xenopus* otic development is described in Paterson (1948), Haddon and Lewis (1991), and Bever et al. (2003). Zebrafish otic morphogenesis is described in Waterman and Bell (1984), Haddon and Lewis (1996), Bang et al. (2001), and Bever and Fekete (2002), and summarized in Whitfield et al. (2002).

In mice, chicks, and frogs, morphogenesis of the roughly spherical otic vesicle initiates with an evagination of the epithelium on the dorsomedial side of the vesicle. This primordium of the nonsensory endolymphatic duct and sac system, which is important for fluid homeostasis in the mature inner ear, elongates dorsally. In zebrafish, endolymphatic duct development is comparatively rudimentary and occurs after semicircular canal formation.

Soon after the appearance of the endolymphatic anlage, the ventral portion of the otic vesicle of mice and chicks begins to elongate in a ventral direction, initiating cochlear development. In mice, the cochlea coils, ultimately attaining 1.75 turns, whereas in chicks it does not coil. Frog and zebrafish ears do not have a cochlea. Zebrafish auditory sensory tissue is housed in the saccule, lagena, and possibly the utricle and frog auditory tissue is housed principally in the amphibian and basilar papilli.

The development of the dorsal (vestibular) portion of the otic vesicle is similar in mice, chicks, and frogs. Shortly after the initiation of cochlear development in the species that have one, a dorsolateral bulge, the vertical canal plate, appears in the otic epithelium. This plate soon develops another groove, separating the vertical plate into distinct regions that will form the anterior and posterior semicircular ducts. Next, the opposing epithelia approach one another, form a fusion plate, and cells begin to leave from the two central regions, forming the anterior and posterior semicircular ducts. At the same time, there is a bulging of the lateral otic epithelium to form the lateral canal plate, in which central cells fuse and are removed to form the lateral semicircular duct. Zebrafish semicircular canal formation differs slightly in that there are no outpouchings of the epithelium from the otic vesicle. Instead, the epithelium protrudes inward at four sites into the vesicle lumen. Opposing protrusions eventually from the equivalent of a fusion plate and central cells are removed to form the canals, as in mice, chicks, and frogs. In all four species, each semicircular duct has a bulge at one base (the ampulla) that houses a rotation-sensing organ (the crista). The utricle and saccule, chambers housing the gravity sensing organs (maculae), then bulge in successive order, ventral to the largely completed semicircular duct system.

It should be emphasized that the morphogenesis of the otic epithelium occurs concomitantly with cellular differentiation of the epithelial cells into sensory versus nonsensory regions and subsequently into numerous functionally distinct, specialized cell types. Indeed, these two processes are likely to be linked, as many of the mutants to be discussed have defects in both morphogenesis and in sensory development. In fact, there is only one reported mutant that has normal nonsensory development (e.g., in semicircular duct formation) accompanied by abnormal sensory development associated with the affected structure (i.e., the cristae). *Hmx3* mutant have apparently normal lateral semicircular ducts that lack cristae (Wang et al. 1998). However, as the focus here is on morphogenesis, issues of cell type specification are not addressed. For complete descriptions of mutant phenotypes the reader is referred to the original citations for the mutants and to Pauley, Matei, Beisel, and Fritzsch, Chapter 4, and Lanford and Kelley, Chapter 5.

4.1 Experimental Approaches to Otic Vesicle Morphogenesis

Three major experimental approaches have been used to dissect the tissues and genes required for morphogenesis of the roughly spherical otic vesicle into the complex labyrinthine structure of the mature otic epithelium. These approaches are to: (1) surgically alter the relationships between the otic vesicle and its surrounding tissues and to observe its subsequent development; (2) culture the otic vesicle in the presence of other tissues or purified factors and follow development in vitro; and (3) characterize the phenotypes associated with mutations of genes expressed in the developing inner ear and in the surrounding tissues. The first approach can be applied easily to all of the species under consideration here, except the mouse, since its embryos can be cultured only for limited periods of time. The second approach is potentially applicable to any species. Genetics, the third approach, is most effectively applied to the zebrafish and mouse. One of the major conclusions from these lines of experimentation is that normal otic vesicle morphogenesis is not an autonomous process: it depends on interactions with the adjacent hindbrain and the surrounding mesenchyme.

4.2 Role of the Hindbrain in Otic Vesicle Morphogenesis

As in the case of otic placode induction (described by Lanford and Kelley, Chapter 5), the hindbrain clearly plays an early role in otic vesicle morphogenesis. In a series of classic experiments, Detwiler and Van Dyke (1950) showed that if the hindbrain of a developing salamander is replaced with neural tissue from the midbrain or spinal cord regions, the ear becomes cystic. In more recent studies, Hutson et al. (1999) unilaterally ablated rhombomeres 5 and 6 from chick hindbrains at stages 11 to 13 and found that the ipsilateral otocysts were smaller than normal and had uniform rather than ventromedially enriched expression of Pax2. If the otocysts were permitted to develop further, severe defects of dorsal (endolymphatic duct and vestibule) development were apparent. In addition, when the otic vesicle is transplanted to other rostrocaudal positions along the axis of the developing CNS, it fails to undergo normal morphogenesis (Detwiler and Van Dyke 1951). These results were extended by Herbrand et al. (1998), who performed transplants of chick otic placode to abnormally rostral positions adjacent to the developing midbrain/hindbrain or to the wing bud. They found that such transplants could form vesicles, but these vesicles did not undergo normal morphogenesis. In addition, they showed that although normal patterns of chick *Nkx5-1* (*Hmx3* in the mouse) expression were maintained in vesicles formed in the midbrain/hindbrain region, such vesicles did not maintain regionalized expression of *Pax2*. The abnormal vesicles that formed in the wing bud did not exhibit normally regionalized expression of either of these marker genes. Taken together, these studies suggest that the hindbrain produces signals that are necessary for normal morphogenesis and patterns of gene expression in the ear. Detwiler's original studies prompted the question, "What is it that the medulla possesses and gives off which so profoundly affects labyrinth differentiation and which is obviously lacking in any other of the primary segments of the embryonic brain?" (Detwiler and van Dyke 1951). Answers to this 50-year-old question are only just beginning to emerge.

The existence of a number of classic and gene-targeted mouse mutants with abnormalities of hindbrain development that are also associated with abnormal otic morphogenesis further cements the role of the hindbrain in otic morphogenesis. In a series of papers describing the classic mouse mutants *kreisler* (Deol 1964a), *dreher* (Deol 1964b), and *Splotch* and *Loop-tail* (Deol 1966), Deol proposed that the malformed inner ears found in these mutants were a consequence of abnormal hindbrain development. This hypothesis was borne out when the responsible genes were identified. The *kreisler* gene encodes a basic helix–loop–helix transcription factor, *Mafb/kr,* that is expressed in the hindbrain adjacent to the developing otic vesicles, but not in otic tissue itself (Cordes and Barsh 1994).

Kreisler mutant ears do not have endolymphatic ducts and they develop as cystic chambers (Deol 1964a). The cystic phenotype is similar to that of zebrafish ears from *valentino* mutants, in which the zebrafish homolog of *Mafb/kr* is mutated (Moens et al. 1998). Similarly, mutations targeted to *Hoxa1*, which encodes a transcription factor expressed in the hindbrain, cause otic phenotypes that are reminiscent of the *kreisler* mutant (Lufkin et al. 1991; Chisaka et al. 1992; Mark et al. 1993). Transcription factors presumably exert their effects on otic morphogenesis by regulating the expression of genes that encode signaling proteins, which in turn activate intracellular signaling pathways in the cells of the otic epithelium. Indeed, there is evidence that both *Mafb/kr* and *Hoxa1* regulate expression of *Fgf3* in hindbrain rhombomeres (r)5 and r6, which are adjacent to the developing otic vesicle (Carpenter et al. 1993; Frohman et al. 1993; McKay et al. 1996).

Fgf3 encodes a fibroblast growth factor (FGF) that is required for normal otic morphogenesis (Mansour et al. 1993). FGFs are secreted molecules that signal through a specific class of transmembrane receptor tyrosine kinases. Binding of an FGF to its specific receptor activates a variety of intracellular growth and differentiation signaling pathways. *Fgf3* mutant ears have variable and incompletely penetrant abnormalities of otic morphogenesis that were traced to a failure of endolymphatic duct formation (Mansour et al. 1993). Consistent with this result, mice that lack the FGF3 receptor, FGFR2b, which is expressed in the dorsomedial (endolymphatic duct-forming) region of the otic vesicle at E9.5, also fail to form an endolymphatic duct and have severe dysmorphogenesis of the otic vesicle, similar to but more penetrant than the phenotype of *Fgf3* mutants (Pirvola et al. 2000). The interpretation of the cause of the *Fgf3* mutant phenotype is complicated by the fact *Fgf3,* while expressed strongly in r5 and r6 during the initial stages of otic morphogenesis (E9.0 to E9.5), is also expressed in the anteroventral otic epithelium starting at E10.0 and continues to be expressed during later stages in the developing sensory patches (Wilkinson et al. 1988, 1989; Pirvola et al. 2000; S.L. Mansour, unpublished data). Thus, loss of *Fgf3* normally expressed in otic epithelium may also contribute to the mutant phenotype. However, as the *Hoxa1* otic phenotype can be rescued by maternal application of retinoic acid, which concomitantly induces *Fgf3* expression in the *Hoxa1* mutant hindbrains, an important role in otic morphogenesis for hindbrain-expressed *Fgf3* is strongly suggested (Pasqualetti et al. 2001).

Retinoic acid signaling to organize the hindbrain is, in fact, the earliest step identified so far in the control of early mouse otic vesicle morphogenesis. Mouse embryos mutant for Retinaldehyde dehydrogenase-2 (*Raldh2*), which is required for retinoic acid biosynthesis, die at E10.5 with small otocysts that have aberrant patterns of gene expression. The otic phenotype seems to be more severe than that of *Hoxa1*, *kreisler*, or *Fgf3* single mutants. Significantly, expression of all three of these genes is absent or greatly diminished in the *Raldh2* mutant hindbrain, suggesting that all three genes are regulated by retinoic acid signaling, and in their absence, otic vesicle development arrests (Niederreither et al. 1999; 2000) The zebrafish *neckless* phenotype is caused by mutations in *Raldh2* (Begemann et al. 2001), but the otic phenotype has not yet been described. The retinoic acid signals required for otic development may be mediated by redundant nuclear receptors, RARα and RARγ, as mice lacking both receptors develop with hypoplastic otic vesicles that fail to form an endolymphatic duct (Romand et al. 2002), similarly to the vesicles seen in *Raldh2* (and *Hoxa1*, *Mafb/kr*, or *Fgf3*) mutants. Significantly, these double mutants also have hindbrain patterning defects in the region adjacent to the developing otocyst (Wendling et al. 2001).

The situation with *Splotch* and *Loop-tail* mice may be somewhat different from that of *Mafb/kr*, *Fgf3*, and *Hoxa1*. The former phenotype is caused by mutations in *Pax3*, which encodes a transcription factor (Epstein et al. 1991; Goulding et al. 1993), and in the latter case by *Ltap/Vangl2*, which encodes a transmembrane protein of unknown biochemical function (Kibar et al. 2001; Murdoch et al. 2001). Both genes are expressed along the anteroposterior axis of the dorsal neural tube and *Pax3* mutants have both neural crest and neural tube closure defects. Despite extensive studies of these mutants with respect to their neural tube closure defects (reviewed in Copp 1994; Juriloff and Harris

2000; Gelineau-van Waes and Finnell 2001; Copp et al. 2003), no recent work on the ear phenotypes of *Pax3* mice has been reported, and on their current genetic background, the *Loop-tail* mutant ears do not appear to exhibit the severe otic vesicle dysmorphology at E11.5 reported by Deol (M. Montcouquiol and M. Kelley, personal communication). It certainly would be interesting to pursue the question first posed by Deol (1966) and determine whether abnormal signaling from the hindbrain causes the otic defects seen in these mutants, or whether the otic defects are a secondary consequence of an open neural tube.

The hindbrain is apparently not required throughout the entire period of otic morphogenesis. When mouse otocysts with adhering mesenchyme are explanted into culture at E10.5 and especially at E11.5, they are able to undergo relatively normal morphogenesis. When similar explants are prepared at E9.5, however, morphogenesis fails (Li et al. 1978; Van De Water et al. 1980). This suggests that the influence of the hindbrain on otic morphoghesis is limited to the period prior to E10.5. Analysis of appropriately designed conditional mutants in the mouse would help to evaluate the validity of this conclusion.

What aspect, then, of otic morphogenesis is the hindbrain required for? One of the most notable characteristics of the phenotypes of the transplanted chick vesicles or of the E9.5 cultured mouse vesicles, as well as of the *kreisler* and *Hoxa1* mutants, and to a variable extent the *Fgf3* mutants mentioned above, is that the very first step of otic vesicle morphogenesis is affected, namely, there is a failure to form the endolymphatic duct. In these cases, the epithelium develops in a cystic fashion, without showing evidence of normal dorsal or ventral morphogenesis. This observation suggests the possibility that the hindbrain sends a signal to the vesicle to initiate endolymphatic duct formation and that further morphogenesis of the vesicle actually depends on proper execution of this command. As zebrafish do not develop an endolymphatic duct until late stages of otic morphogenesis, this scenario would not apply to this species.

The fact that there are no mouse mutants with abnormal or missing endolymphatic ducts in which the rest of otic morphogenesis is normal supports a critical role for the endolymphatic duct in subsequent morphogenesis. In addition, Hendriks and Toerien found that experimental extirpation of the endolymphatic duct from E4 chick otic vesicles in ovo caused abnormal morphogenesis of the vesicle (Hendriks and Toerien 1973). This view was challenged by Van De Water, who removed the endolymphatic duct and sac anlage from E11.5 and E12.5 mouse otocysts, cultured them in vitro, and found that otic morphogenesis was normal (Van De Water 1977; Van De Water et al. 1980). It is possible that these seemingly contradictory results might be resolved if the differences in the timing of the two ablations are significant and there is a requirement for endolymphatic duct formation for the initiation of subsequent morphogenetic steps, but once these steps are initiated, there is no ongoing requirement for the endolymphatic duct/sac. Alternatively, it may be that endolymphatic duct function is not recapitulated in culture.

Studies of the targeted mouse *Foxi1/Fkh10* mutant lend support to the original conclusions from the chick extirpation study suggesting a role for the endolymphatic duct in normal otic morphogenesis. *Foxi1* is expressed in the developing endolymphatic duct/sac. Although the initial formation of the endolymphatic duct appears normal in targeted *Foxi1* null mutants, the duct/sac becomes progressively dilated. Subsequently, the cochlear and vestibular regions of the ear become large irregular cavities, similar to those seen in *Hoxa1*, *kreisler*, and *Fgf3* mutants (Hulander et al. 1998, 2003). It certainly would be of interest to determine whether the initiation of *Foxi1* expression in the endolymphatic duct/ sac depends on signals from the hindbrain. Mutations in the zebrafish *Foxi1* gene lead to an almost complete failure of otic vesicle formation (Nissen et al. 2003; Solomon et al. 2003a) and are thus quite different from the mouse *Foxi1* mutants described above. This could reflect species-specific differences in inner ear development, or more likely, simply reflects an unfortunate assignment of the same name to genes that are not truly orthologous (Solomon et al. 2003b).

Clues to other signaling systems that may play roles in endolymphatic duct formation come from inhibition of DAN protein expression in the chick. Dan is a member of a cysteine knot protein family that is able to inhibit the function of members of the Tgfß superfamily, including members of the Bmp and Gdf subfamilies (Dionne et al. 2001). *Dan* mRNA is expressed in the chick medial otic vesicle and chicks electroporated with morpholinos designed to inhibit Dan expression have enlarged endolymphatic ducts and sacs (Gerlach-Bank et al. 2002, 2004), suggesting that Dan normally limits the function of a signal that promotes endolymphatic duct outgrowth. The source and precise identity of the hypothesized signal is unknown. Dan cannot function in the same way in mouse endolymphatic duct outgrowth as it is not expressed in or near the early otic vesicle and null mutants do not have otic defects (Dionne et al. 2001; Gerlach-Bank et al. 2002). Nevertheless, the results from the chick inhibition study suggest that an expression survey in mouse of genes encoding other members of the same cysteine knot family, as well as of the genes encoding the signaling proteins to which they bind could identify additional signaling systems involved in endolymphatic duct development.

4.3 Role of the Periotic Mesenchyme in Otic Vesicle Morphogenesis

The mesenchyme surrounding the otic vesicle eventually undergoes chondrogenesis to form the bony capsule that surrounds the membranous labyrinth. Many lines of evidence suggest that the morphogenesis of the otic epithelium and the development of the periotic mesenchyme are mutually dependent. An early demonstration of the role of otic epithelium in the normal development of the otic capsule comes from the work of Kaan, who showed that transplantation of salamander otic cup or placode to a site anterior or posterior of its normal position would induce mesenchymal condensation and capsule formation around the transplanted epithelium. Transplantation of the same otic tissues to limb or pronephric mesenchyme, however, did not induce formation of a capsule (Kaan 1930). Conversely, when the otic tissue was removed from its normal location and substituted with either lens, olfactory placode, retina, or hindbrain, these tissues failed to induce the underlying mesenchyme to condense (Kaan 1938). Kaan also noted that in this experimental paradigm, the otic epithelium failed to acquire its normal form.

When E10.5 to E12.3 mouse otocysts are dissected and the adherent mesenchyme is removed enzymatically or mechanically, the otocysts fail to undergo normal morphogenesis in culture, whereas similarly staged otocysts cultured with adherent mesenchyme develop normally (Van De Water et al. 1980). Similar results were reported for E4 chick otocysts (Orr 1976; Orr and Hafft 1980). Furthermore, when such "naked" otocysts prepared from stage 17 to 18 (E3) quail embryos were grafted to chick wing mesenchyme, the grafts failed to undergo morphogenesis, although they did exhibit quite extensive and remarkably normal cellular differentiation (Swanson et al. 1990). These results imply that otic mesenchyme, but not other types of mesenchyme, produces factors that are required for normal otic epithelial morphogenesis.

Given the ability to successfully culture the otic vesicle and achieve relatively normal morphogenesis in the presence but not the absence of the otic mesenchyme, it is disappointing that this approach has not been exploited successfully to identify the morphogenetic signaling molecules expressed by the mesenchyme to control epithelial morphogenesis. Furthermore, there are remarkably few genetic clues as to the identity of the mesenchymal signals that control otic epithelial morphogenesis. Indeed, there are only a few mouse mutants with abnormal ear development that can potentially be traced to abnormalities of the mesenchyme. One of these is the *sex linked fidget* mouse in which the transcription factor *Pou3f4* (*Brn4*), which is widely expressed in otic mesenchyme and not in the epithelium, is mutated (Phippard et al. 2000). *Pou3f4* has also been subjected to targeted mutagenesis (Minowa et al. 1999; Phippard et al. 1999). Whereas the mutant mice have many defects in mesenchymally derived ear structures themselves (Minowa et al. 1999; Phippard et al. 1999, 2000), they also show a failure of normal epithelial morphogenesis. Most notably, there is a failure of cochlear coiling. This phenotype suggests that the mutant mesenchyme fails to send appropriate signals necessary for epithelial morphogenesis (Phippard et al. 1999, 2000) The *Prx1* and *Prx2* genes, which encode members of the aristaless-domain-containing transcription factor family, are required redundantly for normal morphogenesis of the mouse lateral semicircular duct (ten Berge et al. 1998). This epithelial structure and its corresponding semicircular canal are absent from the double mutants, but present in either of the single mutants. Analysis of the developing semicircular duct system in *Prx1/Prx2* double mutant embryos suggests that this phenotype is a result of reduced outgrowth of the duct, which is evident as early as E12.5. Indeed, double mutant embryos showed defects in the outgrowth of all the duct diverticula, but only the lateral duct was absent from the final ear. This phenotype was correlated with both *Prx* genes being expressed in the mesenchyme surrounding the lateral side of the otocyst from E9.5 to E12.5. *Prx2* is also expressed in the primordium of the lateral duct itself, but *Prx1* is not found there and neither single mutant has vestibular abnormalities, so it is likely that mesenchymal abnormalities are the cause of the observed phenotype (ten Berge et al. 1998).

As argued for the hindbrain transcription factors, the mesenchymal transcription factors that play roles in epithelial morphogenesis are likely to do so by controlling the synthesis of signaling molecules that act on the epithelium. Identification of the genes controlled by *Prx1*, *Prx2*, and *Pou3f4* might serve to identify the signaling systems implicated by the respective mutant phenotypes in the control of epithelial outgrowth.

4.4 Midline Signals and Otic Morphogenesis

As noted above in the section on axis formation, Shh signals emanating from the notochord and floorplate seem to play a very early role in determining the mouse otic DV axis. There is also evidence that *Shh* may play later roles in otic morphogenesis through its effects on the otic mesenchyme. Chondrogenesis in the otic mesenchyme of *Shh*−/− mutants is delayed (Liu et al. 2002) and expression of mesenchymal *Tbx1* and *Pou3f4* (*Brn4*) are inhibited (Riccomagno et al. 2002). *Shh* signals are transduced by a receptor complex composed of Patched and Smoothened molecules and ultimately lead to activation of Gli transcription factors by a poorly defined pathway. Most notably, the *Patched1* and *Gli1* genes themselves are transcriptional targets of *Shh* signaling (Ingham and McMahon 2001). Expression studies of these genes in wild type and *Shh* mutants suggest that *Shh* signaling is active in both the otic epithelium and mesenchyme. Detection of *patched* gene expression in *Xenopus* otic vesicles (Koebernick et al. 2001) is consistent with the findings in the mouse. A more precise temporal and spatial dissection of the roles of Shh signals during otic morphogenesis awaits careful application of the conditional *Shh* allele (Dassule et al. 2000). Further information could be gained by examining the otic phenotypes of mice lacking other components of the *Shh* signaling pathway. Indeed, the gene encoding one of these factors, Gli3, is mutated in the *Extra-toes* (*Xt*) mouse (Vortkamp et al. 1992; Hui and Joyner 1993). Homozygous *Xt* mutants have otic defects (Johnson 1967), but neither a complete description of otic morphogenesis in the mutant nor a systematic analysis of *Gli3* expression in the otic region has been reported. Both are needed to help sort out the roles of Shh signaling in epithelial/mesenchymal signaling.

4.5 Epithelial Factors Controlling Otic Morphogenesis

The large majority of genes identified so far as key players in otic epithelial morphogenesis encode transcription factors expressed by the epithelium itself. A small group acts early and affects the morphogenesis of the entire vesicle. A much larger group, discussed below, has effects confined to the cochlea or the vestibule, indicating that at some point the development of these functionally distinct parts of the ear are controlled independently. Of course transcription factors themselves are not likely to be the effectors of morphogenesis. As argued above, they are likely to control the genes encoding signaling molecules, as well as the direct effectors of morphogenesis. Indeed, a few such epithelial genes have now been implicated in otic morphogenesis.

4.5.1 Global Control of Morphogenesis by Genes Expressed in the Epithelium

Eya1-null mice, which serve as a model for branchio-oto-renal syndrome, have otic vesicles that are very small and do not undergo significant morphogenesis, failing even to form an endolymphatic duct (Xu et al. 1999). *Eya1*, which encodes a transcription factor related to the *eyes absent* (*eya*) gene of *Drosophila*, is expressed at least as early as E10.5 in mouse otic epithelium in a ventral domain (Kalatzis et al. 1998). As the effects of the null mutation can be detected at E9.5 by changes in marker gene expression in the vesicle, it seems likely that *Eya1* must also be expressed before E10.5 in the vesicle. If it is similarly restricted to the ventral part of the vesicle, this would imply that signaling from ventral regions of the vesicle is required for normal dorsal development or that *Eya1*-expressing cells migrate and contribute to a much larger part of the later vesicle, as suggested by lineage studies of the chick otic cup (Brigande et al. 2000a). Alternatively, it is possible that *Eya1* is expressed more globally in the vesicle during the early stages of otic morphogenesis. Additional studies of *Eya1* expression in the time period prior to the initiation of otic defects would be very helpful in interpreting the mutant phenotype. At later stages, *Eya1* is expressed both in the otic epithelium and in the periotic mesenchyme, suggesting the possibility that it could play other roles in the later stages of otic development (Kalatzis et al. 1998). Production and analysis of a conditional allele will be necessary to sort out the morphogenetic functions of *Eya1* in different tissues. In addition, cross-species comparisons of ear morphogenesis in zebrafish *dog-eared* (*dog*) mutants, which carry a mutation in *Eya1*, will be informative (Whitfield et al. 2002), although the preliminary descriptions of *dog* mutant ears (Whitfield et al. 1996) suggest that the otic phenotype is less severe that that described for the mouse mutants.

Eya1 is a component of a regulatory cascade including *Pax*, *Six*, and *Dach* genes conserved from *Drosophila* to mammals. *Six1*, which encodes a transcription factor expressed in the otic epithelium, is a downstream target of *Eya1* and mice that lack *Six1* have severely malformed otic vesicles that essentially stop developing at E12.5 (Laclef et al. 2003; Zheng et al. 2003; Ozaki et al. 2004). The mutant phenotype differs from the *Eya1* mutant phenotype in that at least on one genetic background the endolymphatic duct does form, although it is enlarged and there is some evidence that the rest of the vesicle has the character of a developing canal plate (Ozaki et al. 2004). In this case, however, very detailed expression data on *Six1* exist, as do extensive studies of marker gene expression prior to overt appearance of morphological defects. *Six1* is expressed as early as the otic placode and cup stages and in homozygous mutant embryos marker gene expression is altered in such a way as to suggest that the DV axis of the mutant ears may be affected (Zheng et al. 2003; Ozaki et al.

2004), similarly to that of *Shh* mutants (see discussion on axis formation above), although *Six1* is not a target of Shh signals and vice versa (Ozaki et al. 2004). As *Six1* has a dynamically changing pattern of expression in the otic epithelium and is also expressed in periotic mesenchyme, where it seems to be involved in middle ear formation, conditional mutagenesis will be necessary to sort out the temporal and spatial details of *Six1* function. The relationship with *Eya1*, however, strongly suggests that the *Eya1* mutant should be reevaluated with the idea in mind that *Eya1* may be functioning much earlier than previously appreciated. Furthermore, it will be worthwhile to assess otic phenotypes in mutants in other members of the Pax-Six-Dach regulatory cascade.

4.5.2 Global Control of Ear Morphogenesis by Genes That Are Expressed in Multiple Tissues

Mutation of the transcription factor, *Tbx1*, which phenocopies many aspects of human DiGeorge syndrome in mice (Jerome and Papaioannou 2001; Lindsay et al. 2001; Merscher et al. 2001), also has global effects on otic morphogenesis. Unlike the *Eya1* mutants, however, *Tbx1* mutants apparently form an endolymphatic duct, after which point otic epithelial morphogenesis is halted (Jerome and Papaioannou 2001; Vitelli et al. 2003; Raft et al. 2004). Like *Eya1*, *Tbx1* is expressed both in otic epithelium and in otic mesenchyme (Vitelli et al. 2002, 2003; Raft et al. 2004). Analysis of changes in otic epithelial and mesenchymal marker genes, including a *Tbx1^{LacZ}* knockin allele (Vitelli et al. 2003), suggests that *Tbx1* function is required to pattern the epithelium (Raft et al. 2004) and for expansion of cells that contribute to both the cochlear and vestibular regions of the epithelium. Disturbances of gene expression in the mutant periotic mesenchyme suggest that the mutant phenotype may not be strictly autonomous to the epithelium (Vitelli et al. 2003). A conditional *Tbx1* mutant will be required to establish tissue specific contributions to the overall otic phenotype.

The recent identification of a *Tbx1* mutation in the zebrafish *van gogh* (*vgo*) mutant strain may provide additional insight into the role of this gene in inner ear morphogenesis (Piotrowski et al. 2003). Similar to mouse *Tbx1* mutants, *vgo* mutants have small otic vesicles that do not undergo semicircular canal morphogenesis. Interestingly, the primary defect in *vgo* mutants seems to be in the differentiation of pharyngeal endoderm. This defect is postulated to affect patterning of the mesenchyme, which could, in turn, account for the inner ear abnormalities (Piotrowski and Nüsslein-Volhard 2000; Whitfield et al. 2002). It should be noted that mouse *Tbx1* is also expressed and functions in the pharyngeal endoderm (Vitelli et al. 2002; Yamagishi et al. 2003). Together with the zebrafish data, this suggests the possibility that the endoderm may also play an as yet unexplored role in mouse otic morphogenesis.

Gata3 mutants also have major otic morphogenetic defects. Following endolymphatic duct formation, otic development apparently arrests, with occasional ventral extensions of the vesicle (Karis et al. 2001). As this transcription factor is expressed throughout the placodal ectoderm, vesicle epithelium, and in parts of the surrounding mesenchyme (Karis et al. 2001; Lawoko-Kerali et al. 2002), it is not possible to assign a tissue responsibility for the observed defects, again suggesting the utility of a conditional mutant.

As discussed earlier, most of the genes that have been identified to date as having global effects on otic morphogenesis are expressed in the hindbrain rather than in the epithelium. It was suggested that one function of these transcription factors likely is to control secreted signaling molecules that communicate with the epithelium. One would expect that at a minimum, there should be otic epithelial receptors for hindbrain-expressed secreted factors that are required for epithelial morphogenesis. Indeed, selective mutation of the "b" splice variant of Fgf receptor 2, which is expressed in the otic cup and vesicle (Pirvola et al. 2000; Wright and Mansour 2003), and which serves as a receptor for hindbrain-expressed *Fgf3*, leads to a global failure of otic morphogenesis that is similar to but more severe and penetrant than that of *Fgf3* mutants (Pirvola et al. 2000).

4.5.3 Regionalized Morphogenesis at Later Stages—Cochlear Morphogenesis

Only two mouse mutants with relatively specific defects in cochlear morphogenesis have been described, and these both encode transcription factors. Mice homozygous for either a targeted or spontaneous mutation in *Pax2* have relatively normal vestibular development, but fail entirely to form a cochlea (Favor et al. 1996; Torres et al. 1996). This phenotype correlates well with the ventromedial expression of *Pax2* in the epithelium at the otic vesicle stage in mice and chicks (Nornes et al. 1990; Hutson et al. 1999; Hidalgo-Sánchez et al. 2000). As is the case for most of the other transcription factors implicated in otic morphogenesis, the specific effectors that mediate the *Pax2* morphogenetic defect are unknown. Zebrafish *pax2.1* is also expressed in the otic vesicle, but in this species, which does not have a cochlea, its unique role appears to be in hair cell specification rather than morphogenesis (Riley et al. 1999; Whitfield et al. 2002).

Mice that lack the transcription factor Otx1, or lack both copies of the *Otx1* gene and one copy of the related *Otx2* gene, have a range of dysmorphologies affecting many parts of the ear. In particular, the cochlea is not properly separated from the utricle and saccule and there are variable defects in cochlear coiling. Both genes are expressed in the ventrolateral region of the developing otocyst prior to the development of the defects (Morsli et al. 1999). This pattern is consistent with an autonomous role for these transcription factors in cochlear morphogenesis, but the mechanisms by which they act have not been explored in detail.

As mentioned above, null mutations in *Eya1* arrest otic morphogenesis at the otic vesicle stage. A spontaneous mutation in *Eya1* (*Eya1*^{bor}) that reduces, but does not eliminate, *Eya1* transcripts leads to a milder phenotype in which cochlear development is severely affected, but vestibular development is only mildly affected. In particular, the homozygotes are missing most of the cochlea. The development of this phenotype has not been examined in detail, but presumably there is a block to the ventral growth of the otocyst.

Of course there are many other cochlea-expressed genes, but mutations in most of these genes analyzed so far cause defects of cell fate specification or patterning, which are the subject of other chapters in this volume. It is surprising that so few genes affecting cochlear morphogenesis have been described. Perhaps the use of in vitro culture systems, in which the otic vesicle is cultured with extracts of mesenchyme, might allow identification of secreted factors that control cochlear morphogenesis. Systematic preparation and evaluation of targeted mutants in the mouse or morphants in zebrafish will continue to reveal genes important for cochlear morphogenesis.

4.5.4 Semicircular Duct Morphogenesis—Localized Outpocketing of the Otic Epithelium

The first step in forming the semicircular ducts is localized outpocketing of the dorsal otic epithelium—first dorsally to form the vertical canal plate, then laterally to form the lateral (horizontal) canal plate. In vivo inhibitor studies have been used to implicate BMPs (members of the transforming growth factor ß superfamily) in this process. BMPs are secreted signaling molecules that bind to and activate the intracellular serine/threonine kinase activity of heterodimeric transmembrane receptors. Activation of these receptors leads to changes in gene expression that can affect the growth or differentiation of cells. This signaling system can be inhibited by secreted glycoproteins such as Noggin, Chordin, Follistatin, Cerberus, Gremlin, and DAN, which bind to the BMPs and prevent them from binding to their receptors. Bmp2 and Bmp4 bind with particularly strong affinity to Noggin (Zimmerman et al. 1996) and *Bmp4* in particular is expressed by specific domains of the chick otic epithelium that presage the appearance of all of the sensory organs (Wu and Oh 1996). *Bmp4* expression differs somewhat in the mouse, where it marks the sensory tissue of the semicircular canals (the cristae), but is expressed in nonsensory tissue in the cochlea. *Noggin* is expressed in the otic mesenchyme (Chang et al. 1999; Gerlach et al. 2000). This scenario suggests the possibility that localized variations in the active concentration of BMPs might signal particular regions of the epithelium to increase the rate of cell division and thus induce localized outgrowth of the epithelium. When either a retrovirus expressing *Noggin* or agarose beads carrying *Noggin*-expressing cells are applied to the chick inner ear between stages 13 and 20, abnormal morphogenesis results. Specifically, outgrowth of the epithelium to form the semicircular ducts can be prevented in a predictable manner by the localized application of *Noggin* cells. Such effects are not seen with ectopic Bmp4, but Bmp4 applied at the same time as *Noggin* cells can rescue the defects (Chang et al. 1999; Gerlach et al. 2000). Similar results are obtained with the BMP antagonist, DAN, which is normally expressed in restricted areas of the otic epithelium (Gerlach et al. 2000; Gerlach-Bank et al. 2004). While it is tempting to conclude that these particular combinations of signaling molecules and inhibitors play normal roles in otic morphogenesis, it is certainly possible that other BMPs or inhibitors may be operative in vivo. Indeed, both *Bmp5* and *Bmp7* are expressed in the developing chick otic epithelium (Oh et al. 1996). In addition, *bmp2b*, *bmp7*, and *bmp5* are all also expressed in the zebrafish otic vesicle in patterns that are similar to, but not precisely identical to those of *Bmp* genes in the mouse and chick, suggesting that these molecules may have conserved, but not precisely orthologous functions in otic development. It is interesting to note that *Noggin* null mice are reported to have abnormal development of the ear, but this was attributed to severe neural tube closure defects (McMahon et al. 1998), so analysis of a conditional *Noggin* allele that was inactivated in otic mesenchyme might also be revealing with respect to Noggin function in inner ear morphogenesis. The zebrafish *noggin* genes, however, are not expressed in otic mesenchyme, where *follistatin* is detected instead (Mowbray et al. 2001).

Direct genetic tests of the roles of BMP signaling in ear morphogenesis have yet to be reported. Mice that lack *Bmp4* die prior to ear development (Winnier et al. 1995), so a conditional allele will be required before the effects on the ear can be determined. Similarly, zebrafish that lack either *bmp2a* (*swirl* mutants) or *bmp7* (*snailhouse* mutants) are affected too early to assess inner ear development (Kishimoto et al. 1997; Dick et al. 2000). There is, however, indirect genetic evidence for a role for *Bmp4* in the mouse ear. *Dlx5* null mutants have severe hypoplasia of the semicircular canals, the primordia of which are rudimentary, and they also fail to elongate the endolymphatic duct (Acampora et al. 1999; Depew et al. 1999). *Dlx5* is a transcription factor that is normally expressed in these vestibular epithelia and in its absence, vestibular expression of *Bmp4* is reduced or absent, suggesting the possibility that *Bmp4* does have a specific role in ear morphogenesis: namely in the outgrowth of canal primordia (Merlo et al. 2002).

Fgf10 null mutants have severe dysmorphogenesis of the semicircular canals. At E18, no distinct canals can be recognized, although both the anterior and horizontal cristae (both abnormal) can be identified (Pauley et al. 2003). This phenotype is consistent with expression of *Fgf10* initially throughout the otic cup, with gradual restriction to the dorsal aspect of the vesicle and finally to all of the developing sensory patches of the epithelium, adjacent to the nonsensory sites of expression of mRNAs encoding the Fgf10 receptor, Fgfr2b (Pirvola et al. 2000; Pauley et al. 2003). Since there is little or no effect on the morphogenesis of the rest of the otic epithelium in *Fgf10* mutants, it is assumed that these regions express a redundant gene product, perhaps *Fgf3* (Pirvola et al. 2000). Indeed, *Fgfr2b* mutants have more severe and earlier dysmorphogenesis of the otic epithelium, similar to that described for *Fgf3* mutants (Mansour et al. 1993; Pirvola et al. 2000). Additional developmental studies will be needed to determine whether the *Fgf10* mutant phenotype is caused by a failure of canal primordia outgrowth, or some later event in canal morphogenesis, and to assess the relative contributions of epithelial Fgf10 and Fgf3 signals to otic morphogenesis.

Another gene that may play a role in canal plate outgrowth is *Fidgetin* (*Fign*). The original drawings of the inner ears of the spontaneous mutant, *fidget*, show that the lateral canal is entirely missing, but that the vertical canal plate does undergo some rudimentary outgrowth and presumably some fusion and clearing in a very small central region (Truslove 1956). *Fidgetin* (*Fign*) was identified recently as the responsible gene and a *lacZ* knockin allele was generated. *Fign/LacZ* expression is detected at a variety of stages in the otic epithelium, but the details remain to be determined. Figetin is an AAA protein, other types of which are molecular chaparones that mediate a variety of functions including cell-cycle regulation, organelle biogenesis and protein degradation, but the specific biochemical activity of Fidgetin is unknown (Cox et al. 2000). A close examination of the development of the vestibular phenotype and an analysis of otic epithelial proliferation in the *Fign*-null mutant mice might shed some light on the function of this novel gene product.

4.5.5 Fusion Plate Formation and Resorption

After the outpouching of the precursor to each semicircular duct occurs, the two epithelial sheets detach from the surrounding mesenchyme and approach one another prior to fusion in the central region of the outpouching. Inhibitor studies have been used to implicate hyaluronan (also known as hyaluronate or hyaluronic acid) in this process in *Xenopus* semicircular canal formation. Hyaluronan is an extremely large extracellular glycosaminoglycan that is found in many areas of the developing embryo and that has long been postulated to play roles in morphogenesis, particularly of the heart (Bernanke and Markwald 1979). Observations of *Xenopus* semicircular canal formation showed that the developing space between the surrounding mesenchyme and the axial protrusions of the epithelial outpouchings that go on to form the semicircular ducts are filled with hyaluronan. When hyaluronidase was injected into the hyaluronan-rich core of the protrusions, hyaluronan was degraded and development of the injected canal precursor was inhibited. Studies of [^{3}H]glucosamine incorporation into explanted ears showed that this precursor of hyaluronan was incorporated mainly by the epithelium. Thus, it was proposed that the shaping of the epithelium during canal formation is driven in part by localized high-level epithelial production of hyaluronan, which "propels" the epithelium away from the mesenchyme and into the lumen of the otic vesicle (Haddon and Lewis 1991). Hyaluronan is also synthesized by mouse inner ears (McPhee et al. 1987), but nothing is known about the dynamics of its synthesis in the murine inner ear or its potential role in semicircular canal formation.

As hyaluronan is a polysaccharide, there is no gene for it per se, but its production can be inhibited genetically by mutating the gene coding for its synthetic enzyme, hyaluronan synthase. There are such three genes in mammals, only one of which, *Has2*, seems to be required for embryonic development. Unfortunately, *Has2*-null mice die before semicircular canal development (they do, however, have otic vesicles), precluding a simple analysis of canal morphogenesis (Camenisch et al. 2000). As these mutants have heart defects that might be predictable from the original hypotheses about the role of hyaluronan in endocardial cushion development (Markwald et al. 1978), it would be very interesting to develop and analyze a conditional *Has2* mutant that could be inactivated only in the otic epithelium. This would permit a genetic confirmation of the role of hyaluronin in semicircular canal formation in mice.

Other evidence implicating glycosaminoglycans generally in movement of the opposing sheets of the canal primordia toward one another comes from studies of the zebrafish mutant *jekyll* (*jek*), which carries a mutation in the *ugdh1*gene, encoding UDP-glucose dehydrogenase (Walsh and Stainier 2001). This enzyme converts UDP-glucose to UDP-glucuronate, a component of, hyaluronan, as well as of the other glycosaminoglycans, chondroitin sulfate and heparan sulfate. *Jekyll* mutants do not have semicircular canals because although the epithelial projections are formed from the otocyst wall, they fail to extend (Neuhauss et al. 1996). Zebrafish embryos injected with morpholinos directed against the homolog of the human deafness gene, *DFNA5*, have an otic phenotype that is remarkably similar to that of *jekyll* mutants. Indeed, *ugdh1* expression is strongly reduced in these embryos, suggesting that *dfna5* is an upstream component of the pathway leading to hyaluronan synthesis and normal development of the semicircular canals (Busch-Nentwich et al. 2004). It is interesting to note that a chemically induced lesion of mouse *Ugdh* causes a developmental block at gastrulation, similar to that of early inhibition of FGF signaling (García-García and Anderson 2003). Given the importance of FGF signaling to inner ear morphogenesis, it would not be surprising if conditional mutants of mouse *Ugdh* turn out to have defects of otic morphogenesis, although they may not be restricted to semicircular canal formation.

After the two epithelial sides of the semicircular duct primordia approach one another, they become thinner, the underlying basement membrane is disrupted, the cells lose their epithelial morphology, and then the two layers become fused in the center. This may be another aspect of otic morphogenesis that is analogous to neural fold fusion. In mice and zebrafish, these central cells are removed, perhaps by moving back into the duct epithelium (Waterman and Bell 1984; Martin and Swanson 1993). In chicks and frogs, the fused area undergoes apoptosis (Haddon and Lewis 1991; Fekete et al. 1997). One gene that is clearly required for the resorption process in lateral and posterior semicircular duct formation in mice is *netrin1*, which encodes a secreted, laminin-related molecule expressed by the fusion-plate-forming cells. *Netrin1*-null mutants undergo normal development of the epithelial outpocketings for all three semicircular ducts and the epithelial walls of the outpocketings begin to thin. After this point, however, semicircular duct development goes awry. The mutant lateral and posterior epithelia maintain an intact basement membrane and epithelial morphology, and they fail to form a fusion plate. As a consequence, *Netrin1*-null ears lack lateral and posterior canals. Furthermore, the mutants show reduced proliferation of the mesenchymal cells in the vicinity of the *Netrin1*-expressing epithelial cells, suggesting the possibility that proliferation of the mesenchyme may play a role in pushing the epithelia together (Salminen et al. 2000). As the anterior semicircular duct forms normally in *Netrin1* mutants, it is possible that another related molecule may have a redundant function in anterior duct formation. Netrins are best known for their roles in axon guidance and cell migration in the nervous system in which they function by binding to transmembrane receptors that include Dcc, Neogenin, and Unc5h (Livesey 1999). It would be interesting to learn whether Netrin1 function in semicircular duct morphogenesis requires these or other related receptors.

Hmx3 (*Nkx5-1*) mutant mice have otic morphogenetic defects that are remarkably similar to those of the *Netrin1* mutants at the morphological level (Hadrys et al. 1998; Wang et al. 1998). In addition, *Hmx2* (*Nkx5-2*) mutant ears also appear to arrest semicircular canal development at a stage prior to fusion plate formation and in this case, all three canals are affected. As with the *Netrin1* mutation, this phenotype is associated with reduced proliferation of the mesenchyme adjacent to the duct diverticula (Wang et al. 2001). Like *Netrin1*, *Hmx2*, and *Hmx3*, which encode transcription factors, are expressed by the nonsensory, fusion plate-forming cells of the semicircular ducts (Rinkwitz-Brandt et al. 1995, 1996). Neither *Hmx2* nor *Hmx3*, however, are required for *Netrin1* expression, as this gene is expressed in the abnormally small *Hmx2* and *Hmx3* mutant semicircular duct diverticula. Thus, it is thought that these genes may help to specify the region of the epithelium that will form the fusion plate (Salminen et al. 2000; Wang et al. 2001). Future studies of the control of *Netrin1* expression in the ear should eventually reveal additional genes required for semicircular duct morphogenesis.

4.5.6 Cell Death and Otic Morphogenesis

As noted above, after the fusion plates are formed during chick semicircular canal morphogenesis, cell death initiates at the center of the plates and spreads radially. When cell death is blocked by ectopic retrovirally transduced expression of the antiapoptotic gene, *Bcl2*, the number of dying cells is reduced in all three fusion plates, with the posterior fusion plate most severely affected. Significantly, there is a failure of posterior semicircular canal morphogenesis in a high proportion of *Bcl2*-infected ears, suggesting that cell death plays a role in normal morphogenesis of chicken semicircular canals, perhaps by permitting clearing of the center of the fusion plates (Fekete et al. 1997).

Mouse otic morphogenesis is associated with characteristic patterns of cell death that are generally similar to those found during chick otic morphogenesis with the notable exception of the semicircular canal fusion plates, which do not exhibit cell death (Martin and Swanson 1993; Nishikori et al. 1999). Nevertheless, mutation of the proapoptotic genes *Apaf1* or *Caspase9*, which almost completely block apoptosis in the developing mouse inner ear, have profound effects on otic morphogenesis. The endolymphatic duct of *Apaf1*$^{-/-}$ mutants is short and wide and the cochlea is short and poorly coiled. Most notably, the anterior semicircular duct of *Apaf1*$^{-/-}$ mutants was always affected (either reduced or absent), but the other two ducts always formed. Curiously, a mutation in *Bcl2l*, which encodes an antiapoptotic factor and leads to increased numbers of dying cells in normal areas of apoptosis, has remarkably mild effects on otic morphogenesis. The only consistent defect is seen in the posterior semicircular duct, which arrests at the outpocketing stage, without fusing or clearing in the center. Thus, apoptosis is required for normal morphogenesis of the mouse inner ear, but it is not clear yet whether the effects of the mutations in proapoptotic genes are directly related to each individual site of normal cell death or whether an early inhibition of cell death during formation of the endolymphatic duct has secondary consequences to the subsequent stages of morphogenesis. Nor is it understood how increases in cell death can block posterior duct morphogenesis at the outpocketing stage, but compensatory changes in proliferation rates after changes in apoptosis have been proposed to contribute to these phenotypes (Cecconi et al. 2004).

4.5.7 Continued Growth of the Semicircular Ducts

After the semicircular ducts have formed, they continue to grow in overall size as the whole ear and the embryo itself enlarges. One gene that plays a critical role in this process is *Nor1* (*Nr4a3*), which encodes a ligand-independent member of the nuclear receptor superfamily of transcription factors. At E13.5 *Nor1*-deficient mice have normally formed vestibular apparati, but they eventually develop abnormally narrow semicircular ducts. This defect is caused by reduced proliferation of the nonsensory duct epithelia, which normally express *Nor1* in cells of their inner surfaces (Ponnio et al. 2002). The growth signaling pathways by which *Nor1* acts at this late stage of otic morphogenesis have yet to be determined.

5. Summary and Future Directions

There is no aspect of otic morphogenesis that is adequately understood at both the cellular and molecular levels. The knowledge that is available, however, can be used as a basis for further exploration. Clearly, the early stages of morphogenesis have not yet yielded to genetic analysis and the right kinds of mutants, those that block morphogenesis at specific points, would be very valuable. More knowledge of the genes involved in the later stages of morphogenesis is available; however, in some cases, basic characterization of mutant phenotypes is still necessary. In other cases, the responsible gene needs to be identified, and in still other cases, knowledge is more advanced, in that we are beginning to understand genetic pathways involved in particular morphogenetic processes, but their full extent and regulation remain mysterious. Finally, the coordination between morphogenesis and cell-type differentiation remains to be explored. Different experimental approaches utilizing each of the model organisms will be necessary to unravel the remaining mysteries.

Acknowledgments. We are grateful to Michele Bever and Donna Fekete (Purdue University) for working so hard to produce the compilation of paint-filled ears from four species shown in Figures 3.2a and b. Research in the authors' laboratories is supported by grants from the NIH.

References

Acampora D, Merlo GR, Paleari L, Zerega B, Postiglione MP, Mantero S, Bober E, Barbieri O, Simeone A, Levi G (1999) Craniofacial, vestibular and bone defects in mice lacking the *Distal-less*-related gene *Dlx5*. Development 126:3795–3809.

Altmann CR, Brivanlou AH (2001) Neural patterning in the vertebrate embryo. Int Rev Cytol 203:447–482.

Alvarez IS, Navascues J (1990) Shaping, invagination, and closure of the chick embryo otic vesicle: scanning electron microscopic and quantitative study. Anat Rec 228:315–326.

Alvarez IS, Martin-Partido G, Rodriguez-Gallardo L, Gonzalez-Ramos C, Navascues J (1989) Cell proliferation during early development of the chick embryo otic anlage: quantitative comparison of migratory and nonmigratory regions of the otic epithelium. J Comp Neurol 290:278–288.

Bang PI, Sewell WF, Malicki JJ (2001) Morphology and cell type heterogeneities of the inner ear epithelia in adult and juvenile zebrafish (*Danio rerio*). J Comp Neurol 438: 173–190.

Begemann G, Schilling TF, Rauch GJ, Geisler R, Ingham PW (2001) The zebrafish *neckless* mutation reveals a requirement for *raldh2* in mesodermal signals that pattern the hindbrain. Development 128:3081–3094.

Bernanke DH, Markwald RR (1979) Effects of hyaluronic acid on cardiac cushion tissue cells in collagen matrix cultures. Tex Rep Biol Med 39:271–285.

Bever MM, Fekete DM (2002) Atlas of the developing inner ear in zebrafish. Dev Dyn 223:536–543.

Bever MM, Jean YY, Fekete DM (2003) Three-dimensional morphology of inner ear development in *Xenopus laevis*. Dev Dyn 227:422–430.

Bissonnette JP, Fekete DM (1996) Standard atlas of the gross anatomy of the developing inner ear of the chicken. J Comp Neurol 368:620–630.

Bok J, Bronner-Fraser M, Wu DK (2005) Role of the hindbrain in dorsoventral but not anteroposterior axial specification of the inner ear. Development 132:2115–2124.

Brigande JV, Iten LE, Fekete DM (2000a) A fate map of chick otic cup closure reveals lineage boundaries in the dorsal otocyst. Dev Biol 227:256–270.

Brigande JV, Kiernan AE, Gao X, Iten LE, Fekete DM (2000b) Molecular genetics of pattern formation in the inner ear: Do compartment boundaries play a role? Proc Natl Acad Sci USA 97:11700–11706.

Brown JW, Beck-Jefferson E, Hilfer SR (1998) A role for neural cell adhesion molecule in the formation of the avian inner ear. Dev Dyn 213:359–369.

Busch-Nentwich E, Sollner C, Roehl H, Nicolson T (2004) The deafness gene *dfna5* is crucial for *ugdh* expression and HA production in the developing ear in zebrafish. Development 131:943–951.

Camenisch TD, Spicer AP, Brehm-Gibson T, Biesterfeldt J, Augustine ML, Calabro A, Jr., Kubalak S, Klewer SE, McDonald JA (2000) Disruption of hyaluronan synthase-2 abrogates normal cardiac morphogenesis and hyaluronan-mediated transformation of epithelium to mesenchyme. J Clin Invest 106:349–360.

Carpenter EM, Goddard JM, Chisaka O, Manley NR, Capecchi MR (1993) Loss of *Hox-A1 (Hox-1.6)* function results in the reorganization of the murine hindbrain. Development 118:1063–1075.

Cecconi F, Roth KA, Dolgov O, Munarriz E, Anokhin K, Gruss P, Salminen M (2004) Apaf1-dependent programmed cell death is required for inner ear morphogenesis and growth. Development 131:2125–2135.

Chang W, Nunes FD, De Jesus-Escobar JM, Harland R, Wu DK (1999) Ectopic Noggin blocks sensory and nonsensory organ morphogenesis in the chicken inner ear. Dev Biol 216:369–381.

Chang W, Cole L, Cantos R, Wu DK (2002) Molecular genetics of vestibular organ development. In: Highstein SM, Fay RR, Popper AN (eds), The Vestibular Sytem. New York: Springer-Verlag, pp. 11–56.

Chisaka O, Musci TS, Capecchi MR (1992) Developmental defects of the ear, cranial nerves and hindbrain resulting from targeted disruption of the mouse homeobox gene *Hox-1.6*. Nature 355:516–520.

Colas JF, Schoenwolf GC (2001) Towards a cellular and molecular understanding of neurulation. Dev Dyn 221:117–145.

Copp AJ (1994) Genetic models of mammalian neural tube defects. Ciba Found Symp 181:118–134.

Copp AJ, Greene ND, Murdoch JN (2003) The genetic basis of mammalian neurulation. Nat Rev Genet 4:784–793.

Cordes SP, Barsh GS (1994) The mouse segmentation gene *kr* encodes a novel basic domain-leucine zipper transcription factor. Cell 79:1025–1034.

Couly G, Le Douarin NM (1990) Head morphogenesis in embryonic avian chimeras: evidence for a segmental pattern in the ectoderm corresponding to the neuromeres. Development 108:543–558.

Cox GA, Mahaffey CL, Nystuen A, Letts VA, Frankel WN (2000) The mouse fidgetin gene defines a new role for AAA family proteins in mammalian development. Nat Genet 26:198–202.

Cremer H, Lange R, Christoph A, Plomann M, Vopper G, Roes J, Brown R, Baldwin S, Kraemer P, Scheff S, et al. (1994) Inactivation of the N-CAM gene in mice results in size reduction of the olfactory bulb and deficits in spatial learning. Nature 367:455–459.

D'Amico-Martel A, Noden DM (1983) Contributions of placodal and neural crest cells to avian cranial peripheral ganglia. Am J Anat 166:445–468.

Dassule HR, Lewis P, Bei M, Maas R, McMahon AP (2000) Sonic hedgehog regulates growth and morphogenesis of the tooth. Development 127:4775–4785.

Deol MS (1964a) The abnormalities of the inner ear in *kreisler* mice. J Embryol Exp Morphol 12:475–490.

Deol MS (1964b) The origin of the abnormalities of the inner ear in dreher mice. J Embryol Exp Morphol 12:717–733.

Deol MS (1966) Influence of the neural tube on the differentiation of the inner ear in the mammalian embryo. Nature 209:219–220.

Depew MJ, Liu JK, Long JE, Presley R, Meneses JJ, Pedersen RA, Rubenstein JL (1999) *Dlx5* regulates regional development of the branchial arches and sensory capsules. Development 126:3831–3846.

Detwiler SR, Van Dyke RH (1950) The role of the medulla in the differentiation of the otic vesicle. J Exp Zool 113:179–199.

Detwiler SR, Van Dyke RH (1951) Recent experiments on the differentiation of the labyrinth in amblystoma. J Exp Zool 118:389–401.

Dick A, Hild M, Bauer H, Imai Y, Maifeld H, Schier AF, Talbot WS, Bouwmeester T, Hammerschmidt M (2000) Essential role of *Bmp7* (*snailhouse*) and its prodomain in dorsoventral patterning of the zebrafish embryo. Development 127:343–354.

Dionne MS, Skarnes WC, Harland RM (2001) Mutation and analysis of *Dan*, the founding member of the Dan family of transforming growth factor ß antagonists. Mol Cell Biol 21:636–643.

Epstein DJ, Wekemans M, Gros P (1991) *splotch* (*Sp2H*), a mutation affecting development of the mouse neural tube, shows a deletion within the paired homeodomain of *Pax-3*. Cell 67:767–774.

Favor J, Sandulache R, Neuhäuser-Klaus A, Pretsch W, Chatterjee B, Senft E, Wurst W, Blanquet V, Grimes P, Spörle R, Schughart K (1996) The mouse *Pax2*^{1Neu} mutation is identical to a human *PAX2* mutation in a family with renal-coloboma syndrome and results in developmental defects of the brain, ear, eye, and kidney. Proc Natl Acad Sci USA 93:13870–13875.

Fekete DM (1996) Cell fate specification in the inner ear. Curr Opin Neurobiol 6:533–541.

Fekete DM (1999) Development of the vertebrate ear: insights from knockouts and mutants. Trends Neurosci 22:263–269.

Fekete DM, Homburger SA, Waring MT, Riedl AE, Garcia LF (1997) Involvement of programmed cell death in morphogenesis of the vertebrate inner ear. Development 124:2451–2461.

Forsberg E, Hirsch E, Fröhlich L, Meyer M, Ekblom P, Aszodi A, Werner S, Fässler R (1996) Skin wounds and severed nerves heal normally in mice lacking tenascin-C. Proc Natl Acad Sci USA 93:6594–6599.

Fritzsch B, Barald KF, Lomax MI (1998) Early embryology of the vertebrate ear. In: Rubel EW, Popper AN, Fay RR (eds), Development of the Auditory System. New York: Springer-Verlag, pp. 80–145.

Frohman MA, Martin GR, Cordes SP, Halamek LP, Barsh GS (1993) Altered rhombomere-specific gene expression and hyoid bone differentiation in the mouse segmentation mutant, *kreisler (kr)*. Development 117:925–936.

García-García MJ, Anderson KV (2003) Essential role of glycosaminoglycans in Fgf signaling during mouse gastrulation. Cell 114:727–737.

Garcia-Martinez V, Alvarez IS, Schoenwolf GC (1993) Locations of the ectodermal and nonectodermal subdivisions of the epiblast at stages 3 and 4 of avian gastrulation and neurulation. J Exp Zool 267:431–446.

Gelineau-van Waes J, Finnell RH (2001) Genetics of neural tube defects. Semin Pediatr Neurol 8:160–164.

Gerchman E, Hilfer SR, Brown JW (1995) Involvement of extracellular matrix in the formation of the inner ear. Dev Dyn 202:421–432.

Gerlach LM, Hutson MR, Germiller JA, Nguyen-Luu D, Victor JC, Barald KF (2000) Addition of the BMP4 antagonist, noggin, disrupts avian inner ear development. Development 127:45–54.

Gerlach-Bank LM, Ellis AD, Noonen B, Barald KF (2002) Cloning and expression analysis of the chick DAN gene, an antagonist of the BMP family of growth factors. Dev Dyn 224:109–115.

Gerlach-Bank LM, Cleveland AR, Barald KF (2004) DAN directs endolymphatic sac and duct outgrowth in the avian inner ear. Dev Dyn 229:219–230.

Gould SE, Upholt WB, Kosher RA (1995) Characterization of chicken Syndecan-3 as a heparan sulfate proteoglycan and its expression during embryogenesis. Dev Biol 168: 438–451.

Goulding M, Sterrer S, Fleming J, Balling R, Nadeau J, Moore KJ, Brown SD, Steel KP, Gruss P (1993) Analysis of the *Pax-3* gene in the mouse mutant *splotch.* Genomics 17:355–363.

Haddon C, Lewis J (1996) Early ear development in the embryo of the zebrafish, *Danio rerio.* J Comp Neurol 365:113–128.

Haddon CM, Lewis JH (1991) Hyaluronan as a propellant for epithelial movement: the development of semicircular canals in the inner ear of *Xenopus.* Development 112: 541–550.

Hadrys T, Braun T, Rinkwitz-Brandt S, Arnold HH, Bober E (1998) Nkx5-1 controls semicircular canal formation in the mouse inner ear. Development 125:33–39.

Hammond KL, Loynes HE, Folarin AA, Smith J, Whitfield TT (2003) Hedgehog signalling is required for correct anteroposterior patterning of the zebrafish otic vesicle. Development 130:1403–1417.

Harrison RG (1936) Relations of symmetry in the developing ear of amblystoma punctatum. Proc Natl Acad Sci USA 22:238–247.

Hendriks DM, Toerien MJ (1973) Experimental endolymphatic hydrops. S Afr Med J 47:2294–2298.

Herbrand H, Guthrie S, Hadrys T, Hoffmann S, Arnold HH, Rinkwitz-Brandt S, Bober E (1998) Two regulatory genes, *cNkx5-1* and *cPax2*, show different responses to local signals during otic placode and vesicle formation in the chick embryo. Development 125:645–654.

Hidalgo-Sánchez M, Alvarado-Mallart R, Alvarez IS (2000) *Pax2*, *Otx2*, *Gbx2* and *Fgf8* expression in early otic vesicle development. Mech Dev 95:225–229.

Hilfer SR, Randolph GJ (1993) Immunolocalization of basal lamina components during development of chick otic and optic primordia. Anat Rec 235:443–452.

Hilfer SR, Esteves RA, Sanzo JF (1989) Invagination of the otic placode: normal development and experimental manipulation. J Exp Zool 251:253–264.

Hilfer SR, Marrero L, Sheffield JB (1990) Patterns of cell movement in early organ primordia of the chick embryo. Anat Rec 227:508–517.

Hui CC, Joyner AL (1993) A mouse model of Greig cephalopolysyndactyly syndrome: the *extra-toes^J* mutation contains an intragenic deletion of the *Gli3* gene. Nat Genet 3:241–246.

Hulander M, Wurst W, Carlsson P, Enerbäck S (1998) The winged helix transcription factor Fkh10 is required for normal development of the inner ear. Nat Genet 20:374–376.

Hulander M, Kiernan AE, Blomqvist SR, Carlsson P, Samuelsson EJ, Johansson BR, Steel KP, Enerbäck S (2003) Lack of pendrin expression leads to deafness and expansion of the endolymphatic compartment in inner ears of *Foxi1* null mutant mice. Development 130:2013–2025.

Hutson MR, Lewis JE, Nguyen-Luu D, Lindberg KH, Barald KF (1999) Expression of Pax2 and patterning of the chick inner ear. J Neurocytol 28:795–807.

Ingham PW, McMahon AP (2001) Hedgehog signaling in animal development: paradigms and principles. Genes Dev 15:3059–3087.

Jerome LA, Papaioannou VE (2001) DiGeorge syndrome phenotype in mice mutant for the T-box gene, *Tbx1*. Nat Genet 27:286–291.

Johnson DR (1967) *Extra-toes*: a new mutant gene causing multiple abnormalities in the mouse. J Embryol Exp Morphol 17:543–581.

Juriloff DM, Harris MJ (2000) Mouse models for neural tube closure defects. Hum Mol Genet 9:993–1000.

Kaan HW (1930) The relation of the developing auditory vesicle to the formation of the cartilage capsule in Amblystoma punctatum. J Exp Zool 55:263–291.

Kaan HW (1938) Further studies on the auditory vesicle and cartilaginous capsule of Amblystoma punctatum. J Exp Zool 78:159–183.

Kalatzis V, Sahly I, El-Amraoui A, Petit C (1998) *Eya1* expression in the developing ear and kidney: towards the understanding of the pathogenesis of branchio-oto-renal (BOR) syndrome. Dev Dyn 213:486–499.

Karis A, Pata I, van Doorninck JH, Grosveld F, de Zeeuw CI, de Caprona D, Fritzsch B (2001) Transcription factor GATA-3 alters pathway selection of olivocochlear neurons and affects morphogenesis of the ear. J Comp Neurol 429:615–630.

Kaufman MH, Bard JBL (1999) The eye and the ear. In: The Anatomical Basis of Mouse Development. San Diego: Academic Press, pp. 194–208.

Keller R (2002) Shaping the vertebrate body plan by polarized embryonic cell movements. Science 298:1950–1954.

Kibar Z, Vogan KJ, Groulx N, Justice MJ, Underhill DA, Gros P (2001) *Ltap*, a mammalian homolog of *Drosophila Strabismus/Van Gogh*, is altered in the mouse neural tube mutant *Loop-tail*. Nat Genet 28:251–255.

Kiernan AE, Steel KP, Fekete DM (2002) Development of the mouse inner ear. In: Rossant J, Tam PPL (eds), Mouse Development: Patterning, Morphogenesis and Organogenesis. London: Academic Press, pp. 539–566.

Kishimoto Y, Lee KH, Zon L, Hammerschmidt M, Schulte-Merker S (1997) The molecular nature of zebrafish *swirl*: BMP2 function is essential during early dorsoventral patterning. Development 124:4457–4466.

Klüppel M, Vallis KA, Wrana JL (2002) A high-throughput induction gene trap approach defines *C4ST* as a target of BMP signaling. Mech Dev 118:77–89.

Knowlton VY (1967) Correlation of the development of membranous and bony labyrinths, acoustic ganglia, nerves, and brain centers of the chick embryo. J Morphol 121:179–206.

Koebernick K, Hollemann T, Pieler T (2001) Molecular cloning and expression analysis of the Hedgehog receptors *XPtc1* and *XSmo* in *Xenopus laevis*. Mech Dev 100:303–308.

Kozlowski DJ, Murakami T, Ho RK, Weinberg ES (1997) Regional cell movement and tissue patterning in the zebrafish embryo revealed by fate mapping with caged fluorescein. Biochem Cell Biol 75:551–562.

Laclef C, Souil E, Demignon J, Maire P (2003) Thymus, kidney and craniofacial abnormalities in *Six1* deficient mice. Mech Dev 120:669–679.

Lawoko-Kerali G, Rivolta MN, Holley M (2002) Expression of the transcription factors GATA3 and Pax2 during development of the mammalian inner ear. J Comp Neurol 442:378–391.

Li CW, Van De Water TR, Ruben RJ (1978) The fate mapping of the eleventh and twelfth day mouse otocyst: an in vitro study of the sites of origin of the embryonic inner ear sensory structures. J Morphol 157:249–268.

Lindsay EA, Vitelli F, Su H, Morishima M, Huynh T, Pramparo T, Jurecic V, Ogunrinu G, Sutherland HF, Scambler PJ, Bradley A, Baldini A (2001) *Tbx1* haploinsufficieny in the DiGeorge syndrome region causes aortic arch defects in mice. Nature 410:97–101.

Liu W, Li G, Chien J, Raft S, Zhang H, Chiang C, Frenz D (2002) Sonic hedgehog regulates otic capsule chondrogenesis and inner ear development in the mouse embryo. Dev Biol 248:240.

Livesey FJ (1999) Netrins and netrin receptors. Cell Mol Life Sci 56:62–68.

Lufkin T, Dierich A, LeMeur M, Mark M, Chambon P (1991) Disruption of the *Hox-1.6* homeobox gene results in defects in a region corresponding to its rostral domain of expression. Cell 66:1105–1119.

Mansour SL, Goddard JM, Capecchi MR (1993) Mice homozygous for a targeted disruption of the proto-oncogene *int-2* have developmental defects in the tail and inner ear. Development 117:13–28.

Mark M, Lufkin T, Vonesch JL, Ruberte E, Olivo JC, Dollé P, Gorry P, Lumsden A, Chambon P (1993) Two rhombomeres are altered in *Hoxa-1* mutant mice. Development 119:319–338.

Markwald RR, Fitzharris TP, Bank H, Bernanke DH (1978) Structural analyses on the matrical organization of glycosaminoglycans in developing endocardial cushions. Dev Biol 62:292–316.

Martin P, Swanson GJ (1993) Descriptive and experimental analysis of the epithelial remodellings that control semicircular canal formation in the developing mouse inner ear. Dev Biol 159:549–558.

McKay IJ, Lewis J, Lumsden A (1996) The role of FGF-3 in early inner ear development: An analysis in normal and *kreisler* mutant mice. Dev Biol 174:370–378.

McMahon JA, Takada S, Zimmerman LB, Fan CM, Harland RM, McMahon AP (1998) Noggin-mediated antagonism of BMP signaling is required for growth and patterning of the neural tube and somite. Genes Dev 12:1438–1452.

McPhee JR, Van de Water TR, Su HX (1987) Hyaluronate production by the inner ear during otic capsule and perilymphatic space formation. Am J Otolaryngol 8:265–272.

Meier S (1978a) Development of the embryonic chick otic placode. I. Light microscopic analysis. Anat Rec 191:447–458.

Meier S (1978b) Development of the embryonic chick otic placode. II. Electron microscopic analysis. Anat Rec 191:459–477.

Merlo GR, Paleari L, Mantero S, Zerega B, Adamska M, Rinkwitz S, Bober E, Levi G (2002) The *Dlx5* homeobox gene is essential for vestibular morphogenesis in the mouse embryo through a BMP4-mediated pathway. Dev Biol 248:157–169.

Merscher S, Funke B, Epstein JA, Heyer J, Puech A, Lu MM, Xavier RJ, Demay MB, Russell RG, Factor S, Tokooya K, Jore BS, Lopez M, Pandita RK, Lia M, Carrion D, Xu H, Schorle H, Kobler JB, Scambler P, Wynshaw-Boris A, Skoultchi AI, Morrow BE, Kucherlapati R (2001) *TBX1* is responsible for cardiovascular defects in velo-cardio-facial/DiGeorge syndrome. Cell 104:619–629.

Minowa O, Ikeda K, Sugitani Y, Oshima T, Nakai S, Katori Y, Suzuki M, Furukawa M, Kawase T, Zheng Y, Ogura M, Asada Y, Watanabe K, Yamanaka H, Gotoh S, Nishi-Takeshima M, Sugimoto T, Kikuchi T, Takasaka T, Noda T (1999) Altered cochlear fibrocytes in a mouse model of DFN3 nonsyndromic deafness. Science 285:1408–1411.

Moens CB, Cordes SP, Giorgianni MW, Barsh GS, Kimmel CB (1998) Equivalence in the genetic control of hindbrain segmentation in fish and mouse. Development 125: 381–391.

Montcouquiol M, Rachel RA, Lanford PJ, Copeland NG, Jenkins NA, Kelley MW (2003) Identification of *Vangl2* and *Scrb1* as planar polarity genes in mammals. Nature 423: 173–177.

Moro-Balbas JA, Gato A, Alonso MI, Martin P, de la Mano A (2000) Basal lamina heparan sulphate proteoglycan is involved in otic placode invagination in chick embryos. Anat Embryol 202:333–343.

Morsli H, Choo D, Ryan A, Johnson R, Wu DK (1998) Development of the mouse inner ear and origin of its sensory organs. J Neurosci 18:3327–3335.

Morsli H, Tuorto F, Choo D, Postiglione MP, Simeone A, Wu DK (1999) *Otx1* and *Otx2* activities are required for the normal development of the mouse inner ear. Development 126:2335–2343.

Mowbray C, Hammerschmidt M, Whitfield TT (2001) Expression of BMP signalling pathway members in the developing zebrafish inner ear and lateral line. Mech Dev 108:179–184.

Murdoch JN, Doudney K, Paternotte C, Copp AJ, Stanier P (2001) Severe neural tube defects in the *loop-tail* mouse result from mutation of *Lpp1*, a novel gene involved in floor plate specification. Hum Mol Genet 10:2593–2601.

Neuhauss SC, Solnica-Krezel L, Schier AF, Zwartkruis F, Stemple DL, Malicki J, Abdelilah S, Stainier DY, Driever W (1996) Mutations affecting craniofacial development in zebrafish. Development 123:357–367.

Niederreither K, Subbarayan V, Dollé P, Chambon P (1999) Embryonic retinoic acid synthesis is essential for early mouse post-implantation development. Nat Genet 21: 444–448.

Niederreither K, Vermot J, Schuhbaur B, Chambon P, Dollé P (2000) Retinoic acid synthesis and hindbrain patterning in the mouse embryo. Development 127:75–85.

Nishikori T, Hatta T, Kawauchi H, Otani H (1999) Apoptosis during inner ear development in human and mouse embryos: an analysis by computer-assisted three-dimensional reconstruction. Anat Embryol (Berl) 200:19–26.

Nissen RM, Yan J, Amsterdam A, Hopkins N, Burgess SM (2003) Zebrafish *foxi* one modulates cellular responses to Fgf signaling required for the integrity of ear and jaw patterning. Development 130:2543–2554.

Noden DM, Van de Water TR (1992) Genetic analyses of mammalian ear development. Trends Neurosci 15:235–237.

Nornes HO, Dressler GR, Knapik EW, Deutsch U, Gruss P (1990) Spatially and temporally restricted expression of *Pax2* during murine neurogenesis. Development 109: 797–809.

Oh SH, Johnson R, Wu DK (1996) Differential expression of bone morphogenetic proteins in the developing vestibular and auditory sensory organs. J Neurosci 16:6463–6475.

Orr MF (1976) The influence of mesenchyme on the development of the embryonic chick otocyst epithelium. J Cell Biol 70:155A.

Orr MF, Hafft LP (1980) The influence of mesenchyme on the development of the embryonic otocyst. An electron microscopic study. J Cell Biol 87:27A.

Ozaki H, Nakamura K, Funahashi J, Ikeda K, Yamada G, Tokano H, Okamura HO, Kitamura K, Muto S, Kotaki H, Sudo K, Horai R, Iwakura Y, Kawakami K (2004) *Six1* controls patterning of the mouse otic vesicle. Development 131:551–562.

Pasqualetti M, Neun R, Davenne M, Rijli FM (2001) Retinoic acid rescues inner ear defects in *Hoxa1* deficient mice. Nat Genet 29:34–39.

Paterson NF (1948) The development of the inner ear of *Xenopus laevis*. Proc Zool Soc Lond 119:269–291.

Pauley S, Wright TJ, Pirvola U, Ornitz D, Beisel K, Fritzsch B (2003) Expression and function of FGF10 in mammalian inner ear development. Dev Dyn 227:203–215.

Phippard D, Lu L, Lee D, Saunders JC, Crenshaw EB, 3rd (1999) Targeted mutagenesis of the POU-domain gene *Brn4/Pou3f4* causes developmental defects in the inner ear. J Neurosci 19:5980–5989.

Phippard D, Boyd Y, Reed V, Fisher G, Masson WK, Evans EP, Saunders JC, Crenshaw EB, 3rd (2000) The *sex-linked fidget* mutation abolishes *Brn4/Pou3f4* gene expression in the embryonic inner ear. Hum Mol Genet 9:79–85.

Piotrowski T, Nüsslein-Volhard C (2000) The endoderm plays an important role in patterning the segmented pharyngeal region in zebrafish (*Danio rerio*). Dev Biol 225: 339–356.

Piotrowski T, Ahn DG, Schilling TF, Nair S, Ruvinsky I, Geisler R, Rauch GJ, Haffter P, Zon LI, Zhou Y, Foott H, Dawid IB, Ho RK (2003) The zebrafish *van gogh* mutation disrupts *tbx1*, which is involved in the DiGeorge deletion syndrome in humans. Development 130:5043–5052.

Pirvola U, Spencer-Dene B, Xing-Qun L, Kettunen P, Thesleff I, Fritzsch B, Dickson C, Ylikoski J (2000) FGF/FGFR-2(IIIb) signaling is essential for inner ear morphogenesis. J Neurosci 20:6125–6134.

Ponnio T, Burton Q, Pereira FA, Wu DK, Conneely OM (2002) The nuclear receptor Nor-1 is essential for proliferation of the semicircular canals of the mouse inner ear. Mol Cell Biol 22:935–945.

Raft S, Nowotschin S, Liao J, Morrow BE (2004) Suppression of neural fate and control of inner ear morphogenesis by *Tbx1*. Development 131:1801–1812.

Riccomagno MM, Martinu L, Mulheisen M, Wu DK, Epstein DJ (2002) Specification of the mammalian cochlea is dependent on Sonic hedgehog. Genes Dev 16:2365–2378.

Richardson GP, Crossin KL, Chuong CM, Edelman GM (1987) Expression of cell adhesion molecules during embryonic induction. III. Development of the otic placode. Dev Biol 119:217–230.

Riethmacher D, Brinkmann V, Birchmeier C (1995) A targeted mutation in the mouse E-cadherin gene results in defective preimplantation development. Proc Natl Acad Sci USA 92:855–859.

Riley BB, Chiang M, Farmer L, Heck R (1999) The *deltaA* gene of zebrafish mediates lateral inhibition of hair cells in the inner ear and is regulated by *pax2.1*. Development 126:5669–5678.

Rinkwitz-Brandt S, Justus M, Oldenettel I, Arnold HH, Bober E (1995) Distinct temporal expression of mouse *Nkx-5.1* and *Nkx-5.2* homeobox genes during brain and ear development. Mech Dev 52:371–381.

Rinkwitz-Brandt S, Arnold HH, Bober E (1996) Regionalized expression of *Nkx5-1, Nkx5-2, Pax2* and *sek* genes during mouse inner ear development. Hear Res 99:129–138.

Romand R, Hashino E, Dollé P, Vonesch JL, Chambon P, Ghyselinck NB (2002) The retinoic acid receptors RARalpha and RARgamma are required for inner ear development. Mech Dev 119:213–223.

Romberger DJ (1997) Fibronectin. Int J Biochem Cell Biol 29:939–943.

Salminen M, Meyer BI, Bober E, Gruss P (2000) netrin 1 is required for semicircular canal formation in the mouse inner ear. Development 127:13–22.

Schoenwolf GC, Alvarez IS (1989) Roles of neuroepithelial cell rearrangement and division in shaping of the avian neural plate. Development 106:427–439.

Schoenwolf GC, Fisher M (1983) Analysis of the effects of *Streptomyces* hyaluronidase on formation of the neural tube. J Embryol Exp Morphol 73:1–15.

Schoenwolf GC, Smith JL (1990) Mechanisms of neurulation: traditional viewpoint and recent advances. Development 109:243–270.

Sher AE (1971) The embryonic and postnatal development of the inner ear of the mouse. Acta Oto-laryngol 285 (Suppl.):1–77.

Sinning AR, Olson MD (1988) Surface coat material associated with the developing otic placode/vesicle in the chick. Anat Rec 220:198–207.

Smith JL, Schoenwolf GC (1997) Neurulation: coming to closure. Trends Neurosci 20: 510–517.

Solomon KS, Kudoh T, Dawid IB, Fritz A (2003a) Zebrafish *foxi1* mediates otic placode formation and jaw development. Development 130:929–940.

Solomon KS, Logsdon JM, Jr., Fritz A (2003b) Expression and phylogenetic analyses of three zebrafish FoxI class genes. Dev Dyn 228:301–307.

Sulik KK, Cotanche DA (1995) Embryology of the ear. In: Gorlin RJ, Toriello HV, Cohen MM (eds), Hereditary Hearing Loss and Its Syndromes. New York: Oxford University Press, pp. 22–42.

Swanson GJ, Howard M, Lewis J (1990) Epithelial autonomy in the development of the inner ear of a bird embryo. Dev Biol 137:243–257.

ten Berge D, Brouwer A, Korving J, Martin JF, Meijlink F (1998) *Prx1* and *Prx2* in skeletogenesis: roles in the craniofacial region, inner ear and limbs. Development 125: 3831–3842.

Torres M, Giráldez F (1998) The development of the vertebrate inner ear. Mech Dev 71:5–21.

Torres M, Gómez-Pardo E, Gruss P (1996) *Pax2* contributes to inner ear patterning and optic nerve trajectory. Development 122:3381–3391.

Truslove GM (1956) The anatomy and development of the fidget mouse. J Genet 54: 64–86.

Van De Water TR (1977) The effect of the removal of the endolymphatic duct anlage upon organogenesis of the mammalian inner ear "in vitro:" a preliminary report. Arch Otorhinolaryngol 217:297–311.

Van De Water TR, Li CW, Ruben RJ, Shea CA (1980) Ontogenic aspects of mammalian inner ear development. Birth Defects: Orig Article Ser 16:5–45.

Visconti RP, Hilfer SR (2002) Perturbation of extracellular matrix prevents association of the otic primordium with the posterior rhombencephalon and inhibits subsequent invagination. Dev Dyn 223:48–58.

Vitelli F, Morishima M, Taddei I, Lindsay EA, Baldini A (2002) *Tbx1* mutation causes multiple cardiovascular defects and disrupts neural crest and cranial nerve migratory pathways. Hum Mol Genet 11:915–922.

Vitelli F, Viola A, Morishima M, Pramparo T, Baldini A, Lindsay E (2003) *TBX1* is required for inner ear morphogenesis. Hum Mol Genet 12:2041–2048.

Vortkamp A, Franz T, Gessler M, Grzeschik KH (1992) Deletion of *GLI3* supports the homology of the human Greig cephalopolysyndactyly syndrome (GCPS) and the mouse mutant *extra toes* (*Xt*). Mamm Genome 3:461–463.

Wallingford JB, Fraser SE, Harland RM (2002) Convergent extension: the molecular control of polarized cell movement during embryonic development. Dev Cell 2:695–706.

Walsh EC, Stainier DY (2001) UDP-glucose dehydrogenase required for cardiac valve formation in zebrafish. Science 293:1670–1673.

Wang W, Van De Water T, Lufkin T (1998) Inner ear and maternal reproductive defects in mice lacking the *Hmx3* homeobox gene. Development 125:621–634.

Wang W, Chan EK, Baron S, Van De Water T, Lufkin T (2001) *Hmx2* homeobox gene control of murine vestibular morphogenesis. Development 128:5017–5029.

Waterman RE, Bell DH (1984) Epithelial fusion during early semicircular canal formation in the embryonic zebrafish, *Brachydanio rerio.* Anat Rec 210:101–114.

Wendling O, Ghyselinck NB, Chambon P, Mark M (2001) Roles of retinoic acid receptors in early embryonic morphogenesis and hindbrain patterning. Development 128:2031–2038.

Whitfield T, Haddon C, Lewis J (1997) Intercellular signals and cell-fate choices in the developing inner ear: origins of global and of fine-grained pattern. Semin Cell Dev Biol 8:239–247.

Whitfield TT, Granato M, van Eeden FJ, Schach U, Brand M, Furutani-Seiki M, Haffter P, Hammerschmidt M, Heisenberg CP, Jiang YJ, Kane DA, Kelsh RN, Mullins MC, Odenthal J, Nüsslein-Volhard C (1996) Mutations affecting development of the zebrafish inner ear and lateral line. Development 123:241–254.

Whitfield TT, Riley BB, Chiang MY, Phillips B (2002) Development of the zebrafish inner ear. Dev Dyn 223:427–458.

Wilkinson DG, Peters G, Dickson C, McMahon AP (1988) Expression of the FGF-related proto-oncogene *int-2* during gastrulation and neurulation in the mouse. EMBO J 7: 691–695.

Wilkinson DG, Bhatt S, McMahon AP (1989) Expression pattern of the FGF-related proto-oncogene *int-2* suggests multiple roles in fetal development. Development 105: 131–136.

Winnier G, Blessing M, Labosky PA, Hogan BL (1995) Bone morphogenetic protein-4 is required for mesoderm formation and patterning in the mouse. Genes Dev 9:2105–2116.

Wright TJ, Mansour SL (2003) *Fgf3* and *Fgf10* are required for mouse otic placode induction. Development 130:3379–3390.

Wright TJ, Hatch EP, Karabagli H, Karabagli P, Schoenwolf GC, Mansour SL (2003) Expression of mouse fibroblast growth factor and fibroblast growth factor receptor genes during early inner ear development. Dev Dyn 228:267–272.

Wu DK, Oh SH (1996) Sensory organ generation in the chick inner ear. J Neurosci 16: 6454–6462.

Wu DK, Nunes FD, Choo D (1998) Axial specification for sensory organs versus non-sensory structures of the chicken inner ear. Development 125:11–20.

Xu PX, Adams J, Peters H, Brown MC, Heaney S, Maas R (1999) *Eya1*-deficient mice lack ears and kidneys and show abnormal apoptosis of organ primordia. Nat Genet 23:113–117.

Yamagishi H, Maeda J, Hu T, McAnally J, Conway SJ, Kume T, Meyers EN, Yamagishi C, Srivastava D (2003) *Tbx1* is regulated by tissue-specific forkhead proteins through a common Sonic hedgehog-responsive enhancer. Genes Dev 17:269–281.

Zheng W, Huang L, Wei ZB, Silvius D, Tang B, Xu PX (2003) The role of *Six1* in mammalian auditory system development. Development 130:3989–4000.

Zimmerman LB, De Jesus-Escobar JM, Harland RM (1996) The Spemann organizer signal noggin binds and inactivates bone morphogenetic protein 4. Cell 86:599–606.

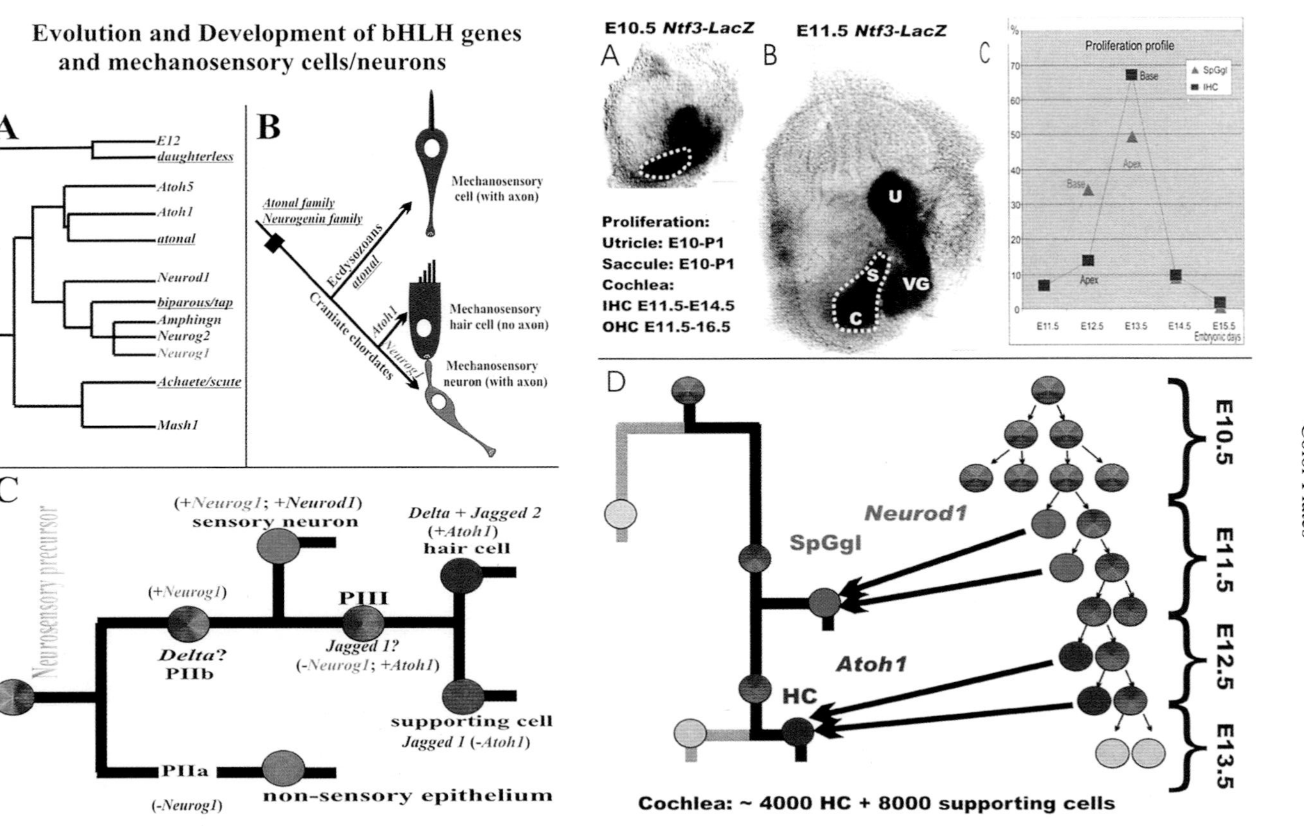

FIGURE 4.1. (*Left*) Basic helix–loop–helix (bHLH) genes. Modified after Fritzsch et al. (2000); Bertrand et al. (2002); Zine (2003); Fritzsch and Beisel (2004). FIGURE 4.2. (*Right*) Possible clonal relationship and clonal expansion of neurosensory precursors. Modified after Ruben (1967); Fritzsch et al. (2000); Farinas et el. (2001); Cai et al. (2002).

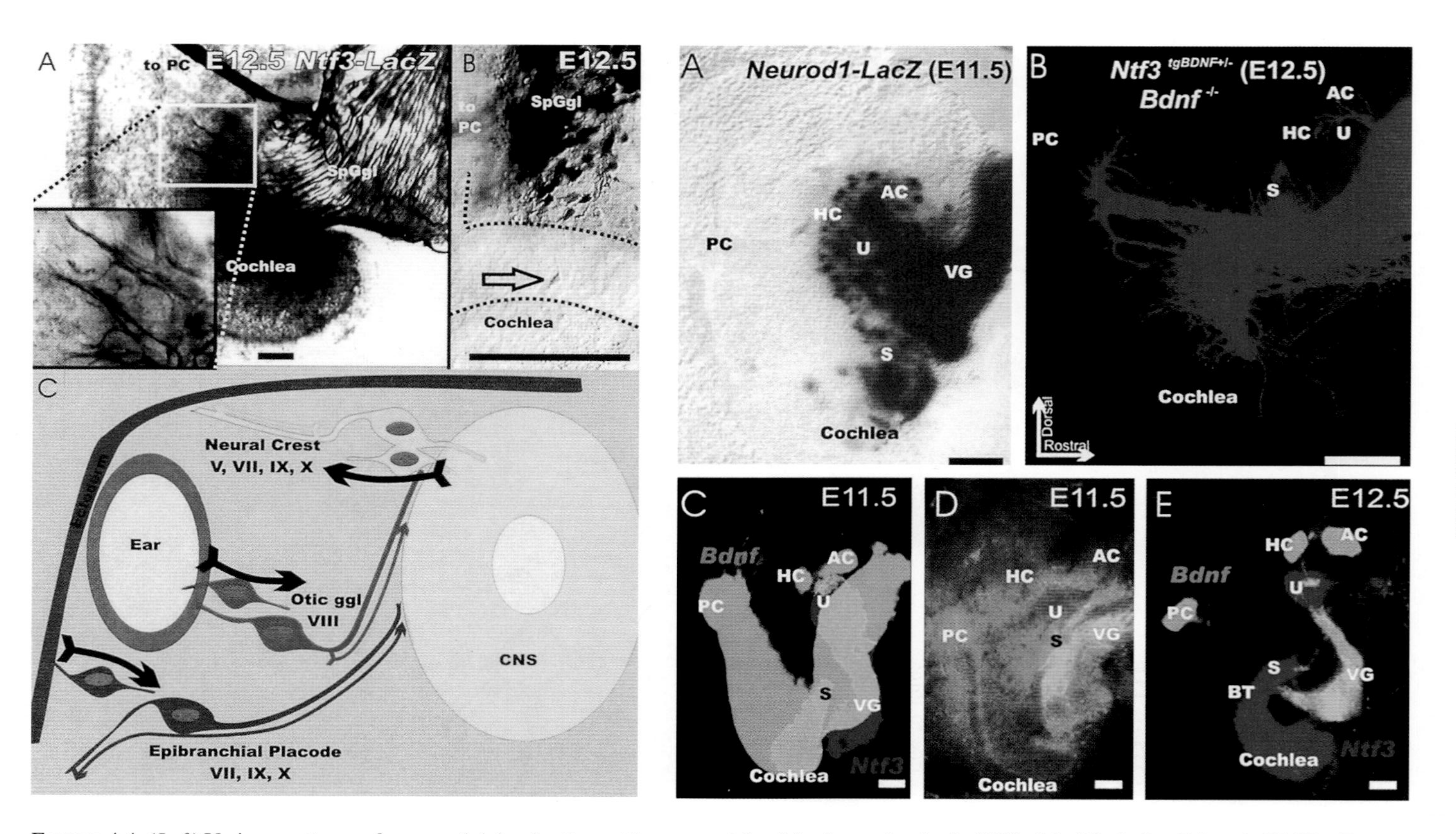

FIGURE 4.4. (*Left*) Various patterns of neuronal delamination with axon and dendrite formation in the PNS. Modified after Fritzsch (2003). FIGURE 4.5. (*Right*) The formation of sensory neurons, expressions of neurotrophins, and effects of neurotrophin misexpression. Modified after Fritzsch et al. (2002); Tessarollo et al. (2004).

4

Wiring the Ear to the Brain: The Molecular Basis of Neurosensory Development, Differentiation, and Survival

SARAH PAULEY, VERONICA MATEI, KIRK W. BEISEL, AND BERND FRITZSCH

Physicists like to think that all you have to do is say:
These are the conditions! Now what happens next? *Richard P. Feynman*

1. Introduction

Affecting both the young, as a result of congenital malformations or deficits acquired in childhood, and the old, as a result of age-related changes, hearing loss is one of the most prevalent chronic disabilities of our time. Hearing reduction or loss can affect up to one in three elderly people. Normal hearing requires proper sound transmission to healthy cochlear hair cells for mechano-electric transduction and sensory neurons to transmit the signal from the hair cells to the brain. In the inner ear, the hair cells are most vulnerable to various drugs and intense sound. The most common solution for neurosensory hearing loss involves the use of cochlear implants to replace the hair cells and stimulate the surviving sensory neurons directly.

Since the first cochlear implants were developed in the late 1970s, nearly 23,000 severely hearing impaired patients have received implants in the United States alone. Cochlear implantation acts in place of hair cells by supplying auditory information, via electrical stimulation, directly to the cochlear nerves. Proper function of such a device depends on its ability to reach and stimulate individual neuronal processes, which have retained their proper connections to the brain. Cochlear implants are largely limited by the number of discrete channels that can be interpreted by the sensory neurons to extract signal from noise for the listener.

An alternative approach for treating hearing loss is to regenerate damaged hair cells. For this, either new hair cells need to be generated out of progenitors and seeded into the ear, or remaining cells need to be converted into a hair cell phenotype. The differentiation of embryonic stem cells into inner ear progenitors has been described. Further, when implanted into the injured sensory epithelia of the inner ear in chick embryos, some of these cells were able to integrate into the developing ear, express hair cell–specific markers, and produce hair bundles (Li et al. 2003 a,b). In addition, transfection with specific transcription factors can lead to transformation of embryonic or adult inner ear cells into hair cells (Gao 2003; Kawamoto et al. 2003). To effectively use this approach to restore hearing, we need to ensure that these hair cells can not only integrate into the proper location in the sensory epithelium but also form specific connections with sensory neurons to transmit the sound signals to the cochlear nuclei in the proper topological (cochleotopic) order. Optimizing such a technology requires an appreciation of the developmental relationships between these essential neurosensory components and how they become specifically connected in order to implement this during regeneration.

The authors aim to develop a complete model of precursor development, their numerical regulation and cell fate determining processes so as to predict and manipulate the generation of both sensory neurons and hair cells in the mouse. Only once this has been accomplished will cochlear neurosensory restoration in humans using a molecular approach be possible. This chapter reviews what has been learned about vestibular and cochlear sensory neuron development, and presents a current model of precursor proliferation and cell fate specification, including the molecular basis of nerve fiber guidance and sensory neuron survival. Throughout this chapter the currently approved genetic nomenclature will be used (Table 4.1).

2. Specification of Areas of Neurosensory Formation

Development of the vertebrate inner ear begins with a thickening of ectoderm adjacent to the hindbrain, referred to as the otic placode (Brown et al. 2003). This placode folds in upon itself and moves inward to become surrounded by mesoderm. This patch of ectoderm, influenced by the mesodermal signals surrounding it, ultimately develops into the otic epithelium, including hair cells, supporting cells, and the primary sensory neurons that innervate the hair cells. The development of this otocyst depends on the pattern of gene expression both within the otic capsule and in the surrounding mesoderm. Global patterning of the otocyst is influenced by mesodermal signals such as sonic hedgehog (*Shh*), and has been described using the compartmental model (Fekete and Wu 2002). Once established, this global patterning gradually gives rise to more elaborate and discrete local patterning as demonstrated by fibroblast growth factor 10 (*Fgf10*) (Pauley et al. 2003). It is this patterning in the inner ear that defines the areas for neurosensory formation.

Primary sensory neuron primordia can be identified either as delaminating cells (Carney and Silver 1983) emigrating through the basal lamina surrounding the otocyst, or as cells that express specific markers such as neurogenin 1 (*Neurog1*; alias *Ngn1*) (Ma et al. 1998), neurogenic differentiation 1 (*Neurod1*; alias *NeuroD*) (Liu et al. 2000; Kim et al. 2001), neurotrophins (Farinas et al. 2001),

TABLE 4.1. Genetic nomenclature.

Gene name	Gene description	Gene aliases
Atoh1	Atonal homolog 1 (*Drosophila*)	*Math1*; *MATH-1*
Neurod1	Neurogenic differentiation 1	*BETA2*; *BHF-1*; *Neurod*
Neurog1	Neurogenin 1	*AKA*; *ngn1*; *Math4C*; *Neurod3*; *neurogenin*
Shh	Sonic hedgehog	*Dsh*; *Hhg1*; *9530036O11Rik*
Fgf10	Fibroblast growth factor 10	*FGF-10*
Gata3	GATA binding protein 3	*Gata-3*
Pou4f1	POU domain, class 4, transcription factor 1	*Brn3*; *Brn-3*; *Brn3a*; *Brn-3.0*
Pou4f3	POU domain, class 4, transcription factor 3	*ddl*; *Brn3c*; *Brn3.1*; *Brn-3.1*; *dreidel*
Pou3f4	POU domain, class 3, transcription factor 4	*Slf*; *Brn4*; *Otf9*; *Brn-4*
Gas1	Growth arrest specific 1	*Gas-1*
Eya1	Eyes absent 1 homolog (*Drosophila*)	*bor*
Six1	Sine oculis-related homeobox 1 homolog (*Drosophila*)	—
Dach1	Dachshund 1 (*Drosophila*)	*Dac*; *Dach*; *E130112M23Rik*
Dach2	Dachshund 2 (*Drosophila*)	—
Pax2	Paired box gene 2	*Pax-2*
Tbx1	T-box 1	—
Myod1	Myogenic differentiation 1	*MYF3*; *MyoD*; *Myod-1*
Eya4	Eyes absent 4 homolog (Drosophila)	—
Bmp4	Bone morphogenetic protein 4	*Bmp2b*; *Bmp2b1*; *Bmp2b-1*
Gdf11	Growth differentiation factor 11	*Bmp11*
Cdkn1b	Cyclin-dependent kinase inhibitor 1B (P27)	*p27*; *Kip1*; *p27Kip1*
Btg2	B-cell translocation gene 2, anti-proliferative	*Pc3*; *TIS21*
Otx1	Orthodenticle homolog 1 (Drosophila	*jv*; *A730044F23Rik*
Lfng	Lunatic fringe gene homolog (Drosophila)	—
Ntf3	Neurotrophin 3	*NT3*; *NT-3*; *Ntf-3*
Ephb2	Eph receptor B2	*Drt*; *Erk*; *Nuk*; *Cek5*; *Hek5*; *Qek5*; *Sek3*; *ETECK*; *Prkm5*; *Tyro5*
Erbb2	v-*erb*-b2 erythroblastic leukemia viral Oncogene homolog 2, neuro/glioblastoma derived oncogene homolog (avian)	*Neu*; *HER2*; *HER-2*; *c-neu*; *Erbb-2*; *c-erbB2*; *mKIAA3023*
Sema3a	Semaphorin 3A	*SemD*; *SEMA1*; *Semad*; *coll-1*; *Hsema-I*
Bdnf	Brain derived neurotrophic factor	—
Ntrk2	Neurotrophic tyrosine kinase, receptor, type 2	*Tkrb*; *trkB*

or other genes (Fekete and Wu 2002). These delaminating cells are first apparent around otocyst formation in mice at embryonic day 9.5 (E9.5; Ma et al. 1998), or even before the otocyst is completely formed in chicken (Adam et al. 1998). Later, primary neuron primordia express many other genes such as GATA binding protein 3 (*Gata3*) (Karis et al. 2001; Lawoko-Kerali et al. 2002), POU domain, class 4, transcription factor 1 (*Pou4f1*; alias *Brn3a*) (Huang et al. 2001) neurotrophic tyrosine kinase receptors (*Ntrks*) (Farinas et al. 2001), and fibroblast growth factors (*Fgfs*) (Pauley et al. 2003). In addition to these more general neuronal genes, subtypes of sensory neurons must be specified through either overlapping expression of general genes or through as yet uncharacterized specific genes.

The possibility for unique identities of early primary sensory neuron precursors is underscored by differential expression of several genes in early delaminating cells (Lawoko-Kerali et al. 2002). Further, expression of specific genes appears to be critical for survival of certain neuron populations, for example, data from *Neurod1* mutants demonstrate that most cochlear neurons die while more of the vestibular neurons survive (Kim et al. 2001). In general, the already known diversity of gene expression at these early stages indicates that various areas of the newly formed otocyst could provide unique identities to delaminating precursors based on overlapping and discrete regions of transcription factor expression. Despite this interesting start, it remains to be seen how differential areas of origin in the otocyst relate to differential gene expression and, ultimately, differential projection of primary neurons to specific sensory epithelia of the ear and specific areas of the brain (Maklad and Fritzsch 2002). It is possible, given sufficient nested expression patterns of various transcription factors within the ear, that primary neuron precursors acquire a unique cell fate assignment in the ear by analogy to that of neural crest derived primary sensory neurons and motoneurons in the central nervous system (CNS) (Gowan et al. 2001; Qian et al. 2001; Brunet and Pattyn 2002). If these initial data in the ear can be confirmed and extended by future work, development of distinct peripheral and central projections could be predicted as a consequence of molecularly acquired cell fates already predetermined in the otocyst.

In this context it is important to realize that proliferation, delamination, migration to the final position, and development of central and peripheral projections is a prolonged phase in mammals and birds that lasts for several days (Ruben 1967; Rubel and Fritzsch 2002). As previously noted by Carney and Silver (Carney and Silver 1983) and recently confirmed by Farinas et al. (Farinas et al. 2001) and Fritzsch et al. (Fritzsch et al. 2002), delaminating cells (which are likely neuronal precursors based on *Neurod1* expression) apparently migrate away from the otocyst, sending their lagging process, the dendrite, along fibers of more differentiated neurons that project toward the future sensory epithelia. It appears that spatiotemporally distinct populations of primary sensory neuron precursors specifically extend along the existing neuronal fibers that reach toward the future primary sensory epithelium. Thus, it is possible that fate acquisition, as specified through the gene expression mosaic in the otocyst, results in restricted areas of primary sensory neuron delamination with specific, predetermined fates.

Primary sensory neurons with acquired identities may subsequently project back to the area from which they originated, using other delaminating cells as substrate to extend their peripheral processes. Such a scenario would allow primary sensory neurons to be randomly distributed in the ganglia and nevertheless project specifically to the ear, using distinct and unique guidance cues to navigate to their various peripheral targets. In fact, recent data clearly show that most primary neurons projecting to distinct sensory epithelia are mixed in their distribution rather than completely sorted within the ganglion (Maklad and Fritzsch 1999). Among the ear sensory epithelia, the cochlea of mammals is an exception with its highly organized peripheral and central projection (Lorente de No 1933). However, even here, topologically mismatched primary sensory neurons can comingle (Fritzsch 2003). In contrast, the distribution of primary sensory neurons in the vestibular ganglion is much more random and the peripheral, exclusive projection to distinct endorgans contrasts with the highly overlapping but nevertheless distinct topology in the central auditory nuclei representation of individual sensory epithelia (Maklad and Fritzsch 1999, 2002). Furthermore, tracing of early primary neurons shows that some have already extended an axon toward the brain before they migrate out of the otocyst wall.

Numerous candidate genes exist that may be, at least in part, responsible for the precise localization of the delaminating sensory neurons. It is possible that the entire invaginating otocyst has the capacity to form neuroblasts and that several genes are utilized to reduce this capacity. The reduction in expression in *Fgf10*, possibly owing to decreasing levels of growth arrest specific gene 1 (*Gas1*) (Lee and Fan 2001; Liu et al. 2002), might relate to this, as does the change in expression of T-box 1 (*Tbx-1*) (Raft et al. 2004). Other genes such as *Gata3*, eye absent 1 (*Eya1*), sine occulis 1 (*Six1*), dachshund (*Dach1* and *Dach2*), and paired-box 2 (*Pax2*) may be important for the maintenance of the neurogenic capacity.

Recent analysis indicates a critical role for *Shh* in sensory neuron formation (Riccomagno et al. 2002). As mentioned earlier, signaling from the mesoderm during early otocyst formation is essential for establishment of the initial, global patterning. Specifically, SHH is a small, highly diffusible molecule that is prominently expressed in the notochord and floorplate. The diffusion gradient set up by SHH defines the "midline" of the embryo and plays a role in the symmetric development of paraxial structures such as the somites (Brent et al. 2003). Strong parallels can be drawn here between somite and otocyst development. Like the somites, the otocyst lies within this SHH gradient with the highest levels of SHH at the ventral (cochlear) aspect. The basic helix–loop–helix (bHLH) gene, myogenic differentiation 1 (*Myod1*; alias *MyoD*), depends on *Shh* expression during somite development. *Neurog1* and *Neurod1* (also bHLH genes) important in otic development are also SHH dependent. Given this apparent requirement for *Shh* in otic dorsoventral patterning, we would expect alterations in the concentration of mesodermal SHH protein expression to disrupt the DV pattering, and therefore the specification of cochlear (ventral) and vestibular (dorsal) areas. Studies of the *Shh* null mice revealed morphogenetic defects in both the vestibular organs and the cochlea. Further, *Shh* signaling was determined to be necessary for specification of the most ventral cells of the developing otocyst. In addition, *Shh* mutants failed to form the cochleovestibular ganglion (Riccomagno et al. 2002). No analysis of *Gata3* (described below) was performed in these mutants. This leaves the question of whether or not a specific effect on spiral neurons occurs in the absence of *Shh* unanswered.

While under the influence of these mesodermal genes, the global patterning of the ear is set up. Three critical genes, expressed early in otic development, are *Gata3*, *Fgf3*, and *Fgf10*. *Gata3* is the only early marker that specifically identifies delaminating spiral sensory neurons (Karis et al. 2001; Lawoko-Kerali et al. 2002), but its role in spiral ganglion development is unclear. Based on pathfinding errors in inner ear efferents that also express this gene (Karis et al. 2001), it is possible that *Gata3* is also involved in pathfinding (a topic that is discussed, shortly). Important to this discussion is that in *Gata3* null mice, otic development is arrested at the otocyst stage. Similarly, although both *Fgf10* (Pauley et al. 2003) and *Fgf3* mutant mice have ears that develop, to varying degrees, past the otocyst stage, it is clear that in *Fgf3*/*Fgf10* double mutants, otic development is also arrested at the otic placode stage (Alvarez et al. 2003; Wright and Mansour 2003). The early expression of these genes can be considered a first tier of patterning. We have good reason to believe that these patterns are maintained over time and are congruent with later, local patterning. One such example is the late expression gradients of *Fgf10* in the end organs of the developing ear (Pauley et al. 2003). This is the second, local tier of *Fgf10* patterning.

Gas1 expression has been noted in the developing otocyst at E9.5 (Lee and Fan 2001), and has been shown to up-regulate late *Fgf10* expression (a gene that is also clearly involved in proliferation as its major in vitro assay) in the developing limb. *Gas1* mutants show a decrease in late *Fgf10* expression accompanied by a down-regulation of *Fgf8* and developmental abnormalities in the limb (Liu et al. 2002). Further characterization of *Gas1* mutants needs to be completed to determine the relationship between *Gas1* and the first tier of *Fgf10* expression in early inner ear compartmentalization. Conditional mutations are required for the selected study of later stage effects.

At slightly later stages in the formation of the embryonic ear, involvement of four gene families that are conserved across phyla is observed in the otic vesicle (Noramly and Grainger 2002). The interaction of these genes has been best documented in their role in *Drosophila* eye development and demonstrates an evolutionarily conserved gene network composed of the *Pax*, *Eya*, *Six*, and *Dach* genes (Hanson 2001). This gene network is also observed in zebrafish (*Danio rerio*), chicken, and mouse. These genes may play a fundamental role in the process of invagination. For example, this network is observed in the involuting hypoblast cells at gastrulation and in or during the formation of otic and optic vesicles. It has been suggested that these genes permit cells to migrate without altering their cell fate commitment (Streit 2002). By variation in family member usage and in coexpression patterns, this gene network can be coopted into a wide variety of different morphogenetic contexts. Thus, these regulatory proteins are usually coexpressed throughout embryogenesis in a wide variety of cell types, tissues, and organs. Their expression in a given tissue does not necessarily imply homology of such tissue, which has to be established by other means.

To understand their individual roles and predict their impact on downstream expression patterns, the interactions and functions of these proteins must be understood. EYA and SIX proteins interact with one another to form a single, *composite* transcription factor. This complex can be functionally modified by binding of DACH with EYA. The DNA binding site is contained within SIX proteins, while EYA mediates transcriptional transactivation and contains SIX and DACH binding domains. Additional regulatory complexity is provided by DACH, which appears to function as a cofactor, interacting directly with EYA. Further, it is speculated that these genes function downstream of the *Pax* genes. However, *Dach* expression does not depend solely on a single *Pax* or *Eya* gene in otic or optic development. A conserved expression pattern is found in the otic vesicle where genes representing these four gene families are found. In the mouse otocyst, these genes are *Pax2*, *Eya1* and *4*, *Six1* and *4*, and both *Dach* genes (Davis et al. 1999, et al. 2001; Noramly and Grainger 2002).

With these genes, it is important to note that, in contrast to *Gata3* and *Fgf3/Fgf10* mutants, the otic development of these mutants begins normally and defects are observed only later in development. This suggests that the *Eya* and *Six* genes are not required for the initiation of global patterning, but are needed to keep the patterns "up and running" for normal morphogenesis. Like other genes involved in ear morphogenesis, these regulatory elements also play a role in histogenesis as indicated by their expression patterns in the developing ear. Recent work on *Eya1* and *Six1* shows reduction to complete loss of sensory neuron formation (Xu et al. 1999, 2003).

Another example is the *Eya4* gene which is initially expressed in the otic vesicle (Borsani et al. 1999; Wayne et al. 2001). This gene is present primarily in the upper epithelium of the cochlear duct, in the region corresponding to the presumptive Reissner's membrane and the stria vascularis of the cochlear duct. At E18.5 of the mouse, *Eya4* is expressed in areas of the cochlear duct destined to become spiral limbus, organ of Corti, and spiral prominence. The highest level of expression occurs in the basal turn and in the early external auditory meatus. Diminishing levels of expression are found at later stages in these tissues and in the developing cochlear capsule during the period of ossification after birth to post natal day 14 (P14). In the vestibular system, *Eya4* is observed in the developing sensory epithelia. Interestingly, mutant mice lacking *Eya4* exhibit late-onset deafness similar to the associated deafness in *DFNA10* (mutated *EYA4*) patients (Wayne et al. 2001; Pfister et al. 2002). In the cases of both *Eya1* and *Eya4* it is unclear whether this relates directly to neuronal capacity regulation or to the size reduction of the otocyst. Given that these genes play a role in morphogenesis and histogenesis of the inner ear as well as a survival role in the mature system, the complete spatiotemporal expression patterns must be ascertained before their function is fully understood. Categorization of their components in ear development must be approached by using conditional mutant mouse lines to understand the contextual role these regulatory genes are playing.

Another gene family involved late in ear morphogenesis is the bone morphogenetic protein genes (*Bmps*). These genes belong to the transforming growth factor-B (*TGFB*) superfamily and have been shown to play an essential role in semicircular canal formation (Chang et al. 1999, et al. 2002; Gerlach et al. 2000; Bober et al. 2003). Indeed, most recent overexpression experiments of *Bmp4* in chicken embryos shows their interaction with other BMPs and FGFs (Chang et al. 2002) (see also Kelley and Wu, Chapter 2, and Mansour and Schoenwolf, Chapter 3).

3. Proliferation of Progenitors and Terminal Mitosis of Sensorineural Cells

While the genes described in the previous section are important for defining areas of neuronal development, many of them play an even more critical role in proliferation. The discussion that follows considers how global and local patterning in the ear relates to the specification and proliferation of individual progenitor cells that become specific sensory epithelia and their innervating neurons.

Progenitor/stem cells are defined by common cellular properties: they proliferate, self-renew, and give rise to progeny that subsequently differentiate. In the simple single-cell sensory organs of the fruit fly, the sequence of events as well as the genes regulating them have been specified in great detail. It suffices here to state that many of the genes involved in these processes are conserved, with the caveat that the mammalian counterparts typically have multiplied and undergone some functional diversification. The remarkable conservation across phyla has been used to homologize individual cell types (Fritzsch et al. 2000) and to demonstrate functional conservation of genes across phyla (Wang et al. 2002). Although interesting from an evolutionary perspective (Fig. 4.1, the con-

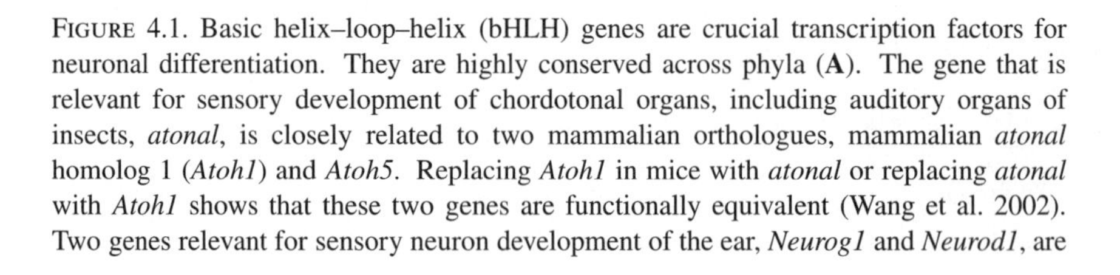

FIGURE 4.1. Basic helix–loop–helix (bHLH) genes are crucial transcription factors for neuronal differentiation. They are highly conserved across phyla (**A**). The gene that is relevant for sensory development of chordotonal organs, including auditory organs of insects, *atonal*, is closely related to two mammalian orthologues, mammalian *atonal* homolog 1 (*Atoh1*) and *Atoh5*. Replacing *Atoh1* in mice with *atonal* or replacing *atonal* with *Atoh1* shows that these two genes are functionally equivalent (Wang et al. 2002). Two genes relevant for sensory neuron development of the ear, *Neurog1* and *Neurod1*, are

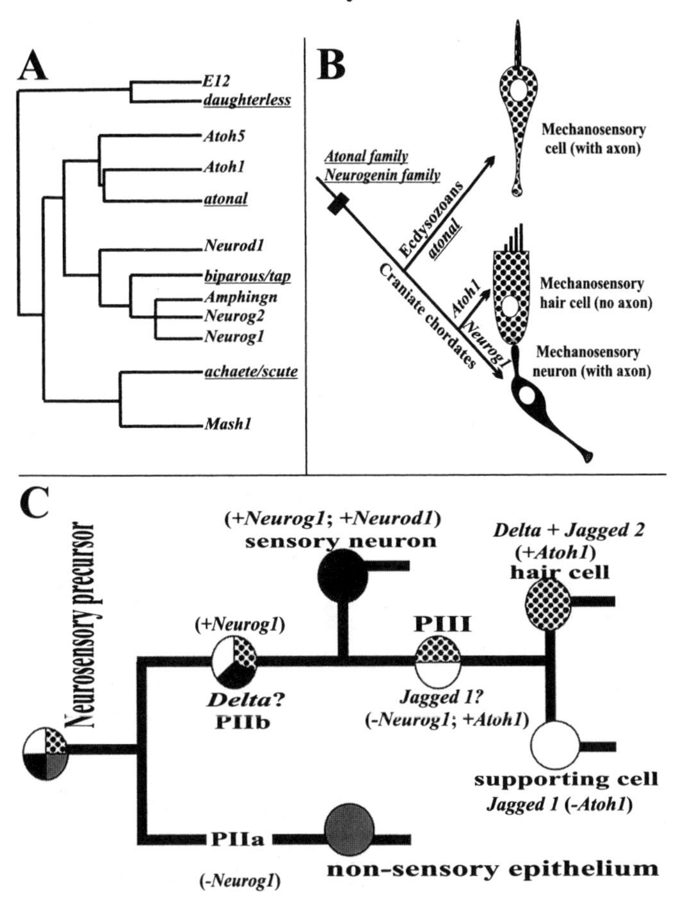

FIGURE 4.1. (*Continued*) each related to a single insect gene, *biparous*. However, in insects *biparous* plays no role in chordotonal organ development. It has therefore been hypothesized that *Neurog1* and *Neurod1* were implemented in vertebrate ear development to accommodate the evolutionarily new formation of a sensory neuron (**B**). Implementing this addition into the cell fate scheme of an insect mechanosensory development (**C**) suggests a stepwise restriction of the cell fate capacity of neurosensory precursors. The initial precursor would likely be omnipotent and able to form all cells of the ear. The immediate neuronal precursor (PIIb) would express *Neurog1* and would have a restricted capacity for formation of sensory neurons, hair cells and supporting cells. Formation of sensory neurons further restricts the capacity of the hair cell precursors (PIII) to hair cells and supporting cells. The ligand *Jagged1* acting through the receptor *Notch1* will activate the downstream factors *Hes1* and *Hes5* to stabilize the fate selection in supporting cells. Modified after Fritzsch et al. (2000); Bertrand et al. (2002); Zine (2003); Fritzsch and Beisel (2004). (See color insert.)

servation of molecular cell fate determining mechanisms might ultimately help us to understand the more complex interplay of proliferation and cell fate determination in the mammalian ear.

These cellular properties of self-renewable stem cells correlate with the expression of general molecular markers [such as members of the SRY-box containing gene (*Sox*) and forkhead box gene (*Fox*) gene families], supporting the likelihood of conserved signaling pathways that maintain these generic and definitive properties (Graham et al. 2003). If and where those genes are expressed in the developing and adult ear remains unexplored. Several models of progenitor proliferation have been proposed in the literature (Cai et al. 2002). These models describe different proportions of asymmetric versus symmetric and terminal versus non-terminal cell divisions. The most widely accepted model states that progenitors must first undergo rounds of multiplicative division to increase the population of progenitor cells (Fig. 4.2). This growth phase is followed by divisions that begin to generate undifferentiated and/or differentiated progeny. Progeny generation could be accomplished by asymmetric divisions that generate both a differentiated cell and another mitotically active progenitor, or by symmetric, terminal divisions that generate two differentiated cells. The fastest mechanism to generate the greatest number of differentiated cells in the smallest number of divisions is by employing symmetric progenitor division followed by terminal divisions (Fig. 4.2). For example, a progenitor that divides in an asymmetric stem cell mode would require four rounds of division to make the four differentiated cells and a progenitor that can continue to proliferate. In contrast, expansion proliferation would require only two rounds of division of a given progenitor cell to generate four differentiated cells without a progenitor.

Another proliferation strategy is a variation of the model of asymmetric division of progenitors, where both a differentiated cell and another proliferative progenitor are generated without an expansion through symmetric division (Fig. 4.2D). Compared to the fast mechanism detailed above, this strategy would require three rounds of division to make the same number of differentiated cells. Consequently, the generation of progenies is extended over a longer period of time and requires the presence of differentiation factors earlier than the simple clonal expansion of the progenitor pool by symmetric division.

These strategies may be employed differentially or in combination in inner ear development (Fig. 4.2) where studies (Ruben 1967) have shown distinct temporal patterns of proliferation for cochlear and vestibular neurosensory components. The cochlea displays a sharp peak of terminal mitoses (between E11.5 and E13.5) whereas the vestibular hair cells undergo the bulk of their terminal mitosis over an extended 5-day period. The fast (symmetric) mechanism provides the type of fast growth needed for cochlea, whereas a mix of symmetric and asymmetric divisions appears to be suitable for the extended proliferation mode of vestibular epithelia. The fast mechanism allows a delayed expression of differentiation factors, after the population is expanded and now is ready for

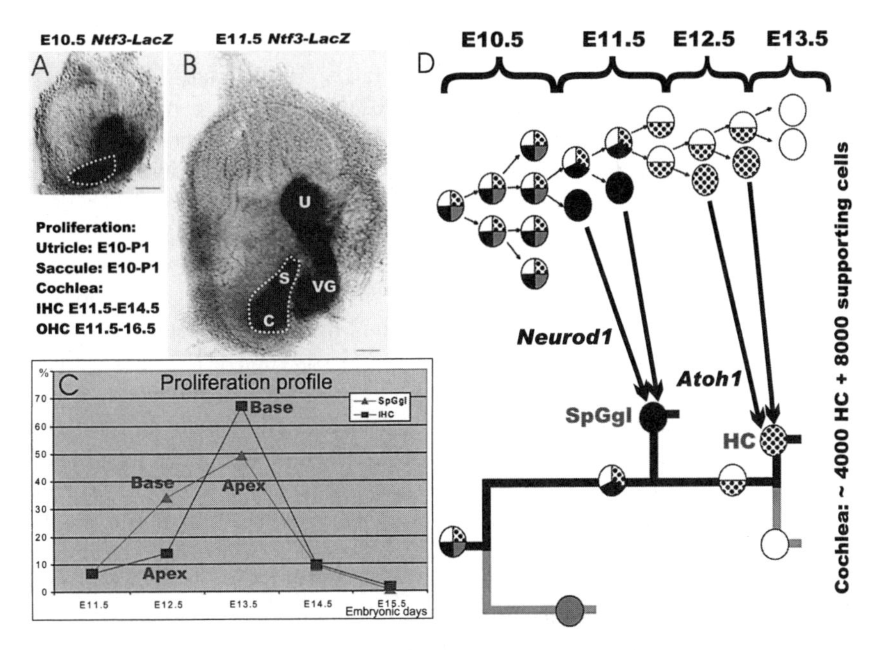

FIGURE 4.2. Possible clonal relationship and clonal expansion of neurosensory precursors are depicted. Formation of neurosensory cells starts with the patterning of areas in the developing ear that are specified to form patches of precursors for hair cells and sensory neurons (**A**, **B**). Several genes have been characterized that highlight the development of those areas (Morsli et al. 1998; Farinas et al. 2001). The neurotrophin *Ntf3*, which in later development is expressed in the supporting cells of the utricle, saccule, and cochlea (Farinas et al. 2001), shows the progressive doubling in size and separation of the proneurosensory area into two patches (**A**, **B**). In addition, sensory neuron precursors delaminate apparently from these patches and form the vestibular and later the cochlear sensory neurons (VG in **A**, **B**). Analysis of the terminal mitosis using [^{3}H] thymidine suggests that formation of sensory neurons is slightly ahead of hair cells in a spatially distinct fashion in the cochlea (**C**). Together these data suggest that there may be a clonal expansion through symmetric division of about three cycles per 24 hr from E10.5 to E11.5 followed by a coordinated transition of precursors with progressively more restricted differentiation capacity (**D**) and a lengthened cell cycle. The present authors hypothesize that some general precursors transform into specific neuronal precursors after three rounds (approximately 1 day) of division. The immediate neurosensory precursors will form, through asymmetric division, sensory neurons that may exit the cell cycle by expressing *Neurod1*. Later, those precursors may transform into immediate precursors for hair cells that, through asymmetric or symmetric division, give rise to hair cells, possibly exiting the cell cycle through up-regulation of *Atoh1* (black arrows in **D**). Note that only the neurosensory part of the possible cell lineage relationships is depicted. Modified after Ruben (1967); Fritzsch et al. (2000), Farinas et al. (2001), Cai et al. (2002). C, Cochlea; HC, hair cells; IHC, inner hair cells; SpGgl, spiral ganglion; U, utricle; VG, vestibular ganglion. Bar = 100 μm in **A**, **B**. (See color insert.)

differentiation. This is in accordance with recent data that have shown expression of differentiation factors such as *Atoh1* (alias, *Math1*) starting at E13.5 in cochlea, in postmitotic cells (Chen et al. 2002). The *Atoh1* expression in the vestibular component is earlier (at least E11.5, if not E10.5) when some of the expressing cells are still proliferating. These findings in the vestibular epithelia suggest an asymmetric proliferation mode.

It is important to have a precise knowledge of the temporal sequence of progenitor divisions in different sensory epithelia because the sequential expression of transcription factors leads to different fates in the resulting progeny. It appears that the progression of the cell cycle is required to advance a "clock" that progenitors use to drive histogenesis (Ohnuma and Harris 2003). The time at which the cell exits the cell cycle is considered its "birth date." In most systems studied, there is a correlation between birth date and cell fate, giving rise to the process known as histogenesis (Caviness et al. 2003). Indeed, mere lengthening of the cell cycle can change the cell fate as these cells then develop in a different "environment" (Calegari and Huttner 2003). One of the most obvious examples of histogenesis in the vertebrate brain is the generation of neurons before glial cells out of the same precursors (Bertrand et al. 2002). A somewhat comparable example might exist in the inner ear. Past research has revealed that neurosensory cells are generated over a prolonged period, with the sensory neurons predating hair cells in cochlear and vestibular systems in a topographically specific fashion (Ruben 1967), and recent experimental work supports the notion of a clonal relationship between sensory neurons and hair cells (Ma et al. 2000).

Tight control of proliferation is important in tissue size regulation. Factors involved in regulating proliferation in other developing systems, such as muscle fibers and olfactory epithelium, are only now becoming apparent (Wu et al. 2003). How these regulatory elements are utilized in the ear and what specific molecules are performing these actions in ear development has yet to be determined; also, most of the genes described in other tissue appear to be present in the ear. Experiments in the olfactory system support the idea that differentiated neurons produce signals that feed back to inhibit the generation of new neurons by neuronal progenitors (Calof et al. 2002). While the molecules that mediate such effects in vivo have not been identified, the bHLH gene *Neurod1* is a good candidate for this and other systems (Canzoniere et al. 2004). The signal produced by differentiated neurons acts upon cells at a very early progenitor stage, causing a fraction of them either to die or to stop producing the downstream neuronal transit amplifying cells that generate olfactory neurons. BMPs and in particular *Bmp4* may provide such a signal. Another candidate is growth differentiation factor 11, *Gdf11*, which reversibly blocks progenitor divisions through a mechanism involving increased expression of cyclin-dependent kinase inhibitor 1B (P27), *Cdkn1b* (alias, *p27Kip1*) (Wu et al. 2003), a factor that is also important in ear development (Chen et al. 2003).

In the vertebrate nervous system, the gradual slowing followed by cessation of progenitor cell proliferation suggests that neurogenesis is also under some form of negative control, similar to that of the olfactory system (Cai et al. 2002).

Gradual slowing of progenitor cell proliferation has also been documented in the inner ear (Fekete and Wu 2002), but further data are needed to uncover the mechanisms that underlie feedback regulation of cell production. Mechanisms of slowing or stopping cell cycles seem to be related to genes expressed inside the cells such as retinoblastoma (Ferguson et al. 2002) and the recently discovered B-cell translocation gene 2, antiproliferative, *Btg2* (*Pc3*) gene (Canzoniere et al. 2004). The latter is particularly interesting as it shows that *Btg2* is upstream of *Atoh1* and may relate to cell differentiation be exerting control of rather than ongoing cycling of precursors through action on cyclin D1 (*Ccnd1*) as well as initiating differentiation through regulation of *Atoh1*. It is possible that the coordinated progression of different bHLH genes in sensory neuron and hair cell precursors as well as up-regulation of specific bHLH genes in differentiating sensory neurons (Kim et al. 2001) and hair cells (Chen et al. 2002) may directly regulate to the negative feedback loop, potentially via the ubiquitous Delta–Notch system.

Superimposed on the generic characteristic of proliferation is the regional patterning that was discussed in the previous section. For example, in the spinal cord, neural progenitors take on distinct dorsoventral identities in response to opposing diffusion gradients of SHH and BMPs and in cooperation with wingless oncogene (WNT) gradients (Maklad and Fritzsch 2003), which also influence regional proliferation (Megason and McMahon 2002). These are reflected by region-specific expression of homeodomain transcription factors (Gowan et al. 2001). This transcriptional regionalization of neural progenitors has been linked with a general program of neurogenesis under the regulation of proneural and neurogenic bHLH transcription factors (Bertrand et al. 2002). As noted above, some bHLH proteins can direct progenitors' exit from the cell cycle and promote differentiation in a coordinated and regulated fashion (Cau et al. 2002).

In general, neurosensory development and evolution of the ear are based on the multiplication of existing sensory patches and their respective innervation (Fritzsch and Wake 1988; Fritzsch et al. 2002), largely through the proliferative expansion of variously committed precursor populations. From a theoretical perspective, larger sensory organs, such as the ear, can be formalized as a multistep expansion of clonal relationship of simple single-cell sensory organs such as those found in the fruit fly (Figs. 4.1 and 4.2). If so, the question then becomes how long the different progenitors remain capable of committing to all, many, or few cell lines, and how this restriction relates to the progression in cell cycles. In addition, topological information needs to be integrated to provide unique identities and directions of differentiation for precursors destined for the cochlea or a semicircular crista. Sensory epithelia modifications are likely followed by functional diversification through creation of modified, unique acellular covering structures (Goodyear and Richardson 2002) that allow transduction of a previously unexplored property of the mechanical energy that reaches the ear.

The simplest ear to be found in extant vertebrates is the hagfish ear. This ear has only three sensory epithelia, one macula communis and two crista organs in a single canal (Fritzsch 2001a, b; Lewis et al. 1985). The largest number of sensory patches in the ear is found in certain species of limbless amphibians, which have nine different sensory patches (three canal cristae, utricle, saccule, lagena, papilla neglecta, papilla basilaris, and papilla amphibiorum) (Sarasin and Sarasin 1892; Fritzsch and Wake 1988). Descriptive developmental evidence has long suggested that the evolution of multiple sensory epithelia comes about through developmental splitting of a single sensory anlage (Norris 1892; Fritzsch et al. 1998) which has to grow by increase in cell number.

Beyond generating the "raw material" by increasing neurosensory cell formation, forming a sensory epithelium that can access a novel sensory stimulus requires transformation of existing mechanoelectric transducers through morphological alterations to tap into a novel mechanical energy source. Sound pressure reception for hearing is not already accomplished by the mere formation of a novel sensory epithelium that can be dedicated to perceive this energy. It is reasonable to assume, however, that once such an uncommitted receptor is available, changes in the ear morphology using some of the genes outlined above may achieve changes upon which further refinement can act in the slow process of selection of appropriate function-based modification. It is conceivable that such alterations will take place as soon as new receptors form simply because of the invariable alteration of FGF and BMP interaction that comes with the formation of a new sensory epithelium. Indeed comparing just *Bmp4* expression in chicken and mice (Wu and Oh 1996; Morsli et al. 1998) shows differences in the expression patterns that need to be further explored by comparing the expression of *Bmp4* with those of FGF ligands and receptors (Pirvola et al. 2002).

It is noteworthy that incomplete segregation of sensory patches has been reported in mutations with altered morphologies and reduced proliferation (Fritzsch et al. 2001a; Pauley et al. 2003). This implies that morphogenesis and segregation of sensory patches are linked, but not necessarily causal, to clonal expansion through proliferation of precursors. The following section discusses, in detail, the role of bHLH genes in cell fate determination as well as in maintaining and terminating the proliferative precursors.

4. Molecular Basis of Neurosensory Cell Fate Specification

Studies in recent years have revealed the molecular basis for the formation of sensory neurons (Ma et al. 1998, 2000) and hair cells (Bermingham et al. 1999) in the ear, mainly using the mouse as a model system. Data from these studies suggest the following molecules and their interactions in determining the fate specification of these cells.

In vertebrates, all neurons derive from ectodermal cells. Theses cells are transformed, via a cascade of genes, into neuronal precursors. Several genes have been identified that appear crucial for this designation of phenotype fate. Owing to their apparent capacity to transform ectodermal cells into neurons (Lee

1997), these genes are referred to as "proneural genes." They belong to the family of genes that encode an ancient protein family, with a basic helix–loop–helix (bHLH) structure (Fig. 4.1), that has a highly conserved DNA binding domain (Bertrand et al. 2002). Proneural bHLH proteins form heterodimers with the ubiquitous E2A proteins (i.e., the insect daughterless proteins) that enable them to bind to DNA and exert their function. These proteins not only have the unique capacity to turn ectodermal cells into neurons in gain-of-function experiments (Ma et al. 1996; Lee 1997), but can also specify cell fate in unrelated tissues such as the pancreas (Liu et al. 2000) or the gut (Yang et al. 2001). In addition to transformation of ectodermal cells into neuronal precursors, they can also generate neuron-like cells, such as Merkel cells (Bermingham et al. 2001).

Given this unique capacity of the proneural genes, it is not surprising that these genes are tightly regulated in their spatiotemporal expression through a number of other transcription-regulating factors. Some of these factors interfere with the heterodimerization of bHLH proteins by binding to the E2A proteins and thus interfering with neuronal differentiation. These genes are therefore referred to as "inhibitors of differentiation" or the bHLH inhibitor of DNA binding (*Id*) genes. Others, such as the vertebrate *hairy* and *enhancer of split* paralogs (*Hes/Hey/Tle*), act as classic DNA-binding repressors of proneural gene transcription. The activation of this gene family appears to be regulated by the ubiquitous Delta–Notch system, which down-regulates the proneural gene expression in neighboring cells. This is called lateral inhibition, and requires the up-regulation of bHLH genes in a limited number of cells to prompt the activation of the Delta–Notch system. Important factors for the Delta–Notch system in the ear appear to be the downstream factors *Hes1* (mutants form extra rows of inner hair cells) and *Hes 5* (mutants form extra rows of outer hair cells (Zine et al. 2001). Eliminating the *Delta/Jagged* receptor *Notch1* causes effects comparable to a double null for both *Hes1* and *Hes5*: double rows of inner and multiple rows of outer hair cells (Zine 2003). Likewise, *Jagged1* null mice have multiple rows of inner and outer hair cells (Zine 2003). Most of the factors that drive this initial up-regulation of proneural bHLH genes are still unknown, but the zinc finger protein family member (*Zic*) genes are good candidates.

In general, the bHLH genes can be divided into three functional groups: true proneural bHLH genes that generate a neural lineage, bHLH genes that drive neural differentiation, and bHLH genes that drive the switch from neural to glial cell lineage (Bertrand et al. 2002; Zhou and Anderson 2002).

Loss-of-function (targeted mutations of a gene or genes) experiments have clarified some of the proneural genes critical for inner ear primary sensory neuron development. Inner ear primary sensory neuron formation requires the vertebrate bHLH gene, *Neurog1* (Ma et al. 1998). A follow-up study showed that no primary sensory neurons ever form in these mutants (Ma et al. 2000). Owing to the absence of primary sensory neuron formation in *Neurog1* mutants, the ear is completely isolated from direct brainstem connections during development, as afferents do not form and neither efferents nor autonomic fibers appear to reach the ear in these mutants (Ma et al. 2000). Nevertheless, these ears develop a fairly normal overall histology. This suggests that ear formation and development, including that of many hair cells, is largely independent of innervation. Although hair cell numbers are reduced to varying degrees in *Neurog1* mutant mice, those hair cells that do form develop normally in the absence of innervation (except for some minor disorientation). Interestingly, the cochlea is shortened and the saccule is almost completely lost (Fig. 4.3). In addition, extra rows of hair cells form in the shortened cochlea (Fig. 4.3). Also, the pattern of terminal mitosis is altered in *Neurog1* mutant. It appears that in the absence of *Neurog1* there is not only a complete loss of neuronal progenitors, but also the hair cell progenitors exit the cell cycle at an earlier embryonic stage than expected (Matei et al. 2005). These data suggest a significant interaction between progenitor cells that form primary neurons and progenitor cells that give rise to hair cells, supporting cells, and other inner ear epithelial cells. Although other possible interactions cannot be excluded (Fritzsch et al. 2002), the most simple explanation would be a clonal relationship between primary sensory neurons and some hair cells/supporting cells (Fritzsch and Beisel 2001; Fekete and Wu 2002).

One model that can explain these data hypothesizes the coexpression of *Atoh1* and *Neurog1* in the same progenitor cell and the cross-inhibitory regulation between bHLH transcription factors. In accordance with this model, in the absence of *Neurog1*, *Atoh1* is up-regulated at an earlier stage and this leads to an accelerated pattern of proliferation and, consequently, the exhaustion of the progenitor pool. Such suggestions are in line with recent experimental data in the cerebellum (Canzoniere et al. 2004). In the mammalian ear, a direct clonal analysis using retrovirus infections and other experiments are needed to support this notion of a clonal relationship between neuron and hair cell progenitors and the interaction between bHLH genes.

Another bHLH gene that is immediately downstream of, and mostly regulated by, *Neurog1* is *Neurod1* (also known as Beta) (Ma et al. 1998). As is true for other proneural genes, *Neurog1* is only transiently up-regulated in primary neuron precursors. As primary sensory neuron precursors delaminate from the otocyst wall, they down-regulate *Neurog1* and up-regulate *Neurod1* (Liu et al. 2000; Ma et al. 2000; Kim et al. 2001). The presence of either *Neurog1* or *Neurod1* seems to be involved in the continued proliferation of neuroblasts through interference with the cell cycle. This is based on data in the CNS (Bertrand et al. 2002). As with *Neurog1* mutants, mutations of *Neurod1* have been analyzed and show severe reduction (Kim et al. 2001) or even complete loss of sensory neurons (Liu et al. 2000). The effect of NEUROD1 on cochlear primary neurons (almost completely lost) is different than that on the vestibular primary neurons (many more survive but they are dislocated and have aberrant projections). Further, it appears that the surviving vestibular and cochlear sensory neurons may be in an unusual position and project in an abnormal pattern to only a part of the cochlea and some of the vestibular sensory epithelia (Kim et al. 2001; Fritzsch and Beisel 2003). Based on these data, *Neurod1* appears to play a role

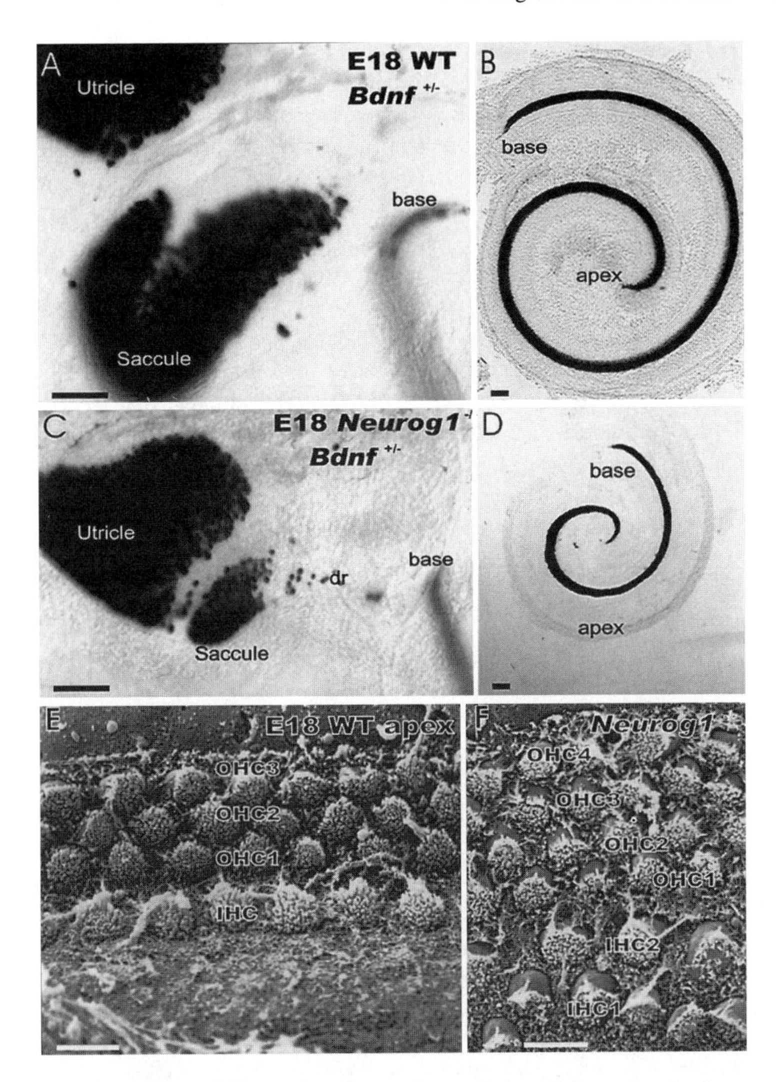

FIGURE 4.3. The effect of *Neurog1* null mutation on hair cell formation is shown using the *BDNF-LacZ* expression as a marker for differentiated hair cells. In contrast to the wild type littermates, *Neurog1* null mutants have almost completely lost the saccule (**A**, **C**) but show individual hair cells scattered along the ductus reuniens (dr) toward the base of the cochlea (C). The cochlea is shortened and the numbers of hair cells are reduced to approximately 60%. SEM details of the cochlea show that in the apex there are up to five rows of disorganized hair cells in *Neurog1* null mutants (**E**, **F**). This implies effects of *Neurog1* on hair cell proliferation, cochlear elongation and hair cell orientation, possibly mediated through the clonal relationship proposed in Figs. 4.1 and 4.2. Modified after Ma et al. (2000); Fritzsch and Beisel (2003): IHC, Inner hair cell; OHC, outer hair cell. Bar = 100 μm (**A–D**) and 10 μm (**E**, **F**).

in neuronal differentiation, survival, migration to appropriate areas, and target selection of peripheral projections. These mutants also display a reduction and/or absence of certain neurotrophin receptor genes known to be essential for neuronal survival (Liu et al. 2000; Kim et al. 2001). Moreover, it is possible that the effects on survival of sensory neurons in *Neurod1* mutants are, in part, mediated by the reduction of expression of a *Pou4f1*. *Pou4f1* appears to have a somewhat similar phenotype, demonstrating lack of innervation of certain sensory epithelia (Huang et al. 2001). Like *Neurod1*, *Pou4f1* affects up-regulation of certain neurotrophin receptors and thus may be only indirectly affecting neuronal development. While no promoter analyses of these genes have been done, their later onset of expression relative to *Neurog1*, and the less severe phenotypes in the mutants of *Neurod* and *Pou34f1*, suggest that *Neurod1* and *Pou4f1* might influence survival via regulation of neurotrophin receptors.

Another gene, the T-box gene, *Tbx1*, is implicated in the DiGeorge syndrome (Merscher et al. 2001). Expression of *Tbx1* has been described in the posterior part of the early otocyst and later in the anterior part (Riccomagno et al. 2002; Vitelli et al. 2003). Specifically, *Tbx1* expression largely borders on *Neurog1* and *Neurod1* expression throughout otocyst development. *Tbx1* expression overlaps with that of *Bmp4* and orthodenticle homolog 1 (*Otx1*), markers for the presumptive cristae and nonsensory cochlea, respectively. Further, *Tbx1* shows expression complementary to that of lunatic fringe homolog (*Lfng*)/*Neurog1*/*Neurod1* (Morsli et al. 1998; Kim et al. 2001).

Recent work with *Tbx1* mutant and *Tbx1* gain-of-function mice demonstrates a role for *Tbx1* in *Neurog1* and *Neurod1* regulation (Raft et al. 2004). *Tbx1* has been shown to suppress expression of the bHLH gene *Neurog1*. In the *Tbx1* mutant, *Neurog1* and *Neurod1* expression is increased and the expected increase in neurogenesis is seen. At E10 in these animals, the VIIIth ganglion, as identified by *Neurod1* expression, is increased by more than 80%. However, by E15, many of these neurons have died, suggesting a role for *Tbx1* in neuron survival at later stages. Conversely, gain of function *Tbx1* mice show little or no neurogenesis in the region of *Tbx1* expression, suggesting a down-regulation of *Neurog1* and, therefore, *Neurod1*. These mice demonstrate marked reduction in expression of *Neurog1* and *Neurod1* as well as a ganglion volume that is 39% of that of the wild type, as determined by *Neurod1* expression (Raft et al. 2004).

Absence of proneural genes, in insects, leads to a complete collapse of sensory organ formation (Caldwell and Eberl 2002). Double-null mutants for both *Neurog1* and *Atoh1* (a hair cell expressed bHLH gene) were therefore analyzed. If these were the only proneural bHLH genes that were independently expressed in the ear, one would expect severe consequences on ear morphology. Preliminary results on a double-null, however, show that even in the absence of both *Neurog1* and *Atoh1*, ear development occurs (Fritzsch and Beisel 2003). Either a third, as yet unidentified, bHLH gene is present in the mammalian ear, or the morphogenesis of the ear is nearly completely independent of the neurogenesis of sensory neurons and the formation of hair cells. The fact that retinoic acid treatment can entirely block the morphogenesis but still allows neurogenesis of the ear placode to proceed (Fritzsch et al. 1998) supports the latter suggestion. Further studies on this subject are clearly warranted.

In contrast to vestibular primary sensory neurons, cochlea (spiral) primary sensory neurons express *Gata3* (Karis et al. 2001; Lawoko-Kerali et al. 2002). GATA factors, including *grain* and, presumably, its vertebrate homolog *GATA3*, (Brown and Castelli-Gair Hombria 2000; Bertrand et al. 2002; Karis et al. 2001) can interact with bHLH dimers for transcriptional regulation. These genes may be directly involved in specific aspects of cochlear and vestibular fate determination via regulation of bHLH gene transcription. Unfortunately, most *Gata3* mutations result in early lethality before hair cells differentiate. Nevertheless it is important to note that a human *GATA3* mutation exists and causes deafness (Van Esch et al. 2000).

In summary, the genes that regulate formation of the neurosensory aspects of ear development are beginning to emerge and their functions have been tested in specific mutant mice. While this line of research has provided dramatic breakthroughs in our understanding of this process, numerous questions remain before this knowledge can be utilized to guide aspects of neurosensory regeneration that would benefit individuals with neurosensory hearing loss.

5. Fiber Growth to Distinct Peripheral and Central Targets

Bipolar sensory neuron fiber growth can occur in one of three different patterns. Not only do these patterns show unique methods of neuron growth, but the dependency on innervation for formation and survival of the targets of these neurons varies. One pattern, exemplified by neurons in the epibranchial placode, is for the neuron to migrate away from the ectoderm, and then send out an axon toward the CNS and a dendrite toward the sensory area. One example is innervation of the taste buds in the tongue (Fig. 4.4). In this case, both axon and dendrite project in a direction different from the course of cell migration. Furthermore, cells innervated by neurons from the epibranchial placode, such as the tastebuds (which are not derived from the epibranchial placode), require innervation for their *survival*, but not for initial development. These sensory cells will die following nerve transection and reappear once nerve fibers have regenerated (Fritzsch et al. 1997b).

In contrast, as neural crest cells migrate away from the CNS, their axons trail behind, causing the final course of the axonal fibers to follow the path of the neuronal delamination. In this way, the migration of the cell body determines the course of the axon whereas the dendritic pathway is determined by other factors related to pathfinding and potentially shared with the dendrite of the epibranchial placode derived neurons. As in the epibranchial placode derived neurons, some target sensors such as muscle spindles critically depend on innervation and will not even develop in the absence of innervation (Farinas et al. 1994).

In the ear, it is apparent that this pattern is reversed and the dendrites are on

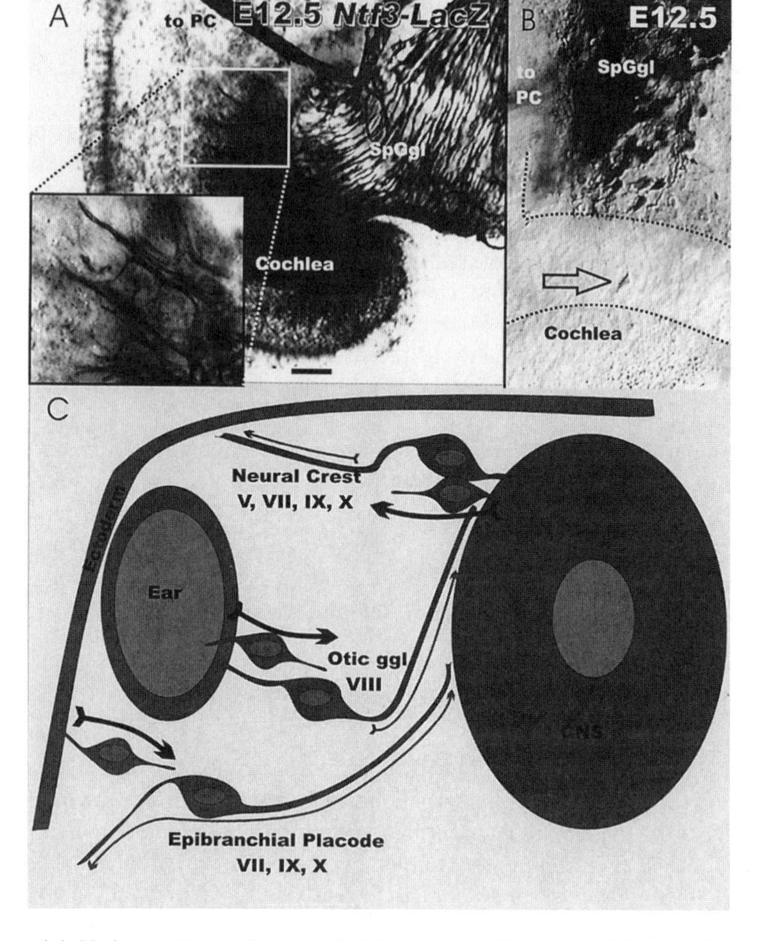

FIGURE 4.4. Various patterns of neuronal delamination with axon and dendrite formation in the PNS are shown. Expression of *Ntf3-LacZ* in the E12.5 cochlea and spiral ganglion (blue staining in **A**) combined with tubulin immunocytochemistry (black in **A**) shows the distribution of nerve fibers selectively to the area of *Ntf3* expression in cells that likely are related to the future supporting cells of the cochlea. Notice B-galactosidase (*Lac-Z*) reaction of both the cochlear sensory epithelia and the delaminating spiral ganglion (SpGgl) neurons. **Inset**: Dendrites in the cochlea will require little additional guidance to reach their hair cell targets and reach the sensory epithelia prior to hair cell formation in the base. Filling inner ear sensory neurons with biotinylated dextran amines shows multiple, delaminated neurons outside the cochlea (SpGgl) as well as single neurons still inside the expanding cochlear duct with an axon already reaching the central nervous system (CNS in **C**). These data suggest an important difference between otic (VIIIth nerve) ganglion cells compared to other neurons: Epibranchial placode derived neurons send axons and dendrites on courses (indicated by *lilac birectional arrow*) different from that of the cell's migration (indicated by the *black arrow*). Neural crest cells migrate away from the CNS (*black arrow*). The axon trails behind and the dendrite extends out to the peripheral target (*yellow arrow*). In the ear, the dendrite trails behind and the axon extends toward the CNS (*green arrow*). Modified after Fritzsch (2003). (See color insert.)

the trailing end of the migrating cell, leaving the dendrite within the ear as the cell body migrates away. In contrast to the above fiber growth patterns, sensory cells (hair cells) innervated in this manner do not require innervation for their formation or initial survival, as demonstrated by neurotrophin mutants. This developmental pattern provides these peripheral processes with a close approximation to their future sensory target(s), that is, the area from which they delaminated. This can be demonstrated in the cochlea using neurotrophin 3 (*Ntf3*; alias, *NT-3*) expression to label both spiral ganglion and cochlear epithelial precursor cells. At E.11.5, *Ntf3* is clearly expressed along the inner radius of the cochlea, labeling future neurons and sensory epithelia (Fig. 4.2). By E12.5, the *Ntf3*-positive spiral ganglion neurons have migrated away from the cochlea, leaving their dendrites in the inner curve of the cochlea and sending their axons toward the CNS (Fig. 4.4). Indeed, occasional neurons may be backfilled via their axons from the brain while the perikaryon is still in the cochlear epithelial wall (Fig. 4.4). In this manner, the path of the dendrites is determined by the migration of the neuronal cell bodies and there is little room for error in overall dendritic pathfinding. Another example is seen in *Fgf10* mutant mice in which the entire sensory epithelia formation of the posterior crista is abolished. In these mutants, the initial projection of afferent fibers is rather targeted (Pauley et al. 2003), perhaps because of delamination of sensory neurons near the posterior crista epithelium. These data can be used to support the idea that the neural sensory cells are originating at or near their future sensory epithelium and the cell bodies move away from them and the ear. The specificity of the allocation of sensory neuron delamination to specific sensory epithelia and the precision in targeting future innervated sensory epithelium requires further analysis.

Despite being generated in such close proximity, the dendrites of these sensory neurons of the inner ear must still undergo extensive guidance to find the proper peripheral sensory patch and the specific hair cell(s) to innervate. They must also follow signals that tell them to stop growing once they have reached their specific target. In addition, the axon must find its path to the CNS by following, as yet undiscovered, molecular cues. Furthermore, it is clear that the dendritic pathfinding is completely uncoupled from that of the axon in the ear, as all of the mutants with defects in dendritic pathfinding show virtually no abnormalities in axonal guidance (Fritzsch et al. 1997a; Xiang et al. 2003).

Recent molecular data suggest some genes that are candidates for regulation of peripheral process development. These genes belong to the ephrin ligand and receptor families, and the semaphorin ligand and receptor families, both of which are known to be important in other developing systems (Tessier-Lavigne and Goodman 1996; Bianchi and Liu 1999). Mutant mice for the ephrin receptor B2, *EphB2*, show circling behavior and altered axonal guidance of midline crossing efferents, but no data on peripheral innervation have been provided (Cowan et al. 2000).

It has been suggested that glia–axon interactions may play a role in proper pathfinding in the lateral line system (Gilmour et al. 2002). Such a role could also be played by glial cells in the developing ear, in particular for the outgrowth of fibers toward the brain (Begbie and Graham 2001). These issues could be studied in v-*erb*-b2 erythroblastic leukemia viral oncogene homolog receptor 2 (*Erbb2*) mutant mice which have no glial cells (Morris et al. 1999). Preliminary data on *Erbb2* mutant mice strongly support this notion (Morris and Fritzsch, unpublished data).

The semaphorins and their receptors, the neuropilins and plexins, are important for their roles in neuronal pathfinding and regeneration (Pasterkamp and Verhaagen 2001; Cloutier et al. 2002). Plexins and semaphorins have been described in the developing ear (Miyazaki et al. 1999; Murakami et al. 2001), but their potential function has not yet been explored in the existing mutant mice (Cloutier et al. 2002). Recently, a role for one semaphorin 3a, *Sema3a*, in providing a stop signal for growing afferents has been described. In this mutant, growing dendrites do not stop at the vestibular sensory epithelia but continue to grow and extend outside the ear and as far as the skin above the ear (Gu et al. 2003). In addition, work on semaphorin receptor mutants indicates that these receptors may play a role in guiding afferents (Gu et al. 2003). Given that all 16 or more semaphorins signal via only two receptors, a large degree of redundancy can be expected. Therefore, understanding the functions of these receptors and their ligands in the ear (Suto et al. 2003) may take some time.

Other factors associated with guidance are the large members of cell adhesion molecules, some of which are expressed in intricate patterns in the developing ear (Davies and Holley 2002). Again, no experimental studies exist that support their function in fiber guidance in vivo.

Another issue that needs to be discussed is the extent of the hair cell's role in fiber guidance. The most direct way to determine how important hair cells are to fiber guidance is to study mutants that lack hair cells. Mice lacking the POU-domain, class 4, transcription factor 3 gene, *Pou4f3* (alias, *Brn3c*), develop only a limited complement of undifferentiated hair cells. Theses cells are identified as hair cells by their hair cell–specific molecular marker expression (Xiang et al. 1998). Examination of afferent innervation showed no correlation between fiber loss and the formation of these morphologically undifferentiated hair cells (Xiang et al. 2003). Specifically, a robust sensory innervation persists through embryogenesis into early neonatal life. Some profound conclusions about the role of hair cells in neuronal pathfinding can be drawn from these mutants. Data from *Pou4f3* mutants suggest that hair cells are *not* needed for the initial neuronal guidance and pathfinding and that at least some of the pathfinding properties must reside in the interaction between growing neurons and the otic cells surrounding them and perhaps the sensory patch per se even without hair cells.

Given that *Pou4f3* null mice still form hair cells that subsequently undergo cell death (Xiang et al. 1998), another mutation was needed that would not form any hair cells, such as *Atoh1* (Bermingham et al. 1999). Examination of mutants lacking the bHLH transcription factor *Atoh1* was expected to help test the functions of guidance and survival factors, released from hair cells, on sensory afferent neurons. Analysis shows that *Atoh1* is required for hair cell differentiation and probably acts upstream of *Pou4f3* (Bermingham et al. 1999; Fritzsch et al. 2000). Surprisingly, there is very little effect on the initial fiber growth of the sensory neurons in this mutant (Fritzsch et al. 2005). Older embryos, however, show a severe reduction of afferents that does not correspond to the pattern of loss observed in neurotrophin mutations. Closer examination of the expression of brain derived neurotrophic factor (*Bdnf*), using the *Bdnf-LacZ* reporter, showed that even in *Atoh1* mutants, some undifferentiated hair cell precursors form and express *Bdnf*. Thus, at least in some hair cell precursors, BDNF expression does not require *Atoh1*-mediated hair-cell differentiation.

Once the neurons reach their target hair cell(s) by whatever means, their survival is known to be mediated, at least in part, by the neurotrophins BDNF and NTF3 as described in the following section. The survival of these neurons in *Pou4f3* mutants throughout embryogenesis is apparently mediated by the limited expression of both neurotrophins, *Bdnf* and *Ntf3*, in undifferentiated sensory epithelia as revealed using in situ hybridization. Even animals that were several months old had considerable innervation of the apical turn of the cochlea. This long-term retention of cochlear innervation may also be mediated by neurotrophins as these neurotrophis are still detectable in neonatal animals (Stankovic and Corfas 2003).

While these data indicate some progress in the molecular network of peripheral neuron guidance, many interesting questions remain to be addressed. It also highlights how rudimentary this research is (Fritzsch 2003). For example, both qualitative and quantitative expression patterns of FGFs may further complicate this picture. Moreover, virtually nothing is known about the molecular guidance of developing vestibular and cochlear afferents into the brain (Rubel and Fritzsch 2002; Maklad and Fritzsch 2003).

6. Cell Survival and Death

In contrast to this apparent scarcity of data on molecular guidance, extensive knowledge exists on the molecular basis of sensory neuron survival (Fritzsch et al. 2005). Numerous observations suggest that hair cells attract fibers to innervate them (Bianchi and Cohan 1991, 1993) and some neurotrophic effects might, in part, mediate these attractions (Cajal 1919). One way of exploring neurotrophin effects in the ear would be to eliminate neurotrophins in the target of the inner ear afferents, the hair cells, by eliminating hair cells or preventing their differentiation. This is expected to eliminate late expression of *Bdnf* because older embryos express *Bdnf* exclusively in hair cells within the ear. In addition, if no hair cells form, it is likely that supporting cells will not form normally. This is so because the differentiation of both hair cells and supporting cells is apparently linked, as the interactions between them are mediated by the Delta–Notch regulatory system (Zine et al. 2001). Recent investigations of two mutations that result in undifferentiated or absent hair cells show limited sensory neuron loss as a result of neurotrophin deficiency (Xiang et al. 2003; Fritzsch et al. 2005). Alternatively, other mutants provide examples of sensory neuron loss that appears to be caused by abnormally low expression of the neurotrophic tyrosine kinase receptors for these neurotrophins (Liu et al. 2000; Huang et al. 2001; Kim et al. 2001). The existence of neurotrophins with lox-P sites as well as the rapid generation of several specific Cre-expressing promoter lines soon will allow investigation of this issue in an even more targeted way.

In 1919, Ramon y Cajal proposed that hair cells secrete neurotrophic substance(s) that attract sensory afferents (Cajal 1919). The data described above show that, despite our attempts to eliminate neurotrophin expression in the ear by causing mutations in essential transcription factors, even to this date, we have been unable to test this proposal satisfactorily. Unfortunately, the recent finding of neurotrophin expression in delaminating sensory neurons has made the interpretation of neurotrophin effects even more complicated. The limited expression of *Bdnf* in the undifferentiated hair cell precursors of *Atoh1* mutant mice is apparently enough to support many afferents throughout embryonic life. Consequently, we cannot exclude a biologically significant effect of the limited expression of neurotrophins within delaminating sensory neurons. Thus, none of the mutations described above has been able to critically test the role exclusively played by hair cells in attraction and maintenance of inner ear sensory neurons. Indeed, recent work on transgenic misexpression of neurotrophins suggests projections to nonsensory areas that lack hair cells (Tessarollo et al. 2004), indicating that the neurotrophin BDNF can override whatever attraction is provided by hair cells.

Within the past 10 years, two neurotrophic factors, BDNF and NTF3, and their high-affinity neurotrophic tyrosine kinase receptors, *Ntrk2* (alias, *trkB*) and *Ntrk3* (alias, *trkC*), as well as the low-affinity receptor *p75*, have been identified in the ear (Pirvola et al. 1992; Wheeler et al. 1994; Fritzsch et al. 1999; Farinas et al. 2001). Remaining issues center on the uniqueness of function of each neurotrophin receptor combination. Since most other vertebrates express only one neurotrophin in the ear, it remains unclear why mammals have two. In addition, the expression patterns described show a highly dynamic change along the cochlea as well as alterations in the vestibular system. We have yet to understand fully the functional significance of these changes (Pirvola et al. 1992; Farinas et al. 2001). Targeted mutations of each of these neurotrophins and receptors have clarified their relative contributions to the survival of different sensory neurons in the ear (Fritzsch et al. 1999). These data have shown that there is the dramatic loss of 85% of cochlear (spiral) sensory neurons in *Ntf3* mutants (Farinas et al. 1994; 2001; Fritzsch et al. 1997a) and of 80% to 85% of vestibular neurons in the *Bdnf* mutant (Jones et al. 1994; Ernfors et al. 1995; Schimmang et al. 1995; Bianchi et al. 1996). Somewhat similar effects have been described in *Ntrk* mutants (Fritzsch et al. 1995, 2001b). Detailed counting has shown that neuronal loss happens within 2 to 3 days after the fibers have first extended toward the sensory epithelia (Bianchi et al. 1996; Farinas et al. 2001). Together these data suggest that vestibular and cochlear sensory neurons have unique but complementary neurotrophin requirements (Ernfors et al. 1995).

The relative distribution of neurotrophins with more prominent expression of *Bdnf* in the vestibular system and of *Ntf3* in the cochlea supports the evidence for complementary roles of these neurotrophins in the vestibular and cochlear sensory epithelia, respectively (Pirvola et al. 1992; Farinas et al. 2001). In the context of the hypothesized role of neurotrophin function in numerical matching of pre- and postsynaptic targets, it needs to be pointed out that in the ear there is no uniform relationship between afferents and hair cells which can vary from 30 to 1 (convergence on a single inner hair cell) to 1 to 30 (divergence on outer hair cells and some vestibular fibers). It is questionable that a single neurotrophin, such as BDNF, distributed fairly uniformly in all hair cells, would be able to mediate these differences. Clearly, quantitative data on specific amounts of BDNF expressed in different types of hair cells are needed to evaluate this aspect of ear innervation.

While specific mutations showed significant effects on sensory neuron survival to distinct endorgans of the ear, it remained unclear whether more neurotrophins or other neurotrophic factors might add to the survival of inner ear sensory neurons. However, double mutant mice, lacking both the neurotrophin receptors *Ntrk2* and *Ntrk3*, or both neurotrophins *Bdnf* and *Ntf3*, have no surviving sensory neurons in the inner ear at birth (Ernfors et al. 1995; Liebl et al. 1997; Silos-Santiago et al. 1997). This dramatic effect of double mutations on ear innervation puts to rest any speculations about the requirement of additional neurotrophins or neurotrophin receptors. These data on double mutants also show that even if other ligands and receptors are present, their function for the development of the inner ear sensory neurons is far less critical than *Bdnf/Ntf3* and their receptors *Ntrk2/Ntrk3*.

Single neurotrophin and neurotrophin receptor mutant mice demonstrate specific losses of distinct cochlear and vestibular afferents. These losses appear to be related to a highly dynamic pattern of expression of neurotrophins but not of neurotrophin receptors in the ear (Pirvola et al. 1992; Farinas et al. 2001). During development and in the adult sensory neurons there appears to be a uniform expression of both *Ntrk2* and *Ntrk3* in all of the sensory neurons in the ear (Fritzsch et al. 1999; Farinas et al. 2001). In addition, primary sensory neurons express neurotrophins soon after delamination from the placodal epithelium is initiated (Farinas et al. 2001). Furthermore, a given primary sensory neuron expresses the same neurotrophin that is present in the area of the otocyst from which it delaminated (Fritzsch et al. 2002). For example, delaminating sensory neurons from the cochlea express the neurotrophin *Ntf3* as revealed in *Ntf3* heterozygotic animals (Figs. 4.2 and 4.4). Comparison of *Bdnf*- and *Ntf3-LacZ* positive cells with delaminating neurons marked by *Neurod1-LacZ* also suggests that the delaminating *Bdnf*- or *Ntf3*-positive precursors are in fact *Neurod1*-expressing neuronal precursors (Liu et al. 2000; Kim et al. Fritzsch et al. 2002). Proof of this, however, will require colabeling for *Neurod1* and each of the neurotrophins.

These data indicate that initial fiber growth occurs normally in the absence of neurotrophins. Indeed, initial fiber growth may use the same molecular cues recognized by the delaminating primary sensory neuron precursors. Subsequently, there is a critical period in which specific neurotrophins are required for sensory neurons to reach their targets. The partially overlapping expression of neurotrophins reported in the developing mammalian ear (Pirvola et al. 1992; Farinas et al. 2001), seems to translate into a spatiotemporal loss of primary sensory neurons in specific mutants (Farinas et al. 2001). Particularly, in the cochlea there is a delayed up-regulation of *Bdnf* expression in the basal turn, leaving all basal turn neurons solely dependent on *Ntf3* for a brief but critical period of embryogenesis (Fig. 4.5). Thus, if *Ntf3* is absent, there is a progressive loss of spiral neurons, especially in the basal turn, where *Bdnf* is not present to compensate for the absence of *Ntf3*. Overall, sensory neuron loss will occur in an embryo with a targeted mutation in *Bdnf* or *Ntf3* only where the other is not present to compensate for its absence.

This suggestion has led to the prediction that in the ear, *Ntf3* and *Bdnf* are functionally equivalent and can be substituted for each other without compromising the development and survival of sensory neurons. Supporting this suggestion, data show that the topological loss of sensory neurons in the basal turn in the *Ntf3* mutant can be rescued by transgenic expression of *Bdnf* under the control of *Ntf3* gene regulatory elements (Coppola et al. 2001). Selective tracer injection in these animals shows reorganization of the peripheral projection pattern of vestibular neurons into the basal turn of the cochlea (Tessarollo et al. 2004). These projections developed at the time the neurons are known to become susceptible to the neurotrophins for their survival, suggesting that they become both neurotrophically as well as neurotropically dependent on neurotrophins at the same time (Fig. 4.5). Interestingly, the central projection remained unchanged in these animals, suggesting little if any influence of neurotrophins on the patterning of the central projections. Overall, these data suggest that the molecular processes that guide peripheral and central projection are distinct. Likewise, the transgenic animal in which the *Ntf3* coding region is inserted into the *Bdnf* gene is equally effective in rescuing the *Bdnf* phenotype in the cochlea, but not in the vestibular system (Agerman et al. 2003). Together these data suggest that the differential expression of neurotrophins in the ear plays a significant role in patterning the ear innervation.

Further, these data support a role for a very early onset of elimination of exuberant or unconnected afferents and the primary sensory neurons that generate these axons. This verification of proper connections occurs immediately after the fibers have reached and start to invade their target organs (as early as E11 in the canal epithelia and E12.5 in the basal turn of the cochlea; Fig. 4.4). The most interesting effect of the single neurotrophin mutation is the striking dependence of the basal turn cochlear neurons on *Ntf3*. This is due to the fact that *Bdnf* shows a delayed expression in the basal turn (Fig. 4.5). Clearly, *Bdnf* not only can compensate for *Ntf3* and rescue the basal turn neurons (Coppola et al. 2001) but can also attract vestibular fibers from the nearby nerve to the posterior crista to innervate the basal turn instead (Fig. 4.5). It is conceivable that evolutionary pressures have resulted in the delayed expression of *Bdnf* in

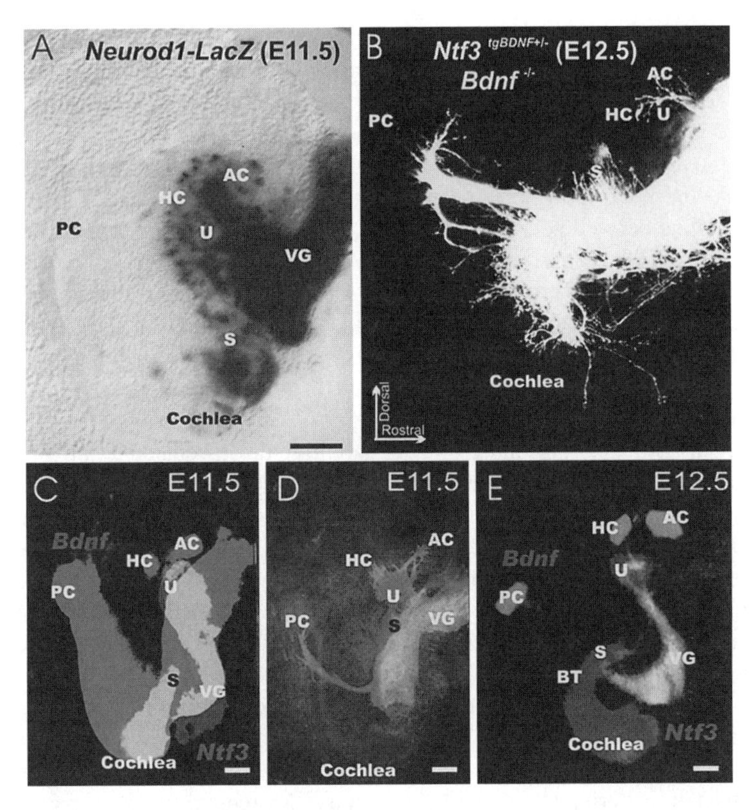

FIGURE 4.5. The formation of sensory neurons, expression of neurotrophins, and effects of neurotrophin misexpression are shown. *Neurod1-LacZ* positive cells exist in the otocyst and indicate the neurosensory area. Note that the area of *Neurod1* expression (**A**) overlaps with the area of *Ntf3* expression (*red* in **C**) as well as the anterior and horizontal crista (HC, AC) that are *Bdnf-LacZ* positive (green in **C**, **E**). Labling of afferents with DiI shows dendrites extending to all patches highlighted by *BDNF* and *Ntf3* expression as early as E 11.5 (**C**, **D**). *Bdnf* expression becomes more restricted in E12.5 embryos (E) offering predominantly *Ntf3* as a neurotrophin in the developing cochlea (*red* in **E**). Altering expression of *Bdnf* through misexpression of *Bdnf* under *Ntf3* promoter control ($Ntf3^{tgBdnf}$) will alter the pathway of vestibular neuron dendrites that project to the cochlea. This can be enhanced by combining the transgene with *Bdnf* mutation (**B**). Under these circumstances, fibers stop on their way to the posterior crista (PC) and redirect their dendrites to the cochlea. In addition, fibers from the saccule project to the cochlea, as shown here after a cerebellar injection of a lipophilic dye tracer. Modified after Fritzsch et al. (2002); Tessarollo et al. (2004). AC, Anterior crista; BT, basal turn (of cochlea); HC, horizontal crista; PC, posterior crista; S, saccule; U, utricle; VG, vestibular ganglion. Bar = 100 μm. (See color insert.)

the basal turn because of the need to avoid misrouting of vestibular afferents. Misrouting of vestibular fibers to cochlear hair cells, assuming they maintain their normal connections in the CNS, will result in auditory information interfering with perceptions of position and motion.

Recently, spiral ganglion cells from the human cochlea were cultured in vitro. Unpublished data show that under the influence of BDNF and NTF3 and/or glial-cell-line derived neurotrophic factor (GDNF), these cells differentiated into elongated neuons. These neurons demonstrated transient expression of the intermediate filament marker, nestin (Rask-Anderson et al. 2005). Although further investigation of these cultured cells is still to come, such as expression analysis for the spiral ganglion-specific *Gata3* gene, these data suggest some plasticity of sensory neurons and an ability of neurotrophins to stimulate neuronal fiber outgrowth.

Interestingly, neurotrophins are apparently down-regulated in neonates (Fritzsch et al. 1999; Stankovic and Corfas 2003; Wheeler et al. 1994) and they appear to be largely lost in adults, despite the fact that their *Ntrk* genes are still expressed in the sensory neurons (Fritzsch et al. 1999). This has led to the suggestion that other neurotrophic factors may play a role in the neonatal death of sensory neurons (Hashino et al. 1999; Echteler and Nofsinger 2000), a suggestion that requires further experimental verification in mutants with conditional targeting of neurotrophin genes.

In summary, certain molecular aspects of sensory neuron formation, guidance, and survival have been clarified in recent years. Still, numerous open issues remain in this rapidly moving field and virtually no molecular data exist for the basic patterning of central projections. Moreover, the emerging role of neurotrophins in neonatal patterning needs to be further explored (Postigo et al. 2002; Davis 2003; Schimmang et al. 2003). Interesting future work will have to sort out how the apparent relationship between specific sensory neurons and specific types of hair cells works at a molecular level inside a given sensory organ.

7. Summary and Conclusions

Our level of understanding of global patterning processes of the ear, restriction of sensory neuron fate determination, molecular regulation of proliferation to determine clonal expansion, and coordinated transition into the generation of different offspring has increased in the ear almost at the same pace as for neuronal development in general. Future work will show which and how the presently known genes can be used to restart cell proliferation and to direct hair cell formation in animals and humans with neurosensory loss. Likewise, additional work is needed to fully characterize the limited number of presently known pathfinding molecules so that they can be used for fiber regeneration in hair cell regenerated patients or patients with cochlear implants. Unique features of the ear, as compared to development of other sensory systems, are becoming apparent and provide challenges for the continued molecular dissection that is needed to gain access to the molecular machinery for guided regrowth of nerve fibers to regenerated hair cells. Progress has been particularly prominent in the neurotrophin areas, where the ear holds an intermediate position between the hypercomplex CNS and the simple one neurotrophin, one neurotrophin receptor that prevails for almost all other sensory neurons. Such differences allow for unique experiments using transgenic misexpressers to exemplify the two levels of function of neurotrophins, trophic and tropic support of growing dendrites.

Future directions will have to identify progenitor populations, hopefully with the capacity to generate both sensory neurons and hair cells, to restore sensorineural hearing loss. Using the known guidance properties of neurotrophins, in particular of BDNF, could allow for a more directed growth of nerve fibers toward cochlear implants. This way the basic science gain could be translated into clinical applications beneficial for a growing number of patients suffering hearing loss.

Acknowledgments. This work was supported by a grant from NIDCD (R01 DC005590 and DC05009). We wish to thank Dr. Carl Hopkins for his EndNote style macro.

References

Adam J, Myat A, Le Roux I, Eddison M, Henrique D, Ish-Horowicz D, Lewis J (1998) Cell fate choices and the expression of Notch, Delta and Serrate homologues in the chick inner ear: parallels with *Drosophila* sense-organ development. Development 125:4645–4654.

Agerman K, Hjerling-Leffler J, Blanchard MP, Scarfone E, Canlon B, Nosrat C, Ernfors P (2003) BDNF gene replacement reveals multiple mechanisms for establishing neurotrophin specificity during sensory nervous system development. Development 130: 1479–1491.

Alvarez Y, Alonso MT, Vendrell V, Zelarayan LC, Chamero P, Theil T, Bosl MR, Kato S, Maconochie M, Riethmacher D, Schimmang T (2003) Requirements for FGF3 and FGF10 during inner ear formation. Development 130:6329–6338.

Begbie J, Graham A (2001) Integration between the epibranchial placodes and the hindbrain. Science 294:595–598.

Bermingham NA, Hassan BA, Price SD, Vollrath MA, Ben-Arie N, Eatock RA, Bellen HJ, Lysakowski A, Zoghbi HY (1999) *Math 1*: an essential gene for the generation of inner ear hair cells. Science 284:1837–1841.

Bermingham NA, Hassan BA, Wang VY, Fernandez M, Banfi S, Bellen HJ, Fritzsch B, Zoghbi HY (2001) Proprioceptor pathway development is dependent on *Math1*. Neuron 30:411–422.

Bertrand N, Castro DS, Guillemot F (2002) Proneural genes and the specification of neural cell types. Nat Rev Neurosci 3:517–530.

Bianchi LM, Cohan CS (1991) Developmental regulation of a neurite-promoting factor influencing statoacoustic neurons. Brain Res Dev Brain Res 64:167–174.

Bianchi LM, Cohan CS (1993) Effects of the neurotrophins and CNTF on developing statoacoustic neurons: comparison with an otocyst-derived factor. Dev Biol 159:353–365.

Bianchi LM, Liu H (1999) Comparison of ephrin-A ligand and EphA receptor distribution in the developing inner ear. Anat Rec 254:127–134.

Bianchi LM, Conover JC, Fritzsch B, DeChiara T, Lindsay RM, Yancopoulos GD (1996) Degeneration of vestibular neurons in late embryogenesis of both heterozygous and homozygous BDNF null mutant mice. Development 122:1965–1973.

Bober E, Rinkwitz S, Herbrand H (2003) Molecular basis of otic commitment and morphogenesis: a role for homeodomain-containing transcription factors and signaling molecules. Curr Top Dev Biol 57:151–175.

Borsani G, DeGrandi A, Ballabio A, Bulfone A, Bernard L, Banfi S, Gattuso C, Mariani M, Dixon M, Donnai D, Metcalfe K, Winter R, Robertson M, Axton R, Brown A, van Heyningen V, Hanson I (1999) *EYA4,* a novel vertebrate gene related to *Drosophila* eyes absent. Hum Mol Genet 8:11–23.

Brent AE, Schweitzer R, Tabin CJ (2003). A somitic compartment of tendon progenitors. Cell 113:235–248.

Brown S, Castelli-Gair Hombria J (2000) *Drosophila* grain encodes a GATA transcription factor required for cell rearrangement during morphogenesis. Development 127:4867–4876.

Brown ST, Martin K, Groves AK (2003) Molecular basis of inner ear induction. Curr Top Dev Biol 57:115–149.

Brunet JF, Pattyn A (2002) *Phox2* genes—from patterning to connectivity. Curr Opin Genet Dev 12:435–440.

Cai L, Hayes NL, Takahashi T, Caviness VS, Jr., Nowakowski RS (2002) Size distribution of retrovirally marked lineages matches prediction from population measurements of cell cycle behavior. J Neurosci Res 69:731–744.

Cajal SR (1919) Accion neurotropica de los epitelios. Trab Lab Invest Biol 17:1–153.

Caldwell JC, Eberl DF (2002) Towards a molecular understanding of *Drosophila* hearing. J Neurobiol 53:172–189.

Calegari F, Huttner WB (2003) An inhibition of cyclin-dependent kinases that lengthens, but does not arrest, neuroepithelial cell cycle induces premature neurogenesis. J Cell Sci 116:4947–4955.

Calof AL, Bonnin A, Crocker C, Kawauchi S, Murray RC, Shou J, Wu HH (2002) Progenitor cells of the olfactory receptor neuron lineage. Microsc Res Tech 58:176–188.

Canzoniere D, Farioli-Vecchioli S, Conti F, Ciotti MT, Tata AM, Augusti-Tocco G, Mattei E, Lakshmana MK, Krizhanovsky V, Reeves SA, Giovannoni R, Castano F, Servadio A, Ben-Arie N, Tirone F (2004) Dual control of neurogenesis by PC3 through cell cycle inhibition and induction of *Math1*. J Neurosci 24:3355–3369.

Carney PR, Silver J (1983) Studies on cell migration and axon guidance in the developing distal auditory system of the mouse. J Comp Neurol 215:359–369.

Cau E, Casarosa S, Guillemot F (2002) *Mash1* and *Ngn1* control distinct steps of determination and differentiation in the olfactory sensory neuron lineage. Development 129:1871–1880.

Caviness VS, Jr., Goto T, Tarui T, Takahashi T, Bhide PG, Nowakowski RS (2003) Cell output, cell cycle duration and neuronal specification: a model of integrated mechanisms of the neocortical proliferative process. Cereb Cortex 13:592–598.

Chang W, Nunes FD, De Jesus-Escobar JM, Harland R, Wu DK (1999) Ectopic *noggin* blocks sensory and nonsensory organ morphogenesis in the chicken inner ear. Dev Biol 216:369–381.

Chang W, ten Dijke P, Wu DK (2002) BMP pathways are involved in otic capsule formation and epithelial-mesenchymal signaling in the developing chicken inner ear. Dev Biol 251:380–394.

Chen P, Johnson JE, Zoghbi HY, Segil N (2002) The role of *Math1* in inner ear development: Uncoupling the establishment of the sensory primordium from hair cell fate determination. Development 129:2495–2505.

Chen P, Zindy F, Abdala C, Liu F, Li X, Roussel MF, Segil N (2003) Progressive hearing loss in mice lacking the cyclin-dependent kinase inhibitor *Ink4d*. Nat Cell Biol 5:422–426.

Cloutier JF, Giger RJ, Koentges G, Dulac C, Kolodkin AL, Ginty DD (2002) Neuropilin-2 mediates axonal fasciculation, zonal segregation, but not axonal convergence, of primary accessory olfactory neurons. Neuron 33:877–892.

Coppola V, Kucera J, Palko ME, Martinez-De Velasco J, Lyons WE, Fritzsch B, Tessarollo L (2001) Dissection of *NT3* functions in vivo by gene replacement strategy. Development 128:4315–4327.

Cowan CA, Yokoyama N, Bianchi LM, Henkemeyer M, Fritzsch B (2000) EphB2 guides axons at the midline and is necessary for normal vestibular function. Neuron 26:417–430.

Davies D, Holley MC (2002) Differential expression of *alpha3* and *alpha6* integrins in the developing mouse inner ear. J Comp Neurol 445:122–132.

Davis RJ, Shen W, Heanue TA, Mardon G (1999) Mouse Dach, a homologue of *Drosophila* dachshund, is expressed in the developing retina, brain and limbs. Dev Genes Evol 209:526–536.

Davis RJ, Shen W, Sandler YI, Heanue TA, Mardon G (2001) Characterization of mouse *Dach2*, a homologue of *Drosophila* dachshund. Mech Dev 102:169–179.

Davis RL (2003) Gradients of neurotrophins, ion channels, and tuning in the cochlea. Neuroscientist 9:311–316.

Echteler SM, Nofsinger YC (2000) Development of ganglion cell topography in the postnatal cochlea. J Comp Neurol 425:436–446.

Ernfors P, Van De Water T, Loring J, Jaenisch R (1995) Complementary roles of BDNF and NT-3 in vestibular and auditory development. Neuron 14:1153–1164.

Farinas I, Jones KR, Backus C, Wang XY, Reichardt LF (1994) Severe sensory and sympathetic deficits in mice lacking neurotrophin-3. Nature 369:658–6180.

Farinas I, Jones KR, Tessarollo L, Vigers AJ, Huang E, Kirstein M, de Caprona DC, Coppola V, Backus C, Reichardt LF, Fritzsch B (2001) Spatial shaping of cochlear innervation by temporally regulated neurotrophin expression. J Neurosci 21:6170–6180.

Fekete DM, Wu DK (2002) Revisiting cell fate specification in the inner ear. Curr Opin Neurobiol 12:35–42.

Ferguson KL, Vanderluit JL, Hebert JM, McIntosh WC, Tibbo E, MacLaurin JG, Park DS, Wallace VA, Vooijs M, McConnell SK, Slack RS (2002) Telencephalon-specific Rb knockouts reveal enhanced neurogenesis, survival and abnormal cortical development. EMBO J 21:3337–3346.

Fritzsch B (2001a) The morphology and function of fish ears. In: Ostrander G (ed) The Laboratory Fish. Exeter: Academic Press, pp. 250–259.

Fritzsch B (2001b) The cellular organization of the fish ear. In: Ostrander G (ed) The Laboratory Fish. Exeter: Academic Press, pp. 480–487.

Fritzsch B (2003) Development of inner ear afferent connections: forming primary neurons and connecting them to the developing sensory epithelia. Brain Res Bull 60: 423–433.

Fritzsch B, Beisel KW (2001) Evolution and development of the vertebrate ear. Brain Res Bull 55:711–721.

Fritzsch B, Beisel KW (2003) Molecular conservation and novelties in vertebrate ear development. Curr Top Dev Biol 57:1–44.

Fritzsch B, Beisel KW (2004) Keeping sensory cells and evolving neurons to connect them to the brain: molecular conservation and novelties in vertebrate ear development. Brain Behav Evol 64:182–197.

Fritzsch B, Wake MH (1988) The inner ear of gymnophione amphibians and its nerve supply: a comparative study of regressive events in a complex sensory system. Zoomorphol 108:210–217.

Fritzsch B, Silos-Santiago I, Smeyne R, Fagan AM, Barbacid M (1995) Reduction and loss of inner ear innervation in *trkB* and *trkC* receptor knockout mice: a whole mount DiI and scanning electron microscopic analysis. Audit Neurosci 1:401–417.

Fritzsch B, Farinas I, Reichardt LF (1997a) Lack of *neurotrophin 3* causes losses of both classes of spiral ganglion neurons in the cochlea in a region-specific fashion. J Neurosci 17:6213–6225.

Fritzsch B, Sarai PA, Barbacid M, Silos-Santiago I (1997b) Mice with a targeted disruption of the neurotrophin receptor *trkB* lose their gustatory ganglion cells early but do develop taste buds. Int J Dev Neurosci 15:563–576.

Fritzsch B, Barald K, Lomax M (1998) Early embryology of the vertebrate ear. In: Rubel EW, Popper AN, Fay RR (eds), Development of the Auditory System. New York: Springer-Verlag, pp. 80–145.

Fritzsch B, Pirvola U, Ylikoski J (1999) Making and breaking the innervation of the ear: neurotrophic support during ear development and its clinical implications. Cell Tissue Res 295:369–382.

Fritzsch B, Beisel KW, Bermingham NA (2000) Developmental evolutionary biology of the vertebrate ear: conserving mechanoelectric transduction and developmental pathways in diverging morphologies. NeuroReport 11:R35–44.

Fritzsch B, Signore M, Simeone A (2001a) Otx1 null mutant mice show partial segregation of sensory epithelia comparable to lamprey ears. Dev Genes Evol 211:388–396.

Fritzsch B, Silos-Santiago I, Farinas I, Jones K (2001b) Neurotrophins and neurotrophin receptors involved in supporting afferent inner ear innervation. In: Mocchetti I (ed), The Neurotrophins. Johnson City, TN: Salzburger and Graham.

Fritzsch B, Beisel KW, Jones K, Farinas I, Maklad A, Lee J, Reichardt LF (2002) Development and evolution of inner ear sensory epithelia and their innervation. J Neurobiol 53:143–156.

Fritzsch B, Tessarollo L, Coppola E, Reichardt LF (2003) Neurotrophins in the ear: their roles in sensory neuron survival and fiber guidance. Prog Brain Res 146:265–278.

Fritzsch B, Matei VA, Nichols DH, Bermingham N, Jones K, Beisel KW, Wang VY (2005) *Atoh1* null mutants show directed afferent fiber growth to undifferentiated ear sensory epithelia followed by incomplete fiber retention. Dev Dyn 233:570–583.

Gao WQ (2003) Hair cell development in higher vertebrates. Curr Top Dev Biol 57: 293–319.

Gerlach LM, Hutson MR, Germiller JA, Nguyen-Luu D, Victor JC, Barald KF (2000) Addition of the BMP4 antagonist, *noggin*, disrupts avian inner ear development. Development 127:45–54.

Gilmour DT, Maischein HM, Nusslein-Volhard C (2002) Migration and function of a glial subtype in the vertebrate peripheral nervous system. Neuron 34:577–588.

Goodyear RJ, Richardson GP (2002) Extracellular matrices associated with the apical surfaces of sensory epithelia in the inner ear: molecular and structural diversity. J Neurobiol 53:212–227.

Gowan K, Helms AW, Hunsaker TL, Collisson T, Ebert PJ, Odom R, Johnson JE (2001) Crossinhibitory activities of *Ngn1* and *Math1* allow specification of distinct dorsal interneurons. Neuron 31:219–232.

Graham V, Khudyakov J, Ellis P, Pevny L (2003) *SOX2* functions to maintain neural progenitor identity. Neuron 39:749–765.

Gu C, Rodriguez ER, Reimert DV, Shu T, Fritzsch B, Richards LJ, Kolodkin AL, Ginty DD (2003). Neuropilin-1 conveys semaphorin and VEGF signaling during neural and cardiovascular development. Dev Cell 5:45–57.

Hanson IM (2001) Mammalian homologues of the *Drosophila* eye specification genes. Semin Cell Dev Biol 12:475–484.

Hashino E, Dolnick RY, Cohan CS (1999) Developing vestibular ganglion neurons switch trophic sensitivity from BDNF to GDNF after target innervation. J Neurobiol 38:414–427.

Huang EJ, Liu W, Fritzsch B, Bianchi LM, Reichardt LF, Xiang M (2001) *Brn3a* is a transcriptional regulator of soma size, target field innervation and axon pathfinding of inner ear sensory neurons. Development 128:2421–2432.

Jones KR, Farinas I, Backus C, Reichardt LF (1994) Targeted disruption of the BDNF gene perturbs brain and sensory neuron development but not motor neuron development. Cell 76:989–999.

Karis A, Pata I, van Doorninck JH, Grosveld F, de Zeeuw CI, de Caprona D, Fritzsch B (2001) Transcription factor *GATA-3* alters pathway selection of olivocochlear neurons and affects morphogenesis of the ear. J Comp Neurol 429:615–630.

Kawamoto K, Ishimoto S, Minoda R, Brough DE, Raphael Y (2003) Math1 gene transfer generates new cochlear hair cells in mature guinea pigs in vivo. J Neurosci 23:4395–4400.

Kim WY, Fritzsch B, Serls A, Bakel LA, Huang EJ, Reichardt LF, Barth DS, Lee JE (2001). *NeuroD*-null mice are deaf due to a severe loss of the inner ear sensory neurons during development. Development 128:417–426.

Lawoko-Kerali G, Rivolta MN, Holley M (2002) Expression of the transcription factors *GATA3* and *Pax2* during development of the mammalian inner ear. J Comp Neurol 442:378–391.

Lee CS, Fan CM (2001) Embryonic expression patterns of the mouse and chick *Gas1* genes. Mech Dev 101:293–297.

Lee JE (1997) Basic helix-loop-helix genes in neural development. Curr Opin Neurobiol 7:13–20.

Lewis ER, Leverenz EL, Bialek WS (1985) The veretebrate inner ear. Boca Raton: CRC Press.

Li H, Liu H, Heller S (2003a) Pluripotent stem cells from the adult mouse inner ear. Nat Med 9:1293–1299.

Li H, Roblin G, Liu H, Heller S (2003b) Generation of hair cells by stepwise differentiation of embryonic stem cells. Proc Natl Acad Sci USA 100:13495–13500.

Liebl DJ, Tessarollo L, Palko ME, Parada LF (1997) Absence of sensory neurons before target innervation in brain-derived neurotrophic factor-, neurotrophin3-, and trkC-deficient embryonic mice. J Neurosci 17:9113–9127.

Liu M, Pereira FA, Price SD, Chu MJ, Shope C, Himes D, Eatock RA, Brownell WE, Lysakowski A, Tsai MJ (2000) Essential role of BETA2/*NeuroD1* in development of the vestibular and auditory systems. Genes Dev 14:2839–2854.

Liu Y, Liu C, Yamada Y, Fan CM (2002) Growth arrest specific gene 1 acts as a region-specific mediator of the *Fgf10/Fgf8* regulatory loop in the limb. Development 129: 5289–5300.

Lorente de No R (1933) Anatomy of the eighth nerve: the central projections of the nerve endings of the internal ear. Laryngoscope 43:1–38.

Ma Q, Kintner C, Anderson DJ (1996) Identification of *neurogenin*, a vertebrate neuronal determination gene. Cell 87:43–52.

Ma Q, Chen Z, del Barco Barrantes I, de la Pompa JL, Anderson DJ (1998) *neurogenin1* is essential for the determination of neuronal precursors for proximal cranial sensory ganglia. Neuron 20:469–482.

Ma Q, Anderson DJ, Fritzsch B (2000) *Neurogenin 1* null mutant ears develop fewer, morphologically normal hair cells in smaller sensory epithelia devoid of innervation. J Assoc Res Otolaryngol 1:129–143.

Maklad A, Fritzsch B (1999) Incomplete segregation of endorgan-specific vestibular ganglion cells in mice and rats. J Vestib Res 9:387–399.

Maklad A, Fritzsch B (2002) The developmental segregation of posterior crista and saccular vestibular fibers in mice: a carbocyanine tracer study using confocal microscopy. Dev Brain Res 135:1–17.

Maklad A, Fritzsch B (2003) Development of vestibular afferent projections into the hindbrain and their central targets. Brain Res Bull 60:497–510.

Matei V, Pauley S, Kaing S, Rowitch D, Beisel KW, et al. (2005) Smaller inner ear sensory epithelia in Neurog1 null mice are related to earlier hair cell terminal mitosis. Dev Dyn (in revision).

Megason SG, McMahon AP (2002) A mitogen gradient of dorsal midline *Wnts* organizes growth in the CNS. Development 129:2087–2098.

Merscher S, Funke B, Epstein JA, Heyer J, Puech A, Lu MM, Xavier RJ, Demay MB, Russell RG, Factor S, Tokooya K, Jore BS, Lopez M, Pandita RK, Lia M, Carrion D, Xu H, Schorle H, Kobler JB, Scambler P, Wynshaw-Boris A, Skoultchi AI, Morrow BE, Kucherlapati R (2001) *TBX1* is responsible for cardiovascular defects in velo-cardio-facial/DiGeorge syndrome. Cell 104:619–629.

Miyazaki N, Furuyama T, Takeda N, Inoue T, Kubo T, Inagaki S (1999) Expression of mouse *semaphorin H* mRNA in the inner ear of mouse fetuses. Neurosci Lett 261: 127–129.

Morris JK, Lin W, Hauser C, Marchuk Y, Getman D, Lee KF (1999) Rescue of the cardiac defect in *ErbB2* mutant mice reveals essential roles of *ErbB2* in peripheral nervous system development. Neuron 23:273–283.

Morsli H, Choo D, Ryan A, Johnson R, Wu DK (1998) Development of the mouse inner ear and origin of its sensory organs. J Neurosci 18:3327–3335.

Murakami Y, Suto F, Shimizu M, Shinoda T, Kameyama T, Fujisawa H (2001) Differential expression of *plexin-A* subfamily members in the mouse nervous system. Dev Dyn 220:246–258.

Noramly S, Grainger RM (2002) Determination of the embryonic inner ear. J Neurobiol 53:100–128.

Norris HW (1892) Studies on the development of the ear in *Amblystoma*. I. Development of the auditory vesicle. J Morphol 7:23–34.

Ohnuma S, Harris WA (2003) Neurogenesis and the cell cycle. Neuron 40:199–208.

Pasterkamp RJ, Verhaagen J (2001) Emerging roles for semaphorins in neural regeneration. Brain Res Brain Res Rev 35:36–54.

Pauley S, Wright TJ, Pirvola U, Ornitz D, Beisel K, Fritzsch B (2003) Expression and function of FGF10 in mammalian inner ear development. Dev Dyn 227:203–215.

Pfister M, Toth T, Thiele H, Haack B, Blin N, Zenner HP, Sziklai I, Murnberg P, Kupka S (2002) A 4-bp Insertion in the eya-homologous region (*eyaHR*) of *EYA4* causes hearing impairment in a Hungarian family linked to DFNA10. Mol Med 8: 607–611.

Pirvola U, Ylikoski J, Palgi J, Lehtonen E, Arumae U, Saarma M (1992) Brain-derived neurotrophic factor and neurotrophin 3 mRNAs in the peripheral target fields of developing inner ear ganglia. Proc Natl Acad Sci USA 89:9915–9919.

Pirvola U, Ylikoski J, Trokovic R, Hebert J, McConnell S, Partanen J (2002) FGFR1 is required for the development of the auditory sensory epithelium. Neuron 35:671.

Postigo A, Calella AM, Fritzsch B, Knipper M, Katz D, Eilers A, Schimmang T, Lewin GR, Klein R, Minichiello L (2002) Distinct requirements for *TrkB* and *TrkC* signaling in target innervation by sensory neurons. Genes Dev 16:633–645.

Qian Y, Fritzsch B, Shirasawa S, Chen CL, Choi Y, Ma Q (2001) Formation of brainstem (nor)adrenergic centers and first-order relay visceral sensory neurons is dependent on homeodomain protein Rnx/Tlx3. Genes Dev 15:2533–2545.

Raft S, Nowotschin S, Liao J, Morrow BE (2004) Suppression of neural fate and control of inner ear morphogenesis by *Tbx1*. Development 131:1801–1812.

Rask-Andersen H, Boström M, Gerdin B, Kinnefors A, Nyberg G. Engstrand T, Miller JM, Lindholm D (2005) Regeneration of human auditory nerve in vitro/in video demonstration of neural progenitor cells in adult human and guinea pig spiral ganglion. Hear Res 203:180–191.

Riccomagno MM, Martinu L, Mulheisen M, Wu DK, Epstein DJ (2002) Specification of the mammalian cochlea is dependent on Sonic hedgehog. Genes Dev 16:2365–2378.

Rubel EW, Fritzsch B (2002) Auditory system development:primary auditory neurons and their targets. Annu Rev Neurosci 25:51–101.

Ruben RJ (1967) Development of the inner ear of the mouse:a radioautographic study of terminal mitoses. Acta Otolaryngol (Suppl) 220:1–44.

Sarasin P, Sarasin F (1892) Über das Gehörorgan der Caeciliiden. Anat Anz 7:812–815.

Schimmang T, Minichiello L, Vazquez E, San Jose I, Giraldez F, Klein R, Represa J (1995) Developing inner ear sensory neurons require *TrkB* and *TrkC* receptors for innervation of their peripheral targets. Development 121:3381–3389.

Schimmang T, Tan J, Muller M, Zimmermann U, Rohbock K, Kopschall I, Limberger A, Minichiello L, Knipper M (2003) Lack of *Bdnf* and *TrkB* signalling in the postnatal cochlea leads to a spatial reshaping of innervation along the tonotopic axis and hearing loss. Development 130:4741–4750.

Silos-Santiago I, Fagan AM, Garber M, Fritzsch B, Barbacid M (1997) Severe sensory deficits but normal CNS development in newborn mice lacking TrkB and TrkC tyrosine protein kinase receptors. Eur J Neurosci 9:2045–2056.

Stankovic KM, Corfas G (2003) Real-time quantitative RT-PCR for low-abundance transcripts in the inner ear:analysis of neurotrophic factor expression. Hear Res 185:97–108.

Streit A (2002) Extensive cell movements accompany formation of the otic placode. Dev Biol 249:237–254.

Suto F, Murakami Y, Nakamura F, Goshima Y, Fujisawa H (2003) Identification and characterization of a novel mouse plexin, plexin-A4. Mech Dev 120:385–396.

Tessarollo L, Coppola V, Fritzsch B (2004) NT-3 replacement with brain-derived neurotrophic factor redirects vestibular nerve fibers to the cochlea. J Neurosci 24:2575–2584.

Tessier-Lavigne M, Goodman CS (1996) The molecular biology of axon guidance. Science 274:1123–1133.

Van Esch H, Groenen P, Nesbit MA, Schuffenhauer S, Lichtner P, Vanderlinden G, Harding B, Beetz R, Bilous RW, Holdaway I, Shaw NJ, Fryns JP, Van de Ven W, Thakker RV, Devriendt K (2000) *GATA3* haplo-insufficiency causes human HDR syndrome. Nature 406:419–422.

Vitelli F, Viola A, Morishima M, Pramparo T, Baldini A, Lindsay E (2003) *TBX1* is required for inner ear morphogenesis. Hum Mol Genet 12:2041–2048.

Wang VY, Hassan BA, Bellen HJ, Zoghbi HY (2002) *Drosophila* atonal fully rescues the phenotype of Math1 null mice:new functions evolve in new cellular contexts. Curr Biol 12:1611–1616.

Wayne S, Robertson NG, DeClau F, Chen N, Verhoeven K, Prasad S, Tranebjarg L, Morton CC, Ryan AF, Van Camp G, Smith RJ (2001) Mutations in the transcriptional activator *EYA4* cause late-onset deafness at the *DFNA10* locus. Hum Mol Genet 10: 195–200.

Wheeler EF, Bothwell M, Schecterson LC, von Bartheld CS (1994) Expression of BDNF and *NT-3* mRNA in hair cells of the organ of Corti:quantitative analysis in developing rats. Hear Res 73:46–56.

Wright TJ, Mansour SL (2003) *Fgf3* and *Fgf10* are required for mouse otic placode induction. Development 130:3379–3390.

Wu DK, Oh SH (1996) Sensory organ generation in the chick inner ear. J Neurosci 16: 6454–6462.

Wu HH, Ivkovic S, Murray RC, Jaramillo S, Lyons KM, Johnson JE, Calof AL (2003) Autoregulation of neurogenesis by *GDF11*. Neuron 37:197–207.

Xiang M, Gao WQ, Hasson T, Shin JJ (1998) Requirement for *Brn-3c* in maturation and survival, but not in fate determination of inner ear hair cells. Development 125:3935–3946.

Xiang M, Maklad A, Pirvola U, Fritzsch B (2003) *Brn3c* null mutant mice show long-term, incomplete retention of some afferent inner ear innervation. BMC Neurosci 4: 2.

Xu PX, Adams J, Peters H, Brown MC, Heaney S, Maas R (1999) *Eya1*-deficient mice lack ears and kidneys and show abnormal apoptosis of organ primordia. Nat Genet 23:113–117.

Xu PX, Zheng W, Huang L, Maire P, Laclef C, Silvius D (2003) *Six1* is required for the early organogenesis of mammalian kidney. Development 130:3085–3094.

Yang Q, Bermingham NA, Finegold MJ, Zoghbi HY (2001) Requirement of *Math1* for secretory cell lineage commitment in the mouse intestine. Science 294:2155–2158.

Zhou Q, Anderson DJ (2002) The bHLH transcription factors *OLIG2* and *OLIG1* couple neuronal and glial subtype specification. Cell 109:61–73.
Zine A (2003) Molecular mechanisms that regulate auditory hair-cell differentiation in the mammalian cochlea. Mol Neurobiol 27:223–238.
Zine A, Aubert A, Qiu J, Therianos S, Guillemot F, Kageyama R, de Ribaupierre F (2001) *Hes1* and *Hes5* activities are required for the normal development of the hair cells in the mammalian inner ear. J Neurosci 21:4712–4720.

5

Notch Signaling and Cell Fate Determination in the Vertebrate Inner Ear

PAMELA J. LANFORD AND MATTHEW W. KELLEY

1. Introduction and Historical Background

Since Alfonso Corti's first microscopic description of the cochlea in the mid-nineteenth century, biologists have had a deep appreciation for the beautifully ordered arrays of cells within the vertebrate organ of Corti. Sensory hair cells and nonsensory supporting cells are arranged in a regular, alternating pattern that has been compared to the ordered arrays of mosaic tiles (Posakony 1999). Recently, this "hair cell mosaic" has held interest for developmental biologists as well: such highly ordered cellular patterns are often regulated by similar molecular mechanisms and can be manipulated to reveal the specific pathways involved. Based on studies in other systems, it was suggested that a developmental mechanism known as "lateral inhibition" might be involved in the specification of sensory versus nonsensory cell types in the inner ear (Corwin et al. 1991; Lewis 1991). This mechanism, in which one progenitor cell produces a signal that inhibits differentiation in its immediate neighbor, plays an important role in the development of a variety of invertebrate cell types (Fig. 5.1). However, lateral inhibition had not been directly shown to be involved in the development of any vertebrate system.

In 1995, Kelley et al. provided the first experimental evidence that supported a role for lateral inhibition in the formation of the hair cell mosaic. Specifically, this study set out to demonstrate that developing hair cells produce an inhibitory signal that prevents neighboring progenitor cells from adopting the hair cell fate. If this hypothesis is true, a decrease in this inhibitory signal should allow progenitor cells that would not normally develop as hair cells to adopt the sensory fate. To test this hypothesis, individual developing hair cells were identified within explant cultures of the organ of Corti and ablated via laser irradiation. The results demonstrated that progenitor cells immediately adjacent to ablated hair cells could change position and phenotype in order to adopt the hair cell fate. This evidence strongly supported a role for lateral inhibition in the generation of the hair cell mosaic but did not examine the specific molecular pathways that might be responsible for mediating the inhibitory signal.

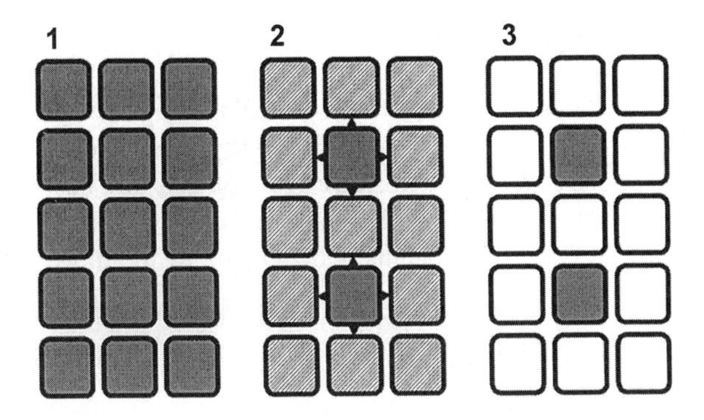

FIGURE 5.1. Lateral inhibition and Notch function in cell fate specification. (**1**) A homogeneous population of progenitor cells (*solid gray*) can be divided into subpopulations via a molecular switching mechanism known as "lateral inhibition." (**2**) A subset of cells within the progenitor pool produces an inhibitory signal, causing a decrease in gene expression in their immediate neighbors. (**3**) The end result of lateral inhibition includes two populations of cells, each with distinct gene expression.

The prime candidate for mediating lateral inhibition in the cochlea was the Notch signaling pathway, a well characterized molecular mechanism known to play an important role in cell fate decisions in many other systems. In the fruit fly, *Drosophila melanogaster*, Lehmann et al. (1983) demonstrated that mutations in primary elements of the Notch pathway result in an overproduction of neural cell types in the developing nervous system, at the expense of nonneural cell types. The results of later studies led to the suggestion that Notch influenced cell fate decisions by lateral inhibition, which delayed or diverted precursor cells from differentiation (Simpson 1990; Coffman et al. 1993; Fortini et al. 1993).

At about the same time, Notch pathway components were identified in mammals and were shown to be expressed in the inner ear (Weinmaster et al. 1991; Williams et al. 1995). Subsequent in situ hybridization studies further defined the specific expression patterns of Notch pathway component genes in auditory development, which, combined with the results of the laser ablation studies, strongly suggested that Notch signaling mediated cell fate decisions in the sensory epithelia. Finally, and perhaps most significantly, experimental studies have demonstrated that manipulation of the Notch pathway has a significant effect on the formation of the organ of Corti and the specification of sensory versus nonsensory cell fates (Lanford et al. 1999, 2000; Zheng et al. 2000; Kiernan et al. 2001; Tsai et al. 2001; Zine et al. 2001). The results of these studies also indicate, however, that the molecular regulation of auditory development is more complex than originally proposed. The goal of this chapter is to provide an overview of Notch function in general and to examine the rapidly expanding body of evidence that implicates Notch as a major player in the development of the inner ear.

2. The Notch Signaling Pathway

The breadth of systems in which Notch signaling plays a significant role is enormous, and our knowledge of the pathway is relatively ancient (in molecular biology terms). Below we provide an overview of both the historical and biological context in which studies of this pathway must be placed, and describe in some detail the various components and function of the molecular cascade.

2.1 Background and Significance

Notch was first identified in *Drosophila* through a random mutation resulting in a malformed or "notched" wingtip (Morh, 1919; reviewed in Artavanis-Tsakonas et al. 1995, 1999). Later studies of this mutation (a partial deletion of the Notch gene) described a variety of gross anatomical phenotypes in Notch-deficient flies, including neural hyperplasia (Poulson 1937, 1940). Based on these initial descriptions, Notch and Notch pathway genes became commonly referred to as "neurogenic" genes. It was not until much later, however, that an understanding of the structure and function of DNA and the development of molecular biological techniques allowed researchers to more fully elaborate the truly widespread role of Notch in cell fate specification.

In the mid-1980s, the genes encoding Notch and one of its ligands, Delta, were first cloned in *Drosophila* and their basic structural components were determined (Wharton et al. 1985; Kidd et al. 1986; reviewed in Artavanis-Tsakonas et al. 1995, 1999). These and other studies indicate that the Notch molecule itself is a transmembrane receptor protein that is expressed on the surface of developing progenitor cells. Activation of the pathway occurs when Notch is bound by one of its ligands, Delta or Serrate, which are also membrane-bound molecules. The structural nature of both receptor and ligand necessarily limits the range of activation to only those cells that are in direct contact with one and other. Since the effects of Notch activation are largely inhibitory, the process by which Notch directs the specification of alternating cell fates in a given system is referred to as "contact-mediated" lateral inhibition (Simpson 1990; Artavanis-Tsakonas 1995).

Since the original cloning experiments, a vast array of studies (several hundred at least) have demonstrated that this signaling pathway is involved in the development of a wide variety of systems. For example, in *Drosophila*, Notch signaling is important for the development of cells arising from each of the three germ layers of the developing embryo. Cell types influenced by Notch signaling in the developing fly include sensory bristles, muscle, heart, midgut, and eye, among others (Hartenstein et al. 1992; Dominguez and de Celis 1998). Interestingly, Notch holds a similar level of importance in cell fate determination across a variety of other species as well. Homologs to *Drosophila* Notch have been identified in nematode (Greenwald 1985), sea urchin (Sherwood and McClay 1997), and a number of vertebrates including frog, zebrafish, chick, mouse, rat, and human (Coffman et al. 1990; Weinmaster et al. 1991; Del Amo et al. 1992; Bierkamp and Campos-Ortega, 1993; Lardelli and Lendahl 1993; Henrique et al. 1995; Myat et al. 1996). See Table 5.1 for an abbreviated list of Notch gene nomenclature across species.

In vertebrates, Notch signaling directs cell fate decisions during myogenesis, hematopoiesis, osteogenesis, endocrine development, vascular morphogenesis, and neurogenesis (Kopan et al. 1994; Shawber et al. 1996; Milner and Bigas 1999; Xue et al. 1999; Jensen et al. 2000; Shindo et al. 2003). Studies have also demonstrated that Notch, its ligands, and its downstream mediators play a part in human genetic disease and certain types of cancer (reviewed, Gridley 1997; see Section 2.3.2). Lastly, we have begun to discover that Notch plays a pivotal role in the development of the inner ear and the mosaic of cells that comprise the sensory epithelia.

2.2 The Structure and Function of the Notch Pathway

Data from a wide range of systems and organisms indicate a strong conservation of the molecular structure of Notch and Notch pathway components. While the bulk of data regarding Notch structure and function has come from the fly model, much of what we know about Notch signaling mechanisms has been demonstrated in vertebrates as well. The information presented here is a compilation of what is now known about the structure and function of this pathway.

The Notch receptor itself consists of an extracellular domain (N^{EC}) containing multiple epidermal growth factor (EGF)-like repeats and three cysteine-rich Notch/Lin-12 repeats (Fig. 5.2). Following N^{EC} is a transmembrane domain (N^{TM}) and an intracellular domain (N^{IC}) that includes six tandem ankyrin repeats and a PEST sequence (proline—P, glutamine—E, serine—S, threonine—T) (reviewed in Artavanis-Tsakonas et al. 1995; Weinmaster 1997). Recent studies have shown that the Notch protein acquires this mature functional form during posttranslational processing, when it is cleaved by a furinlike convertase (Logeat et al. 1998). In addition, some data suggest that Notch receptor function requires the dimerization of the molecule, via interactions between conserved cysteines in the extracellular domain of the molecule. However, direct biochemical data in support of this hypothesis have not yet been presented (reviewed in Weinmaster 1997).

Each Notch ligand is comprised of domains similar to those found in the receptor; an extracellular domain comprised of an N-terminal domain, the unique "DSL" motif (for the ligand names Delta, Serrate, Lag-2) and multiple EGF-like repeats (reviewed in Fleming, 1998; Artavanis-Tsakonas et al. 1999). These regions are followed by a cysteine-rich domain, a transmembrane domain, and a truncated intracellular domain (Fig. 5.2). Binding between the receptor and a ligand occurs between the conserved N-terminal domain of the ligand and one or more of the EGF-like repeats on the receptor (Fehon et al. 1990; Rebay et al. 1991). Interactions between receptor and ligand result (in most cases) in the activation of the inhibitory molecular cascade that underlies lateral inhibition and other cellular events. There is some evidence to suggest, however, that

TABLE 5.1. A sampling of Notch gene nomenclature across species.

Gene	Organism	Reference
Receptors		
Notch	Fly	Mohr 1919
Notch1,2,3,4	Rat/Mouse	Weinmaster et al. 1991, 1992 Lardelli and Lendahl 1993; Lardelli et al. 1994; Uyttendaele et al. 1996
TAN-1	Human	Ellisen et al. 1991
(Truncated allele of Notch-1) Notch 2,3		Sugaya et al. 1994
Xotch	Frog	Coffman et al. 1990
Cnotch	Chicken	Myat et al. 1996
Notch	Zebrafish	Bierkamp and Campos Ortega 1993
Lin-12	Nematode	Greenwald 1985
(Lineage defect-12)		Austin and Kimble 1989
Glp-1		
(Germ line proliferation defective)		
Ligands		
Delta	Fly	Knust et al. 1987; Kopczynski et al. 1988
Serrate		Fleming et al. 1990
Dll1,3,4	Rat/Mouse	Bettenhausen et al. 1995;
(Delta-like 1,3,4)		Dunwoodie et al. 1997; Yoneya et al. 2001
Jagged1, 2		Lindsell et al. 1995, 1996; Lan et al. 1997; Valsecchi et al. 1997
Jagged1, 2	Human	Gray et al. 1999; Yoneya et al. 2001;
Dll1, 4		Luo et al. 1997
(Delta-like 1,4)		
X-Delta-1, 2	Frog	Chitnis et al. 1995; Jen et al. 1997
X-Serrate-1		Kiyota et al. 2001
C-Delta-1	Chicken	Henrique et al. 1995; Myat et al. 1996
C-Serrate-1		
Delta D	Zebrafish	Dornseifer et al. 1997
LAG-2	Nematode	Tax et al. 1994
(LIN-12 and Glp-1)		Mango et al. 1994
APX-1		
Intermediaries and negative regulators		
Su(H)	Fly	Schwiesguth and Posakony 1992;
(Suppressor of Hairless)		Fortini & Artavanis-Tsakonas 1994
E(spl)	Fly	Knust et al. 1987
Enhancer of split		
RBPJk	Mouse/ Human	Schweisguth and Posakony 1992;
(Recombination signal sequence binding protein for Jk genes)		Furukawa et al, 1992; Hsieh et al. 1996
CBF1		
(C-promoter binding factor 1)		
HES1–7	Mouse	Sasai et al. 1992; Akazawa et al. 1992;
(Hairy enhancer of split homolog 1–7)		Ishibashi et al. 1993; Tomita et al. 1996; Nishimura et al. 1998; Bae et al. 2000; Bessho et al. 2001;
TLE	Human	Stifani et al. 1992;
(transducin-like enhancer of split)		Bessho et al. 2001
hHES4,7		

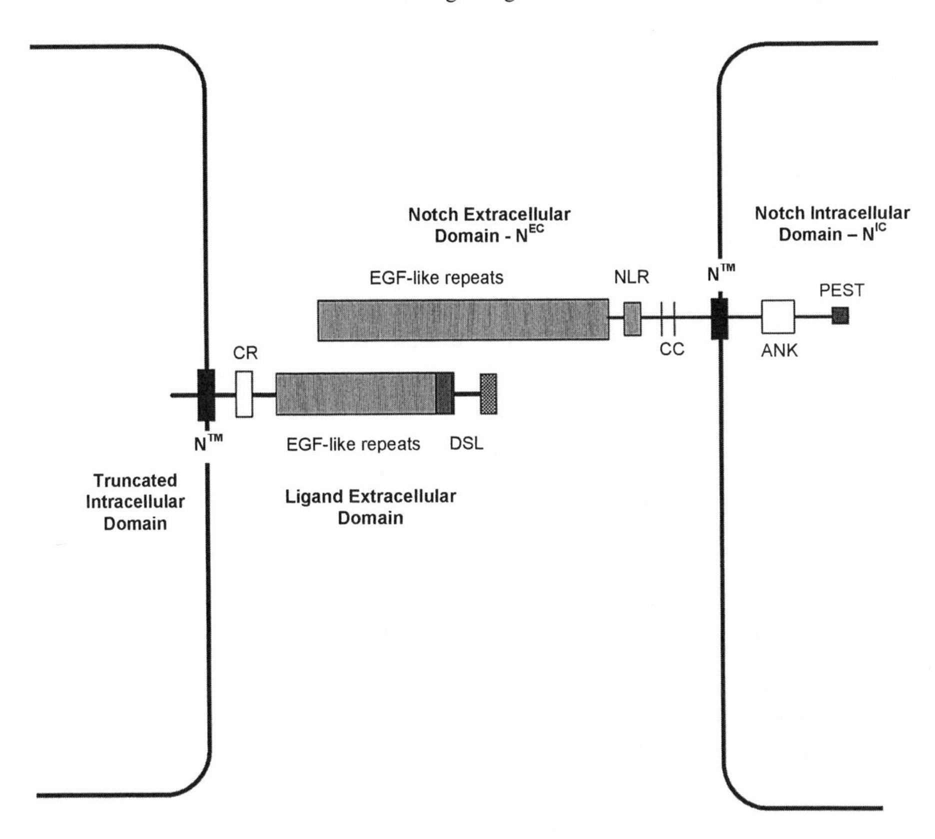

FIGURE 5.2. Diagram representing the structure of the Notch receptor and a generic DSL ligand. The receptor, Notch, is comprised of three major domains, extracellular (N^{EC}), transmembrane (N^{TM}), and the intracellular domain (N^{IC}). N^{EC} consists of a series of epidermal growth factor-like repeats (EGF-like repeats), cysteine-rich Notch/lin-12 repeats (NLR), and two conserved cysteines (CC). Following the transmembrane domain, the intracellular domain consists of six tandem ankyrin repeats (ANK) and a PEST sequence (proline—P; glutamine—E, serine—S, threonine—T). Notch ligands are similar in overall structure to the receptor, with the major structural difference being the truncated intracellular domain of the ligand. The extracellular domain of each DSL ligand is comprised of a unique N-terminal DSL domain, followed by a series of EGF-like repeats. A cysteine-rich (CR) region lies between the EGF-like repeats and the transmembrane domain. Binding between the ligand and the receptor appears to occur between the conserved N-terminal domain of the ligand and the EGF-like repeats of the receptor (reviewed in Artavanis-Tsakonas et al. 1995 and Weinmaster 1997, 2000).

ligand-receptor binding can also result in the negative regulation of the Notch pathway (Weinmaster 1997, 2000).

While DSL ligands are structurally distinct from one another, they can exhibit at least partial functional redundancy (reviewed, Fleming 1998). For example, Gu et al. (1995) demonstrated that the *Drosophila* DSL ligand Serrate can act as a functional replacement for another version of the ligand, Delta. Similar rescue experiments have not been performed in mammals however targeted deletion of specific Notch ligands results in phenotypes that indicate the possibility of functional redundancy between the various ligands (see below).

Once the Notch receptor has been bound by a ligand, the N^{IC} subunit is cleaved via the protease Presenilin. This cleavage results in the translocation of N^{IC} to the nucleus (reviewed in Miele and Osborne 1999; Weinmaster 2000). N^{IC} associates with a complex of proteins that includes Suppressor of Hairless (SuH) or C-promoter binding factor 1 (CBF1), as it is called in vertebrates (Luo et al. 1997; Zhou et al. 2000; Zhou and Hayward 2001; reviewed in Bray and Furriols 2001). In some literature, CBF1 is also referred to as "recombination signal sequence binding protein for Jk genes" or RBP-Jk (Schweisguth and Posakony, 1992; Jarriault et al. 1995). The complex formed by N^{IC}, SuH/CBF1, and other proteins ultimately results in the activation of a family of genes encoding basic helix–loop–helix (bHLH) transcription factors (reviewed in Bray and Furriols 2001; Fig. 5.3). In *Drosophila,* the genes encoding these bHLH transcription factors are collectively referred to as *Enhancer of split* [*E*(*spl*)] genes (Hartley et al. 1987; reviewed in Bray, 1997; Lecourtois and Schweisguth 1997) while the vertebrate homologs of these genes are referred to as *HES* (*Hairy Enhancer of split* homolog) (Akazawa et al. 1992; Sasai et al. 1992). The regulatory relationships between SuH and [*E*(*spl*)] are still not completely fleshed out. It seems clear that in many cases Notch acts directly to activate [*E*(*spl*)] (Bang et al. 1995; Heitzler et al. 1996). However, it appears that (at least in *Drosophila*), SuH may also acts as a repressor of [*E*(*spl*)] prior to Notch activation, and that activation of Notch relieves this repression (reviewed in Bray and Furriols 2001).

Once [*E*(*spl*)] has been activated, the transcription factors encoded by these genes act to negatively regulate the activity of downstream gene sequences. These target sequences include genes that encode bHLH transcription factors such as the genes of the *Acheate–Scute* complex (*AC-SC*) and *atonal* (*ato*) or *MASH* (mammalian acheate–scute homolog) and *Math* (mammalian atonal homolog) as they are known in mammals (Jarman et al. 1993, 1994; Akazawa et al. 1995; Bang et al. 1995; Ishibashi et al. 1995; Heitzler et al. 1996; Nakao and Campos-Ortega 1996; Gupta and Rodrigues, 1997; Jimenez and Ish-Horowicz, 1997; reviewed in Modolell, 1997). These genes are referred to as "proneural" since activation of these genes promotes the development of neural phenotypes in precursor cells. Consequently, activation of Notch in neural precursors results in the suppression of proneural gene activity and the inhibition of neural phenotypes in Notch-bearing precursor cells.

Interestingly, one of the functions of proneural genes appears to be the reg-

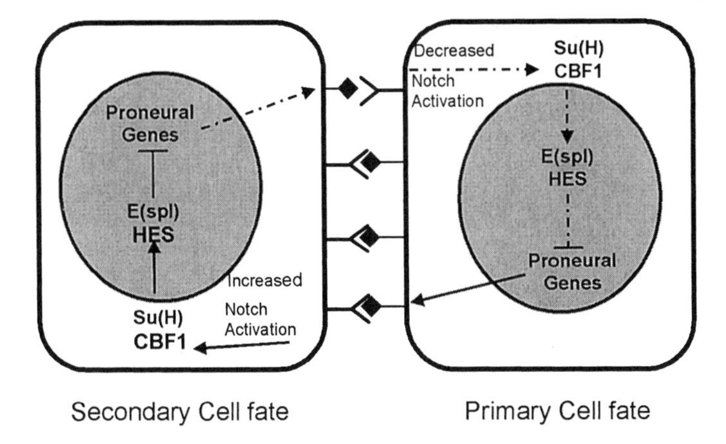

FIGURE 5.3. Diagram of intracellular Notch signaling and proposed feedback mechanism in neural progenitor cells. Cells expressing relatively large amounts of the ligand (*diamond heads*) activate the Notch receptor (*forked heads*) on the surface of an adjacent cell. Receptor-ligand binding results in the activation of Notch and the formation of a Notch-SuH–CBF1 complex. Notch is translocated to the nucleus, where it activates *E(spl)/HES* genes. These target genes act to negatively regulate the expression of proneural genes such as *AC-SC* and *atonal* (*MASH* and *Math* in vertebrates). Decreased proneural gene expression (*dotted line*) results in a reduction in the expression of Notch ligands, which in turns reduces the Notch activation in the adjacent cell (*dotted lines*). See Table 5.1 for gene nomenclature across species.

ulation of DSL ligand expression (Kunisch et al. 1994). The consequence of this particular function is that there is the potential for negative self-regulation of Notch activity in a given cell Fig. 5.3). Specifically, evidence in *Drosophila* has demonstrated that such a feedback loop may act to refine the pattern of receptor and ligand expression in progenitor cells (Kunisch et al. 1994; Heitzler et al. 1996): Elevated expression of proneural genes in a subset of progenitor cells may lead to elevated expression of DSL ligands within those same cells. The increased expression of these ligands may subsequently result in increased activation of Notch in neighboring progenitor cells. Since Notch acts to repress proneural gene expression, increased Notch pathway activation in the neighboring cells ultimately leads to a down-regulation of Notch ligand expression. In this manner, a small initial bias in proneural gene expression may result in the establishment of two distinct progenitor populations: one that expresses DSL ligands and proneural genes and another that expresses Notch but not proneural genes Fig. 5.3).

2.3 Cellular Events Regulated by the Notch Pathway

A complete review of the extremely large body of data regarding Notch and its capacity to regulate cellular events is not within the scope of this chapter. However, a brief overview of this subject is necessary in order to appreciate fully the potential complexity of Notch signaling in the inner ear.

2.3.1 Differentiation, Proliferation, and Apoptosis

Early studies of Notch noted that disruption of this gene pathway frequently resulted in increased proliferation of progenitor cells and the overproduction of certain cell types (Poulson 1937, 1940). Thus, the Notch pathway was widely considered to be a regulator of proliferation. Recently, the question of an active role for Notch in maintaining proliferation has been investigated in greater detail. For example, Carlesso et al. (1999) demonstrated that an increase in Notch signaling in HL-promyelocytic leukemia cells and CD34 bone marrow stem cells accelerates these cells through the G_1 phase (Carlesso et al. 1999). This acceleration pushes progenitor cells through a critical G_1 lag period that is associated with commitment and terminal differentiation. Consequently, in this and other systems, Notch signaling influences cell fate decisions through the direct manipulation and regulation of the cell cycle.

In a large number of cases, however, it is clear that Notch regulates proliferation merely by default. That is, Notch signaling may cause precursor cells to remain in the proliferative state as a result of having repressed differentiation. The distinction between these two types of events is difficult to tease apart, but many of the target genes located downstream in the Notch pathway have now been identified and have been shown to be cell-specific regulators of differentiation (such as the proneural genes) rather than proliferation. Repression of these genes results in a population of precursor cells that remain undifferentiated and thus continue to proliferate.

Interestingly, evidence also suggests that in some cases, Notch may function in the opposite manner. That is, Notch signaling may *promote* differentiation rather than suppress it. This effect has recently been demonstrated in cultured stem cells from the developing vertebrate nervous system (Furukawa et al. 2000; Morrison et al. 2000). Specifically, Morrison et al. (2000) demonstrated that neural crest stem cells (NCSCs) isolated from E14.5 rat sciatic nerve could be pushed toward a glial fate either by transfection with a dominant-negative form of *Notch1*, or the addition of a secreted form of the Notch ligand Delta. NCSCs in culture normally give rise to very large numbers of three cell types—neurons, glia (Schwann cells), and myofibroblasts. Under either treatment condition, however, a significantly reduced number of daughter cells were produced and nearly all of these cells expressed glial fibrillary acidic protein (GFAP), while few to none expressed markers for the neural or myofibroblast cell types. Similar results have been produced in mouse retina, in which increased expression of *Notch1* and *Hes1* resulted in increased numbers of progenitor cells that express markers for Müller glia (Furukawa et al. 2000), and in the mouse cortex, in which transfection with N^{IC} results in an increase in the number of cells that express glial markers (Gaiano et al. 2000). These data indicate that the function of Notch as a suppressor of differentiation may be context dependent.

Finally, Notch signaling has been shown to play an important role in the regulation of programmed cell death (apoptosis). In mammals (mouse), this role has been closely examined during the development of peripheral T cells (Robey et al. 1996; Deftos et al. 1998; Jehn et al. 1999; reviewed in Miele and Osborne, 1999). For example, at one point in development immature thymocytes (the so-called "double-positive" or DP thymocytes, which express both coreceptor molecules, CD4 and CD8) acquire resistance to glucocorticoids, which would otherwise induce apoptosis and cell cycle arrest (Deftos et al. 1998). The acquisition of this resistance appears to be critical in the process of progressive commitment that occurs during the next phase of development—the generation of separate CD4 versus CD8-positive thymocytes (separate lines that will become T helper or T killer cells, respectively). In DP thymocytes that are retrovirally transduced with an activated form of Notch (N^{IC}), there is an increase in resistance to glucocorticoids and an up-regulation of *Bcl-2* (*B-cell leukemia/lymphoma 2*), a gene that inhibits glucocorticoid sensitivity (Deftos et al. 1998). Similar antiapoptosis effects of Notch have been demonstrated in T-cell hybridoma cell lines (Jehn et al. 1999) as well as many other systems (reviewed in Miele and Osborne, 1999). These results strongly implicate Notch in the regulation of cellular development via the inhibition of cell death.

2.3.2 Notch and Disease

As could be suggested from the information presented in Section 2.3.1, Notch signaling has been implicated in a number of types of cancer, including the malignant transformation of T cells. Ellisen et al. (1991) first demonstrated that a constitutively active form of Notch (*TAN-1* or *Truncated Allele of Notch 1*) is apparently responsible for the transformation of T cells and, ultimately, the induction of T-cell acute lymphoblastic leukemias (reviewed in Allman et al. 2002). In addition to its role in T-cell leukemia and thyroid tumors, Notch appears to be involved in the generation of mammary tumors and the immortalization of B cells. In particular, one of the four mammalian Notch genes, *Notch4*, appears to be the site of mammary tumor virus integration (Gallahan and Callahan 1997; reviewed in Callahan and Raafat 2001). Similarly, the oncogenic effects of the Epstein–Barr virus, shown to be involved in B-cell transformation, appear to be mediated via downstream components of the Notch pathway (Hsieh et al. 1996, 1997; reviewed in Smith 2001). It is important to note, however, that much of this research has been performed in mammals other than humans, and that studies in human systems have only begun to elucidate the role of Notch in human cancer (reviewed in Gridley 1997).

Mutations in Notch or its ligands are integral to at least two human genetic diseases: Alagille's syndrome and CADASIL (cerebral autosomal dominant arteriopathy with subcortical infarcts and leukoencephaly) (reviewed in Joutel and Tourner-Lasserve, 1998). Alagille's syndrome appears to be the result of a defect in the gene encoding one of the mammalian Notch ligands, Jagged1 (Li et al. 1997; Oda et al. 1997). Specifically, mutations in the *Jagged1* gene (*Jag1*) result in individuals who are born with an extremely large range of serious birth defects, including kidney, cardiovascular, ocular, and skeletal abnormalities; mental retardation; and other effects (reviewed in Piccoli and Spinner 2001). The widespread nature of this genetic disorder is a reflection of the range of developmental events in which the correct function of the Notch pathway is critical. In fact, complete deletion of mammalian *Notch1* (in mice) causes embryonic death at midgestation as a result of widespread cell death (Swiatek et al. 1994). Deletion of *Jag1* in mice results in embryonic lethality resulting from incomplete vascular development and subsequent hemorrhage (Xue et al. 1999).

CADASIL is a relatively late-onset (approximately 45 years of age) condition that results from a mutation of the Notch3 receptor gene (Joutel et al. 1997). As in Alagille's syndrome, the effects of this Notch mutation are wide ranging, but have particularly strong effects on the central nervous system (CNS). Specifically, CADASIL produces abonormalities within the CNS and cerebral arteries, resulting in a variety of symptoms including subcortical ischemic strokes and migraine headaches (Tournier-Lasserve et al. 1993; Chabriat et al. 1995; reviewed in Joutel and Tournier-Lasserve, 1998). Ultimately, CADASIL results in a progressive neural degeneration and dementia (Chabriat et al. 1995).

2.4 Regulation of Notch Signaling

There are numerous mechanisms by which the activity of the Notch signaling pathway may be regulated and/or modulated, starting with functional diversity among receptor and ligand types. Depending on which molecules are expressed, the specific distribution of this expression, and the tissue/system in which the expression occurs, ligands may act as either positive or negative regulators of Notch activity. For example, in fly neurogenesis, the Notch ligands Delta and Serrate can be interchanged (Gu et al. 1995); however, in the development of the fly imaginal disk, the two ligands initiate entirely different downstream effects in adjacent compartments (Fleming et al. 1997). It is not clear how this difference is mediated, although various mechanisms may be involved.

Some evidence has suggested that certain ligands bind the receptor with greater affinity than do others (Rebay et al. 1991). If true, then variations in ligand distribution may reflect differences in the level of Notch activation. It has also been shown that the activity of a given Notch ligand may be modulated by molecules in the Fringe family of proteins (fringe, in fly; lunatic fringe, manic fringe, radical fringe in mammals) (Cohen et al. 1997; Fleming et al. 1997; Panin et al. 1997; Johnston et al. 1997). Biochemical studies of fringe activity have demonstrated that these molecules modulate receptor–ligand binding through glycosylation of the receptor (Brückner et al. 2000; Moloney et al. 2000; reviewed in Blair, 2000). This glycosylation modifies specific EGF-like repeats within the extracellular domain of Notch and may inhibit or promote pathway activation, depending upon the specific ligand(s) present in the system (Brückner et al. 2000; Moloney et al. 2000). Interestingly, evidence also indicates that fringe activity in mammalian tissue potentiates binding between Notch and its ligand *Delta1* (*Dll1*), while it inhibits binding between Notch and *Jag1* (Hicks et al. 2000).

Regulation of Notch pathway activity may also occur as a result of the simultaneous expression of Notch and its ligands in a given precursor cell. Since the extracellular domains of Notch and its ligands are very similar in structure, interactions may also occur in *cis*—that is, binding between receptor and ligand may occur within the same cell (Fehon et al. 1990; Heitzler and Simpson 1993; Jacobsen et al. 1998). Studies have indicated that dominant-negative effects may result from *cis* interactions between receptor and ligand, or ligand and ligand (Jacobsen et al. 1998; Klueg and Muskavitch 1999). These effects have yet to be fully elucidated, and many other regulatory mechanisms are sure to play major roles in Notch pathway activity. These mechanisms are too numerous and complex to be detailed here; however, an excellent review of Notch regulation has recently been presented by Baron et al. (2002).

3. Notch Signaling in the Developing Inner Ear

In mammals, Notch has been shown to play a significant role in neurogenesis, and has been well studied in such peripheral systems as the olfactory bulb (Lindsell et al. 1996; Kageyama and Nakanishi 1997) and the eye (Ahmad et al. 1995; Ohtsuka et al. 1999). The Notch pathway has also been implicated in regions of the developing CNS (Del Amo et al. 1992; Guillemot and Joyner 1993), including relatively late rounds of cell fate specification in the cerebellum and cerebral cortex (Zhong et al. 1996; Berezovska et al. 1997, 1998; Lutolf et al. 2002). However, prior to 1999, only a few studies had reported the expression of Notch-related genes in the mammalian inner ear (Weinmaster et al. 1991; Williams et al. 1995; Lindsell et al. 1996). Over the last several years, there has been an intense period of increased interest in the molecular aspects of inner ear development and in the ear as a model system for studying the role of Notch in vertebrate systems. As a result, we now have a much greater basis for a discussion on Notch and the development of the hair cell mosaic.

3.1 The Embryology of Notch in the Vertebrate Inner Ear

To date, the expression patterns of Notch-related genes have been studied in many different vertebrate species, however, the most complete set of Notch-related data in any vertebrate system has been collected from mouse models. This is attributable in large part to the increased numbers of mutant mouse strains available to study the effects of Notch signaling and, to a lesser extent, on the structural and functional similarity between rodent and human ears. Consequently, in order to provide the most coherent presentation of the subject, this review focuses on the expression and function of Notch genes during the development of the mouse ear.

The majority of data describing Notch gene expression patterns in the mammalian ear have been derived from in situ hybridization studies, although LacZ reporter constructs and reverse transcriptase-polymerase chain reaction (RT-PCR) methods of detection have also been used. The results of these studies (and the specific methods used in each) are noted in Table 5.2, which outlines the embryology of Notch pathway expression in the ear (see also Mansour and Schoenwolf, Chapter 3, and Pauley, Matei, Beisel, and Fritzsch Chapter 4). An example of this expression is represented in Figure 5.4, which demonstrates the expression of *Jag1* at varying stages of cochlear development. The expression patterns of Notch-related genes in the inner ear can be loosely subdivided into three developmental timeframes including (1) those whose onset occurs very early (otocyst or before), (2) those whose onset occurs at midgestation, and (3) those whose onset occurs during late gestation.

In addition to the chronology of Notch gene expression in the ear, we must also consider the spatial distribution of gene expression patterns. The basic categories of gene expression patterns that can be examined include (1) those expressed very broadly, (2) those that are (or quickly become) restricted to certain areas and subpopulations of cells and that appear to mediate boundary formation, and (3) those that appear to be restricted to a very small, perhaps specific subset of cells at the onset of expression. In the following sections, we attempt to combine both the chronological and spatial aspects of Notch gene expression in the ear into one complete story, and provide an overview of Notch gene expression as it relates to the development of the ear. References to developmental staging are as described in Kaufman (1992).

3.2 Gene Expression Patterns

Genes encoding *Notch1*, *Jag1*, *Dll1*, and *Lfng* are active very early in the development of the inner ear, with expression concentrated in the ventral portion of the otocyst at E9 to E10 (Bettenhausen et al. 1995; Williams et al. 1995; Lewis et al. 1998; Morsli et al. 1998; Morrison et al. 1999). At this time point, Notch gene expression may be related to the delamination of the neurons that will comprise the ganglion of the VIIIth nerve, as well as to the determination of the sensory regions of the otocyst (Morrison et al. 1999; Fritzsch, 2003—see also Pauley, Matei, Beisel, and Fritzch, Chapter 4). By E12, the expression patterns of many Notch-related genes appear to be more specifically related to the development of the vestibular and auditory endorgans. Other than *Notch1*, which continues to be broadly expressed in the ear at this time point, transcripts for Notch-related genes are concentrated in regions of the ear that will develop as sensory epithelia, the so-called “sensory patches” (Morrison et al. 1999). At E12, *Jag1* is expressed in six regions, corresponding to the developing saccule, utricle, the three cristae, and the cochlea (Morrison et al. 1999; see figure 4). In the rudimentary cochlear duct, *Jag1* transcripts are expressed in a broad band that extends the very short distance between its basal and apical turns Fig. 5.4A,D; Morrison et al.1999). At the same time, transcripts for *Dll1* and *HES5* are restricted to the developing cristae, while *Math1* transcripts are only faintly

TABLE 5.2 Embryology of the Notch pathway in the mouse inner ear.

Embryological day	Gene	Distribution of expression	Detection method	Authors
Early gestation (otocyst)				
E8.5–10.5	*Notch1*	Ventral otocyst	ISH	Williams et al. 1995
			ISH	Lewis et al. 1998
E9–E10.25	*Lfn*	Anteroventral otocyst	ISH	Morsli et al. 1998
E10.5	*Jag1*	Ventral otocyst	ISH	Lewis et al. 1998
E9–E10	*Delta1*	Ventral otocyst	LacZ reporter	Morrison et al. 1999
			ISH	Bettenhausen et al. 1995
Mid gestation (E12–E13)				
E12	*Notch1*	Entire ear	ISH	Lanford et al. 1999
E12.5	*Lfn*	Utricle; extends the full length of the cochlear duct	ISH	Morsli et al. 1998
E12.5	*Jag1*	All SE; extends the full length of cochlear duct	ISH	Morrison et al. 1999
E12.5	*HES5*	Cristae	ISH	Shailam et al. 1999
E12.5	*Math1*	All SE; extends partway from base of cochlear duct	LacZ reporter	Bermingham et al. 1999
			ISH	Shailam et al. 1999
E12.5	*Delta1*	Cristae	LacZ reporter	Morrison et al. 1999
E13.5	*Jag2*	Vestibular SE	ISH	Shailam et al. 1999
Mid to late gestation (E14–E15.5)				
E14	*Notch1*	All regions of the ear; sensory and nonsensory	ISH	Lanford et al. 1999
			ISH	Shailam et al. 1999
E14	*Lfn*	All SE; full length of cochlear duct	ISH	Morsli et al. 1998
			LacZ reporter	Zhang et al. 2000
E14	*Jag1*	All SE; full length of cochlear duct	ISH	Morrison et al. 1999
E14	*HES5*	Vestibular SE	ISH	Shailam et al. 1999
		Cochlea	RT-PCR	Zine et al. 2001
E14	*HES1*	Cochlea	RT-PCR	Zine et al. 2001
E14.5	*Jag2*	Expression detected in base of cochlea	ISH	Lanford et al. 1999
E14.5–15.5	*Delta1*	All SE; restricted to HC	LacZ reporter and ISH	Morrison et al. 1999
E15	*Math1*	All SE; restricted to HC	LacZ reporter and ISH	Bermingham et al. 1999
			ISH	Lanford et al. 2000
E15.5	*HES5*	Expression detected in base of cochlea	ISH	Lanford et al. 2000
Late gestation to early postnatal (E16–P0+)				
E16	*Lfn*	All SE; restricted to SC within organ of Corti	ISH	Morsli et al. 1998
E17	*HES5*	All SE; full length of cochlea, restricted to SC within organ of Corti	ISH	Shailam et al. 1999
			ISH	Lanford et al. 2000
			RT-PCR	Zine et al. 2001
E17	*HES1*	Peak expression in cochlea	RT-PCR	Zine et al, 2001

TABLE 5.2 (*Continued*)

Embryological day	Gene	Distribution of expression	Detection method	Authors
E17	*Math1*	All SE; restricted to HC	LacZ reporter ISH	Berminham et al. 1999 Lanford et al. 2000
E17.5	*Jag1*	All SE; restricted to SC in organ of Corti and inner sulcus	ISH	Morrison et al. 1999
E17.5	*Delta1*	All SE; restricted to HC and cells in the endolympatic duct	LacZ reporter	Morrison et al. 1999
E17–18	*Notch1*	All SE; decreased expression in HC?	ISH	Lanford et al. 1999
E18–P0	*Jag2*	Restricted to HC in vestibular SE; full length of cochlea	ISH ISH ISH	Lanford et al. 1999 Shailam et al. 1999 Lanford et al. 2000
E18–P0	*HES1*	Restricted to SC outside organ of Corti	ISH	Zine et al. 2001

E8–E18, Embryological days 8 to 18; HC, hair cells; ISH, in situ hybridization; P0, postnatal day 0 (birth); SC, supporting cells; SE, Sensory end organs.

detectable in the saccule (Morrison et al. 1999; Shailam et al. 1999 Lanford et al. 2000).

By E13-E14, *Notch1*, *Jag1*, *Lfng*, and *Math1* are expressed in all of the sensory endorgans of the ear Fig.4B,E; Morsli et al.1998; Morrison et al.1999; Shailam et al. 1999). *HES5* and *Dll1* expression remains restricted to vestibular epithelia, which are beginning to assume their mature configuration. A third Notch ligand, *Jagged2* (*Jag2*), is detectable in the vestibular epithelia as well. *Notch1* expression continues to be widespread in the ear (Lanford et al. 1999, 2000; Shailam et al. 1999). The cochlear duct has, at this time, formed about half a turn, where *Jag1* expression continues along the length. *Math1* expression is present in the cochlea in a relatively narrow band that begins near the base of the duct and extends a short distance toward the apex (Lanford et al. 2000).

As development progresses through mid to late gestation (E14 to E15), *Lfng* and *Jag1* continue to be expressed in both the vestibular and auditory end organs. The cochlear duct is still expanding to form its mature spiral, and as it does, a subset of Notch-related genes begins to be expressed in the base of the cochlea. A very narrow (possibly restricted to a single cell width) streak of cells, located along the abneural edge of the greater epithelial ridge, begins to express *Dll1* and *Jag2* (Lanford et al. 1999; Morrison et al. 1999). This streak of expression begins near the base of the cochlear duct and extends, over time, toward the apex and appears to be restricted to cells that will develop as hair cells. Transcripts for *Lfng* and *Jag1* continue to be expressed in a relatively broader band, localized to the greater epithelial ridge and inner sulcus regions of the duct Figs. 5.4 and 5.5 (Morsli et al. 1998; Morrison et al. 1999). In contrast, *Notch1* transcripts at this time frame are broadly expressed within the cochlear duct and

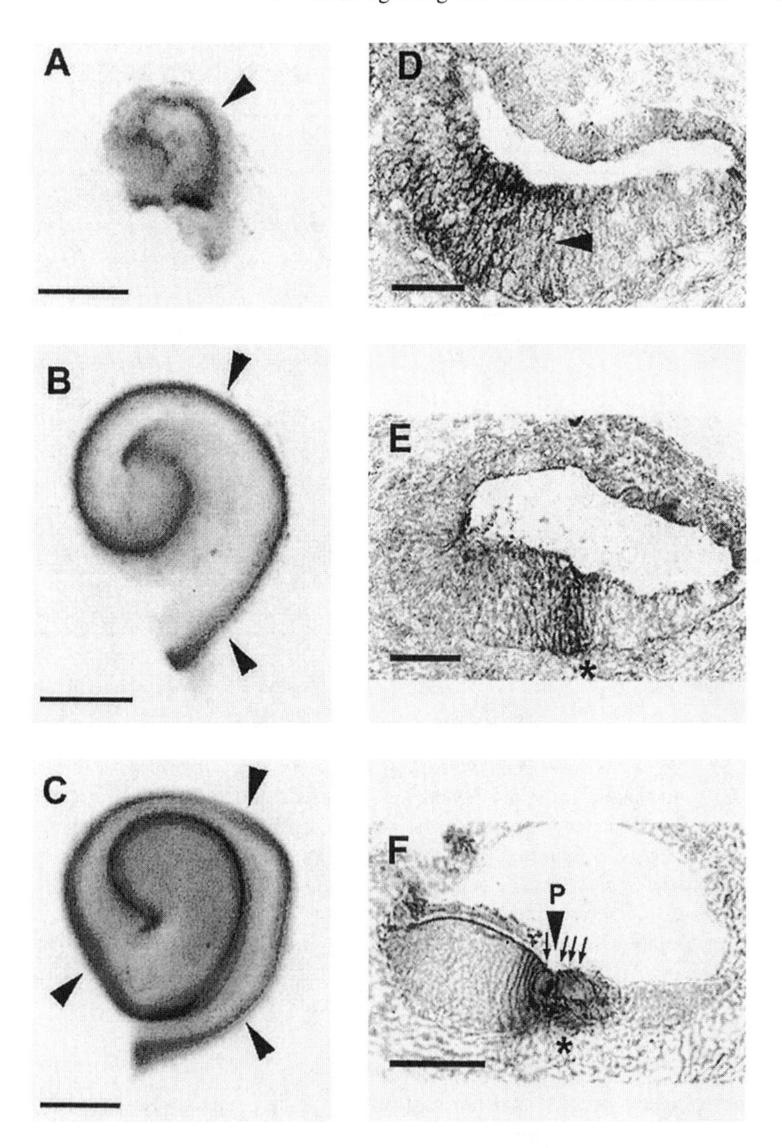

FIGURE 5.4. In situ hybridization and immunohistochemistry of the mammalian Notch ligand, *Jagged1* (*Jag1*), in the developing mouse ear. (**A–C**): *Jag1* gene expression (*arrowheads*) in the cochlear duct extends the full length of the duct at E12 (**A**), E15 (**B**), and E17.5 (**C**). (**D–F**). Immunohistochemistry on sections of the cochlear duct at approximately the same timepoints (E12, E15, E17.5) demonstrates that Jag1 protein expression (*arrowhead*) becomes restricted to supporting cells within the organ of Corti and the inner sulcus. In (**E**) and (**F**), *asterisks* mark the position of the spiral vessel, an anatomical landmark for the position of the developing organ of Corti. In (**F**), P (*arrowhead*) marks the position of pillar cells within the organ of Corti, while *small arrows* show the position of hair cells in the epithelium. Scale bars in **A–C** equal 300 μm. Scale bars in **D**, **E**, and **F** equal 25 μm.

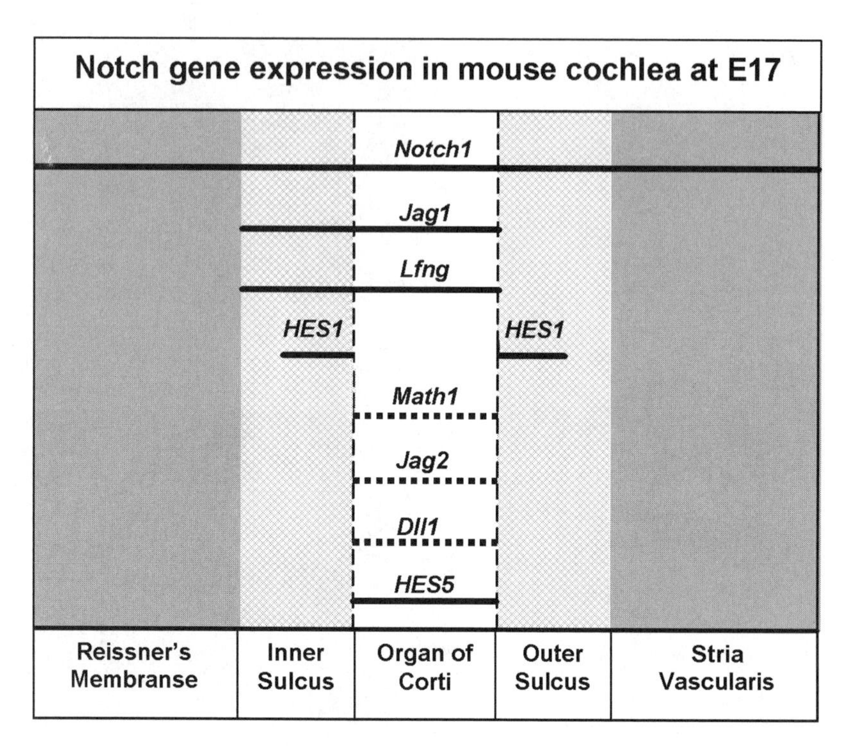

FIGURE 5.5. Notch pathway gene expression in a section near the base of the cochlear duct at E17. In this diagram, the developing Reissner's membrane stria vascularis have been opened and flattened to expose the sensory epithelium (*center*). Homogeneous expression of a gene is indicated by a *solid black line*, while expression that has become restricted to developing hair cells is indicated by a *dotted line*. The diagram was derived from the results of several independent studies (see Table 5.2 for references). Consequently, the locations of gene expression are best estimates, based on their positions relative to anatomical features of the cochlear duct. *Dotted lines* represent gene expression in hair cells.

other sensory end organs, and are still present in nonsensory regions of the ear such as the semicircular canals (Shailam et al. 1999). RT-PCR analysis of the cochlear epithelium has demonstrated that transcripts for *HES1* and *HES5* are expressed in the epithelium on E14, although apparently not at levels detectable via in situ hybridization (Shailam et al. 1999; Lanford et al. 2000; Zine et al. 2001). Consequently, the exact expression patterns of these genes at this timepoint are unknown. Lastly, one study (Lindsell et al. 1996) has demonstrated weak, widespread expression of *Notch3* in the developing ear at E13.5. However, various other studies have concluded that *Notch 2*, *3*, and *4* are not expressed in the ear (Williams et al. 1995; Lewis et al. 1998; P.J. Lanford, personal observation).

The vestibular epithelia are relatively well developed at E14 to E15. Genes that continue to be expressed in the vestibular system include *Notch1*, *Jag1*, *Jag2*, *Lfng*, *Math1*, and *HES5*. Of these, *Jag1*, *Lfng*, and *HES5* are restricted to the supporting cell layer of the epithelia (Morsli et al. 1998; Shailam et al. 1999), while *Math1*, *Dll1*, and *Jag2* are restricted to the luminal half of the epithelium (Morrison et al. 1999; Shailam et al. 1999; Lanford et al. 2000). The expression of these three genes in cells that appear to be developing as hair cells correlates well with results of radiological and morphological studies of ear development that described the birth dates and ultrastructural development of cells in the mammalian ear (Ruben 1967; Lim and Anniko, 1985; Mbiene and Sans 1986). From E15 through postnatal day 0 (about 18 to 19 days postfertilization) the structure of the developing cochlea becomes increasingly recognizable in its mature form (Lim and Anniko, 1985; Kaufman, 1992; Lim and Rueda, 1992). At E15, *HES5* expression begins in the base of the cochlea, and extends partway along the length of the duct. The initial stripe of expression occurs in a narrow band quite similar to the initial pattern of *Jag2* expression (Lanford et al. 2000). Over the next few days, the expression domain for *HES5* expands from base to apex, in a wave of gene activity that mirrors the expression of *Math1* and *Jag2*, which have, at this point, expanded the full length of the duct (Lanford et al. 1999, 2000).

During this same timeframe, the expression of a number of Notch-related genes becomes further restricted to specific cell types. By E17, *Jag1*, *Lfng*, and *HES5* transcripts are restricted to supporting cells in the organ of Corti (Morsli et al. 1998; Morrison et al. 1999; Lanford et al. 2000), while *HES1* is restricted to inner and outer sulcus cells (Zheng et al. 2000; Zine et al. 2001). Finally, between E17 and birth, *Jag2* and *Dll1* expression is turned on in all hair cells within the inner ear (Morrison et al. 1999; Lanford et al. 2000). Interestingly, *Dll1* expression is also present at E17.5 in scattered cells within the endolymphatic duct (Morrison et al. 1999).

As development proceeds through P0 to P3 and beyond, down-regulation of some genes in the cochlea occurs in much the same manner as it began, from base to apex. This progressive down-regulation has been noted in studies using in situ hybridization to localize *Jag2*, *Math1*, and *HES5* (Lanford et al. 1999, 2000; Shailam et al. 1999). RT-PCR products for *HES1* and *HES5* have also begun to decrease by postnatal day 0 (P0) (Zine et al. 2001). The timing of the down-regulation of Notch genes in the vestibular epithelia has not been clearly documented, but appears to occur in a similar time frame (if advanced by 1 day). In contrast, transcripts for *Notch1* and *Jag1* are more persistant, and are present in the cochlea through at least P5 (Lanford et al. 1999; Shailam et al. 1999).

3.3 Gene Expression: Summary

While it is difficult to make a brief and completely accurate overview of Notch gene expression in the mouse inner ear, several general statements can be offered. First, at the early otocyst stage, Notch is expressed broadly throughout the developing ear, while *Dll1* is expressed in cells that probably correlate with delaminating vestibuloacoustic ganglion neuroblasts. Later, broad Notch expression is maintained and *Jag1* and *Lfng* are expressed in regions of the otocyst that appear to correlate with developing sensory endorgans. As development progresses through midgestation and the vestibular regions of the ear begin to differentiate, expression of *HES1*, *HES5*, *Math1*, and *Jag2* begins in the developing epithelia. Expression of *Notch1, Jag1,* and *Lfng* is maintained in the developing sensory epithelia at this timeframe as well. It is not clear whether *Notch2*, *3*, or *4* are expressed at significant levels during ear development, but existing data suggests probably not. In addition, it is not yet known whether other Notch pathway genes such as *Delta3, Manic fringe,* or *radical fringe* are expressed in the ear during development.

As the cochlear duct begins to extend from the otocyst, expression of *Math1*, *Jag1*, and *HES1* is evident in the end organ. About a day later, genes such as *Jag2*, *Dll1*, and *HES5* begin to be expressed in the base of the cochlea. Expression of these genes extends up the cochlear duct as development proceeds, and becomes restricted to specific cell types within the epithelium. *Math1*, *Jag2*, and *Dll1* become restricted to hair cells, while *Jag1*, *HES5*, and *HES1* are restricted to supporting cells within and adjacent to the organ of Corti (Fig. 5.5). Finally, at or about P0, some of these genes (*Jag2*, *HES5*, *Math1*) appear to become down-regulated in a similar base to apex pattern.

3.4 Notch Function in the Inner Ear: Experimental Studies

Based on the function of Notch in other, well-described systems, various hypotheses and models of Notch function in the developing inner ear have been proposed (e.g., Lanford et al. 1999; Eddison et al. 2000). As is often the case in experimental biology, some of these early predictions have been upheld by further studies, while others have not. The goal of this section is to present a basic, unelaborated model of how Notch might function in the mammalian inner ear, and then describe the data that support or refute specific aspects of this model (Fig. 5.6). It should be noted that the model described below is intended merely as a context within which to structure the current accumulation of information in this system. We will then examine the ways in which the data are or are not in agreement with this model. Lastly, we recognize that research in the mouse ear is not performed in a vacuum, and that numerous important studies in other vertebrate classes have been performed and are ongoing. Consequently, where appropriate, we incorporate data and insights from nonmammalian systems.

3.4.1 A Basic Model of Notch in the Mammalian Ear

Step1: An initial round of Notch signaling specifies cells that will delaminate to form the neuroblasts of the vestibuloacoustic ganglion. During early development of the ear (E9 to E10) *Notch1* and *Jag1* are broadly expressed throughout

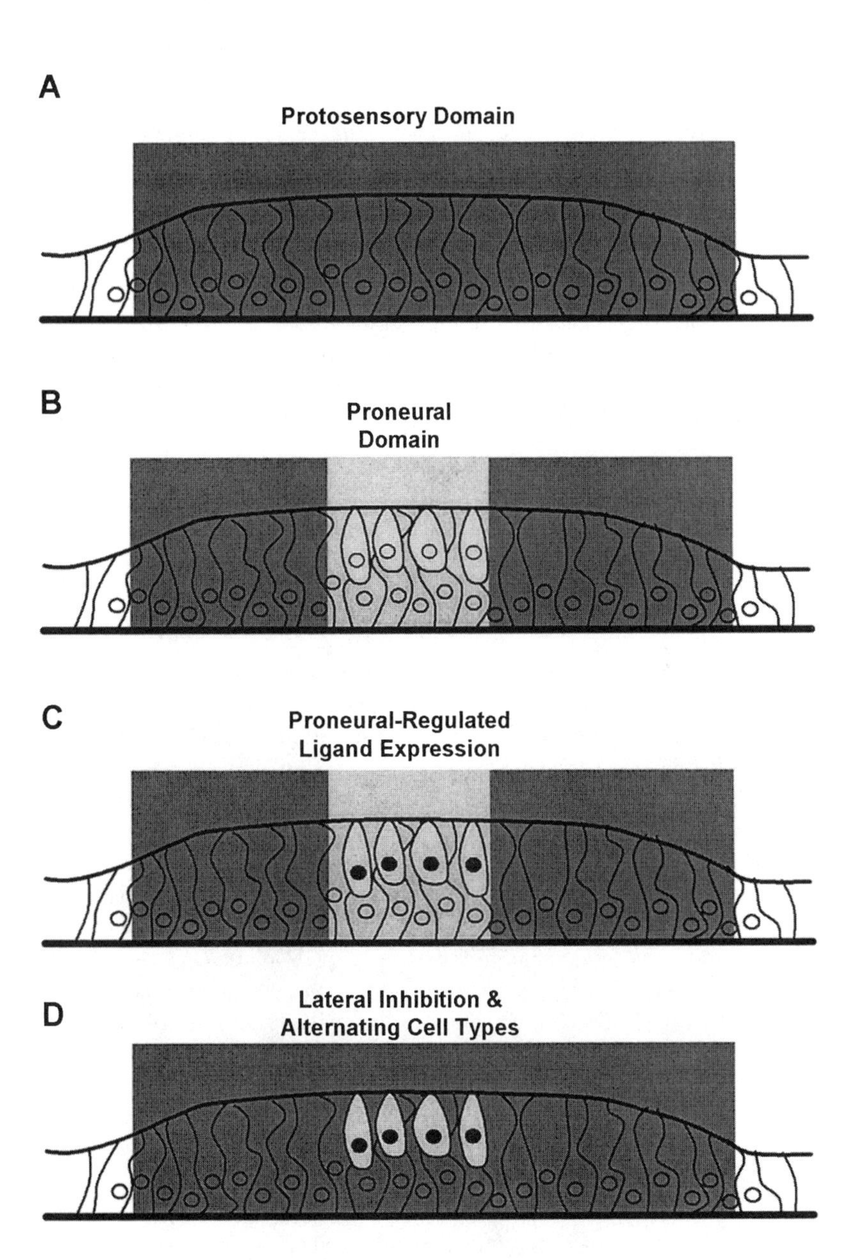

FIGURE 5.6. Diagram of proposed model of Notch pathway function in the developing sensory epithelia of the vertebrate inner ear. (**A**) *Jag1* establishes a broad "protosensory domain" (*gray shading*) in which progenitor cells become determined to develop as cells within the sensory epithelia (hair cells and supporting cells) or cells immediately adjacent to the sensory epithelia (e.g., inner and outer sulcus cells in the cochlea). (**B**) *Math1* expression establishes a "proneural" domain (*thatched shading*) in which, with no further influence, all *Math1*-positive cells will develop as hair cells. (**C**) Within the proneural domain, specific cells begin expressing *Jag2* and *Dll1* (*darkened nuclei*). (**D**) Expression of *Jag2* and *Dll1* in these cells results in increased activation of Notch in their immediate neighbors and the sorting of cells into sensory (hair cell) and nonsensory cell fates.

the otocyst. Expression of proneural genes, *Dll1*, and *Lfng* in the anteroventral portion of the otocyst results in the determination of the neuroblasts that will form the vestibuloacoustic ganglion Fig. 5.6A).

Step 2: *Jag1*–Notch signaling specifies the boundaries of the protosensory domain. During development, discrete regions of the ear become determined to develop as cells within the sensory epithelia (hair cells and supporting cells) or cells immediately adjacent to the sensory epithelia (e.g., inner and outer sulcus cells in the cochlea). Collectively, these cells comprise the "protosensory domain" (also referred to as "the sensory patches" [Haddon et al. 1998]). As development progresses through E12, *Jag1* is expressed in restricted areas of the otocyst that will become the protosensory domains of the cochlea, utricle, saccule, and cristae. The pattern of *Jag1* expression is established via upstream molecules that bias the activation of *Jag1* within this domain, while *Lfng* is expressed the same pool of cells and acts to reinforce the boundary. The cells within protosensory domain are competent to respond to downstream signaling that will determine them as one of several cell types within or adjacent to the organ of Corti or vestibular maculae Fig. 5.6B).

Step 3: The proneural gene *Math1* specifies cells within the protosensory domain as hair cells. As development proceeds (E12^{+}), expression of *Math1* is initiated in a large but restricted number of cells within the protosensory domain. Activation of this gene initiates a molecular program that determines these cells as hair cells. That is, with no further influence, *Math1*-positive cells will develop as hair cells. Conversely, without the expression of *Math1*, cells within the protosensory domain will develop as nonsensory cell types (Fig. 5.6C).

Step 4: A final round of Notch signaling establishes the hair cell/supporting cell mosaic via lateral inhibition. The expression of *Math1* in progenitor cells within the protosensory domain promotes the expression of *Dll1* and *Jag2* in a subset of *Math1*-positive cells. The expression of these ligands initiates the final and definitive round of Notch signaling in the ear—in the form of lateral inhibition,—which reinforces the commitment of protosensory progenitors to hair cell (sensory) or supporting cell (nonsensory) fates, and establishes the alternating pattern of cell types in the epithelia (Fig. 5.6D).

3.4.2 Application of Current Experimental Data to the Hypothetical Model

With this rough model of Notch function and cochlear development in hand, we can now examine the experimental data that have been presented to date, and determine whether these data support or refute the various components of the model.

Step 1: An initial round of Notch signaling specifies cells that will delaminate to form the neuroblasts that will give rise to the vestibuloacoustic ganglion. The first round of Notch signaling that occurs in the developing mouse ear begins at about E9.5, when *Notch1* is already expressed throughout the otocyst. At this timepoint, the proneural bHLH gene, *neurogenin 1* (*ngn1*), is expressed in a subset of cells in the anteroventral portion of the otocyst (Ma et al. 1998; Fritzch

2003). Expression of *ngn1* appears to be spatially coincident with the expression of *Lfng* and later *Dll1* (Ma et al. 1998; Morsli et al. 1998; Morrison et al. 1999; Fritzch 2003). This pattern of expression is consistent with a role for Notch signaling in regulating the number of cells within the otocyst that continue to express *ngn1* and therefore become committed to develop as neuroblasts. The results of two studies strongly support this hypothesis. First, Ma et al. (1998) demonstrated that targeted deletion of *ngn1* results in a loss of primary neuroblasts that form the vestibuloacoustic ganglion. Second, studies in the zebrafish mutant *mindbomb* (*Mib*) demonstrate that inactivation of Notch signaling results in a marked increase in the number of cells within the vestibuloacoustic ganglion (Haddon et al. 1998). The role of *Lfng* in this process is unclear, since deletion of this gene does not result in any discernible phenotype in the inner ear (Zhang et al. 2000; D. Wu, personal communication; P.J. Lanford, personal observation).

Step 2: *Jag1*–Notch signaling specifies the boundaries of the protosensory domain. The role of *Jag1* in the development of the inner ear has been difficult to identify in mammalian model systems primarily because homozygous deletions in the *Jag1* allele result in embryonic lethality at about E10 (Xue et al. 1999). However, the results of at least one study support the hypothesis that there is a relatively broad domain of competent cells in the sensory epithelia that can respond to *Math1* signaling. Specifically, Zheng and Gao (2000) demonstrated that overexpression of *Math1* in the inner sulcus region of cultured cochleae produced supraepithelial hair cells bearing all of the anatomical hallmarks of sensory hair cells, albeit without innervation (see further details below). This study did not determine the limits of this domain, however, nor was it geared toward examining the role of *Jag1* in establishing this competency.

Two recent papers began to address this issue through analysis of animals that are heterozygous for induced mutations of the *Jag1* gene (Kiernan et al. 2001; Tsai et al. 2001). Kiernan et al. (2001) presented anatomical and physiological evidence that mice heterozygous for two different mutant alleles of *Jag1* [*headturner* (*Htu*) and *coloboma* (*Cm*)] have gross anatomical mutations of the inner ear such as missing or truncated sensory epithelia (anterior and posterior cristae). In addition, the cochlear epithelia of these mutant mice contain changes in the number of inner and outer hair cells, although no total counts were performed and it is not possible to say whether the entire number of hair cells in the cochlea was significantly altered. Physiological evidence from this same study suggests that *Htu*/+ animals do not suffer significant auditory deficits. In the study by Tsai et al. (2001), a separate mutation at the *Jag1* locus was identified [*slalom* (*Slm*)]. The inner ears of *Slm* /+ mice have defects similar to, although perhaps not as severe as, *Htu* or *Cm*, including loss of posterior and/or anterior cristae and changes in the number of inner and outer hair cells. In both of these studies, the results show that *Jag1* clearly plays a role in the development of the sensory epithelia, although the specific nature of this role is not yet clear. In particular, reductions in the numbers of outer hair cells suggests that *Jag1* might, indeed, establish the boundaries of the protosensory domain.

However, these results are complicated by the fact that the specific nature of the mutations in each of the mutant *Jag1* alleles has not been determined. Since each allele contains a point mutation in the extraceullular domain of the Jag1 protein, it is not entirely clear whether these represent loss-of-function, hypomorphic, dominant-negative, or even activation mutations. Clearly, the best way to examine the role of *Jag1* in cochlear development would be to generate a targeted mutation in which expression of *Jag1* is specifically deleted in the developing ear only.

Step 3: The proneural gene *Math1* specifies cells within the protosensory domain as hair cells. Experimental studies have now demonstrated that the expression of *Math1* in the developing cochlea is clearly linked to the formation of hair cells in the cochlea and other endorgans. Bermingham et al. (1999) demonstrated that the targeted deletion of *Math1* results in a total lack of sensory cells in the inner ear. The cells remaining in the putative epithelium bear resemblance to supporting cells, but might equally be undifferentiated precursor cells, since in some vertebrates, the two are virtually indistinguishable (Presson et al. 1996). Similarly, as stated above, overexpression of *Math1* in the inner sulcus region of cultured organs of Corti (Zheng and Gao 2000) produces isolated cells bearing all of the anatomical hallmarks of sensory hair cells, albeit without innervation. A recent study by Kawamoto et al. (2003) has also demonstrated that introduction of a *Math1* transgene in vivo could produce new hair cells in the inner sulcus region of the mature mammalian cochlea. These new hair cells also appeared to receive innervation from the nearby epithelium. Although this last study was performed in guinea pigs rather than mice, it seems likely that similar results would be possible across mammalian species. Combined, these data indicate that *Math1* is sufficient to initiate a molecular program that results in the production of hair cell phenotypes in the developing cochlea.

The factors that regulate the spatial expression of *Math1* in the ototcyst are still unknown. However, there are data to suggest that Notch signaling could play a role in this regulation. In the developing mouse cochlea, *Math1* expression occurs in a relatively narrow band of cells near the edge of the inner sulcus. Notch signaling via *Jag1* and *HES1* may act to refine the neural edge of the *Math1* expression band, since *Jag1* and *HES1* mutants have increased numbers of sensory cells in the inner hair cell region. In other words, it appears that deletion of these genes and the subsequent relief of inhibition allows an expansion of the epithelium in only the very limited region of *Math1/Jag1/HES1* coexpression. Careful colocalization studies are necessary to determine the exact relationships between the expression patterns of these and other Notch-related genes.

Step 4: A final round of Notch signaling establishes the hair cell/supporting cell mosaic via lateral inhibition. As discussed in Section 1, the obvious mosaic structure of most hair cell sensory epithelia was one of the initial motivating factors for the study of Notch signaling in the ear. However, it is important to consider that the physical arrangement of cells within the cochlear mosaic is not completely consistent with the model of *Math1/DSL/Notch* function described above. Most obviously, it is clear that the structure of the organ of Corti is *not* a true checkerboard of sensory and nonsensory cell types, but rather is checkerboard pattern that skips a row of sensory cells between the inner and outer pillar cells. In fact, cells that comprise the inner region of the epithelium (inner hair cells, phalangeal cells) are quite different in many ways from those within the outer region of the epithelium (outer hair cells, Deiter's cells, etc.). It is not clear to what we may attribute these differences, or whether or not they are present at the onset of differentiation or arise as a result of downstream "fine tuning" of the cellular population.

With that caveat in mind, the lion's share of experimental data is related to the role of Notch in cochlear development. In general, these data uphold the premise that Notch-mediated lateral inhibition plays a major role in the determination of sensory versus nonsensory cell fates in the organ of Corti. For instance, as previously stated, laser ablation studies performed by Kelley et al. (1995) demonstrated that removal of developing sensory cells allowed other, nonsensory precursor cells to take on the hair cell fate. This result strongly suggests that newly formed hair cells produce some kind of inhibitory signal that represses the development of sensory cell phenotypes in their immediate neighbors.

In 1999, Lanford et al. demonstrated that cochleae from mice containing a targeted deletion of the *Jag2* gene (Jiang et al. 1998) contained significantly higher numbers of hair cells (driven predominantly by an increase in inner hair cells). A subsequent study also showed an increase in the expression of *Math1* in *Jag2* mutants, indicating that the number of cells that continue to express *Math1* and ultimately develop as hair cells is greater in *Jag2* mutants (Lanford et al. 2000). The same study demonstrated a distinct decrease in the expression of *HES5* in *Jag2* mutant cochleae, an apparent reflection of the decrease in Notch pathway activation. Lastly, studies by Zine et al. (2001) demonstrate that targeted deletion of *HES5* also results in an increase in the number of hair cells that develop in the epithelium.

While these results clearly demonstrate that Notch signaling is involved in the formation of the cochlear mosaic, it is important to consider that deletion of what are considered to be major elements of this system (*Jag2*, *HES5*, *Lfn*) does not result in a complete disruption of the cellular mosaic in the organ of Corti. In fact, much of the basic cellular structure of the organ of Corti in *Jag2* and *HES5* mutants remains consistent with controls. For example, the ears of such mutants maintain clear distinctions between inner and outer hair cell regions, and contain pillar cells, Deiter's cells, and inner phalangeal cells.

Most significantly, while the auditory and vestibular epithelia of Notch pathway mutants contain increased numbers of hair cells, they do not appear to contain a concomitant decrease in supporting cells. Such alterations of the cellular mosaic would be predicted in a classic model of lateral inhibition and have been demonstrated in other vertebrate systems. For example, the zebrafish gene, *Mindbomb* (*Mib*), encodes a ubiquitin ligase that is critical for Notch activation in this species (Haddon et al. 1998; Itoh et al. 2003). The inner ears of *Mib* mutants have a dramatic increase in hair cells at the expense of supporting cells in the epithelium. That many Notch mutations in mouse do not result in a similar dramatic disruption in the mosaic raises the distinct possibility that Notch-mediated lateral inhibition (at least in the sense normally thought of in the fly eye) may not occur in the organ of Corti. However, there are a variety plausible explanations for why a complete disruption of the mosaic is not observed in *Jag2* or *HES5* mutants. First, it is possible that Notch signaling is not completely disrupted in the ears of these animals. As discussed, another Notch ligand, *Dll1,* is expressed in the same cells that express *Jag2* and may provide functional redundancy. In addition, it remains possible that *Notch2* or *Notch3* may play a role in the development of the ear, and it is possible that there are as yet unidentified members of the Notch family of genes that are also expressed in the ear. For example, *HES1* and *HES5* are not the only *E(spl)*-related genes expressed in the developing cochlea. A variety of other Notch pathway mediators appear to be present, including *HESR1* and the related *Hey* genes (P.J. Lanford, personal observation). These genes may also provide functional redundancy to the system.

A second possible explanation for the presence of supporting cells in the organ of Corti of *Jag2* and *HES* mutants is the recent demonstration that developing hair cells produce inductive signals that are sufficient to induce surrounding cells to develop as supporting cells (Woods et al. 2004). Moreover, the same study demonstrated that supporting cells can be recruited from cells that would not normally develop as part of the sensory epithelium, suggesting that an overproduction of hair cells, as occurs in Notch pathway mutants, would result in an increase, rather than a decrease, in the number of supporting cells as a result of increased supporting cell recruitment.

An additional sticking point to our classic model of Notch-mediated lateral inhibition in the organ of Corti relates to the function of *Fng* genes. According to studies of *Fng* in the fly, the mammalian homologs to this gene might function as mediators of boundary formation (see Section 2.4). Fng molecules have been shown to regulate Notch-ligand binding and, ultimately, Notch pathway activation. In fact, the expression pattern of *Lfng* in the mouse inner ear appears to be coincident with the expression domain of *Jag1* (Morsli et al. 1998). However, deletion of *Lfng* alone does not seem to have a discernible affect on the development of the cochlea in wild type mice (Zhang et al. 2000). In contrast, deletion of *Lfng* does have an effect on cochlear development in *Jag2* mutant mice. Specifically, deletion of *Lfn* in *Jag2*$^{-/-}$ mice results in the partial rescue of the wild type phenotype—a reduction in the number of extra hair cells produced in the inner hair cell region of the epithelium (Zhang et al. 2000).

These interesting and perhaps puzzling results may be clarified by the results of a ligand-binding study performed in C2C12 myoblasts (Hicks et al. 2000). This study demonstrated that Lfng modulates Notch-ligand binding in a ligand-dependent manner. Specifically, in the myoblast system, Lfng was shown to potentiate interactions between Notch1 and Dll1 (by at least two times), but inhibit Notch1–Jag1 interactions (by at least a 4.4 times). This may occur through the glycosylation of O-linked fucose residues on the extracellular EGF-like repeats region of the Notch receptor molecule (Brückner et al. 2000; Moloney et al. 2000).

Assuming that this information is transferable to the mouse cochlea, the following scenario can be proposed:

If Lfng potentiates the binding activity of Dll1 but inhibits the activity of Jag1, then it is possible that deletion of *Lfng* alone results in no net gain or loss of Notch activation in the cells that express both of theses genes and *Jag2*. Thus, there would be no additions or subtractions of cells in the organ of Corti. In contrast, when both *Lfng* and *Jag2* are deleted we would see a significantly different result. Under this scenario, the removal of Lfng would result in up to a 4.4 times increase in Jag1 signaling in the cochlea. The corresponding decrease in Dll1 signaling would be less than half of the level of the Jag1 increase. Combined, one might see a retightening of the inner boundary of the epithelium due to increased Jag1 activity, but perhaps not a perfect one, due to the decreased Dll1 signaling.

Lastly, it is important to consider that there are several other possible explanations for the limited phenotype observed in *Lfng* mutants. First, two other *Fng* genes, *manic fringe* and *radical fringe*, are present in mammals. At present, data regarding the patterns of expression for these genes in the ear are limited, but the ears of animals with targeted deletions of either gene or both genes appear grossly normal (D. Wu, personal communication). Moreover, the pattern of expression for *Lfng*, at least within the cochlea, is not as simplistic as in other hair cell sensory epithelia. In particular, while *Lfng* is expressed in most supporting cell types within the organ of Corti, it is clearly absent from inner pillar cells (Zhang et al. 2000), suggesting additional possible roles for *Fng* and Notch signaling in cochlear development.

3.4.3 Future Directions

Over the past 8 to 10 years, the role of Notch signaling in inner ear development has progressed from an intriguing hypothesis to a well-documented phenomenon. However, over the same time period, ongoing studies of the Notch signaling pathway have demonstrated that we are just beginning to understand how this molecule functions. As a result, our understanding of its role in the ear must be considered to be largely incomplete and in need of further experimental study: In particular, a more complete characterization of the patterns of expression and the effects of deletion of other members of the pathway must be determined. The potential for functional redundancy probably means that compound mutants containing deletions of multiple pathway components will also be required. In addition, the recent demonstration that Notch activation can be instructive for the formation of certain types of glia raises the question of the potential instructive role of Notch in supporting cell formation. Finally, the potentially intriguing role of *Fringe* genes in the inner ear remains virtually unknown with only a handful of studies examining this very important member of the Notch signaling pathway in the ear.

Given the vast molecular influence that Notch clearly has in biology, it is not likely that the complexities of this pathway will be fully understood anytime soon. The task of understanding the role of Notch in the developing inner ear is only minutely less daunting, and will require diligent and innovative efforts on a number of investigative fronts. The greatest challenge throughout this process, however, will not be in obtaining new and interesting data, which will certainly be gained now with increasing speed. Rather, our greatest challenges lie in the integration of new information across many systems and approaches, in the thoughtful construction of context for these data, and in the cooperative process of experimental design among the numerous laboratories now focused on this problem.

4. Summary

The development of the inner ear can be considered to be a series of progressive decisions that initially delineate VIIIth nerve neuroblasts from developing otic epithelial cells followed by subsequent sorting of the otic epithelium into sensory and nonsensory regions and ultimately sorting of cells within sensory regions into hair cells or supporting cells. Surprisingly or perhaps, not so surprisingly considering evolutionary conservation of function, the Notch signaling pathway appears to play a role in each of these decisions. The Notch signaling pathway refers to a family of related receptors (Notch) and ligands (Delta and Jagged) that are all membrane bound. Activation of Notch has profound implications for individual cell fates, including the maintenance of cells in an uncommitted state. Within the developing ear, Notch signaling is apparently invoked each time a subset of cells must be selected for a specific cell fate. However, beyond its most basic effects, Notch also appears to play additional, more complex, roles in inner ear development that are not yet fully understood. In addition, the simplified interactions described above cannot account for the high degree of regularity that is present in many aspects of inner ear anatomy. Therefore, a better integration of Notch with the other molecular signaling pathways that regulate inner ear development is clearly required.

References

Ahmad I, Zaqouras P, Artavanis-Tsakonas S (1995) Involvement of Notch-1 in mammalian retinal neurogenesis: association of Notch-1 activity with both immature and terminally differentiated cells.Mech Dev 53:73–85.

Akazawa C, Sasai Y, Nakanishi S, Kageyama R (1992) Molecular characterization of a rat negative regulator with a basic helix-loop-helix structure predominantly expressed in the developing nervous system. J Biol Chem 267:21879–21885.

Akazawa C, Ishibashi M, Shimizu C, Nakanishi S, Kageyama R (1995) A mammalian helix-loop-helix factor structurally related to the product of *Drosophila* proneural gene atonal is a positive transcriptional regulator expressed in the developing nervous system. J Biol Chem 270:8730–8738.

Allman D, Punt JA, Izon DJ, Aster JC, Pear WS (2002) An invitation to T and more: notch signaling in lymphopoiesis. Cell 109 (Suppl):S1–11.

Artavanis-Tsakonas S, Matsuno K, Fortini ME (1995) Notch signaling. Science 268: 225–268.

Artavanis-Tsakonas S, Rand MD, Lake RJ (1999) Notch signaling: cell fate control and signal integration in development. Science 284:770–776.

Austin J, Kimble J (1989) Transcript analysis of *glp-1* and *lin-12*, homologous genes required for cell interactions during development of *C. elegans*. Cell 58:565–571.

Bae S, Bessho Y, Hojo M, Kageyama R (2000) The bHLH gene *Hes6*, an inhibitor of *Hes1*, promotes neuronal differentiation. Development 127:2933–2943.

Bang AG, Bailey AM, Posakony JW (1995) *Hairless* promotes stable commitment to the sensory organ precursor cell fate by negatively regulating the activity of the *Notch* signaling pathway. Dev Biol 172:479–494.

Baron M, Aslam H, Flasza M, Fostier M, Higgs JE, Mazaleyrat SL, Wilkin MB (2002) Mol Multiple levels of Notch signal regulation (review). Membr Biol 19:27–38.

Berezovska O, Xia MQ, Page K, Wasco W, Tanzi RE, Hyman BT (1997) Developmental regulation of presenilin mRNA expression parallels *Notch* expression. J Neuropathol Exp Neurol 56:40–44.

Berezovska O, Xia MQ, Hyman BT (1998) *Notch* is expressed in adult brain, is coexpressed with *presenilin-1*, and is altered in Alzheimer disease. J Neuropathol Exp Neurol 57:738–745.

Bermingham NA, Hassan BA, Price SD, Vollrath MA, Ben-Arie N, Eatock RA, Bellen HJ, Lysakowski A, Zoghbi HY (1999) *Math1*: an essential gene for the generation of inner ear hair cells. Science 284:1837–1841.

Bessho Y, Miyoshi G, Sakata R, Kageyama R (2001) *Hes7*: a bHLH-type repressor gene regulated by Notch and expressed in the presomitic mesoderm. Genes Cells 6:175–185.

Bettenhausen B, Hrabe de Angelis M, Simon D, Guenet JL, Gossler A (1995) Transient and restricted expression during mouse embryogenesis of *Dll1*, a murine gene closely related to Drosophila Delta. Development 121:2407–2418.

Bierkamp C, Campos-Ortega JA (1993) A zebrafish homologue of the *Drosophila* neurogenic gene *Notch* and its pattern of transcription during early embryogenesis. Mech Dev 43:87–100.

Blair S (2000) Notch signaling: fringe really is a glycosyltransferase. Curr Biol 10:608–612.

Bray SJ (1997) Expression and function of enhancer of split bHLH proteins during *Drosophila* neurogenesis. Perspect Dev Neurobiol 4:313–323.

Bray S, Furriols M (2001) Notch pathway: making sense of *Suppressor of Hairless*. Curr Biol 11:R217–221.

Brüchner K, Perez L, Clausen H, Cohen S (2000) Glycosytransferase activity of Fringe modulates Notch–Delta interactions. Nature 406:411–415.

Callahan R, Raafat A (2001) Notch signaling in mammary gland tumorigenesis. J Mammary Gland Biol Neoplasia 6:23–36.

Carlesso N, Aster JC, Sklar J, Scadden DT (1999) Notch1-induced delay of human hematopoietic progenitor cell differentiation is associated with altered cell cycle kinetics. Blood 93:838–848.

Chabriat H, Vahedi K, Iba-Zizen MT, Joutel A, Nibbio A, Nagy TG, Krebs MO, Julien J, Dubois B, Ducrocq X, et al. (1995) Clinical spectrum of CADASIL: a study of 7 families. Cerebral autosomal dominant arteriopathy with subcortical infarcts and leukoencephalopathy. Lancet 346:934–939.

Chitnis A, Henrique D, Lewis J, Ish-Horowicz D, Kintner C. (1995) Primary neurogenesis in *Xenopus* embryos regulated by a homologue of the *Drosophila* neurogenic gene *Delta*. Nature 375:761–766.

Coffman C, Harris W, Kintner C (1990) *Xotch*, the *Xenopus* homolog of *Drosophila Notch*. Science 249:1438–1441.

Coffman CR, Skoglund P, Harris WA, Kintner CR (1993) Expression of an extracellular deletion of *Xotch* diverts cell fate in *Xenopus* embryos. Cell 73:659–671.

Cohen B, Bashirullah A, Dagnino L, Campbell C, Fisher WW, Leow CC, Whiting E, Ryan D, Zinyk D, Boulianne G, Hui CC, Gallie B, Phillips RA, Lipshitz HD, Egan SE (1997) Fringe boundaries coincide with Notch-dependent patterning centres in mammals and alter Notch-dependent development in *Drosophila*. Nat Genet 16:283–288.

Corwin JT, Jones JE, Katayama A, Kelley MW, Warchol ME (1991) Hair cell regeneration: the identities of progenitor cells, potential triggers and instructive cues. Ciba Found Symp 160:103–120.

Deftos ML, He YW, Ojala EW, Bevan MJ (1998) Correlating notch signaling with thymocyte maturation. Immunity 9:777–786.

Del Amo FF, Smith DE, Swiatek PJ, Gendron-Maguire M, Greenspan RJ, McMahon AP, Gridley TE (1992) Expression pattern of *Motch*, a mouse homolog of *Drosophila Notch*, suggests an Important role in early postimplantation mouse development. Development 115:737–744.

Dominguez M, de Celis JF (1998) A dorsal/ventral boundary established by Notch controls growth and polarity in the *Drosophila* eye. Nature 396:276–278.

Dornseifer P, Takke C, Campos-Ortega JA (1997) Overexpression of a zebrafish homologue of the *Drosophila* neurogenic gene *Delta* perturbs differentiation of primary neurons and somite development. Mech Dev 63:159–171.

Dunwoodie SL, Henrique D, Harrison SM, Beddington RS. (1997) Mouse *Dll3*: a novel divergent *Delta* gene which may complement the function of other *Delta* homologues during early pattern formation in the mouse embryo. Development 124:3065–3076.

Eddison M, Le Roux I, Lewis J (2000) Notch signaling in the development of the inner ear: lessons from *Drosophila*. Proc Natl Acad Sci USA 97:11692–11699.

Ellisen LW, Bird J, West DC, Soreng AL, Reynolds TC, Smith SD, Sklar J (1991) *TAN-1*, the human homolog of the *Drosophila Notch* gene, is broken by chromosomal translocations in T lymphoblastic neoplasms. Cell 66:649–661.

Fehon RG, Kooh PJ, Rebay I, Regan CL, Xu T, Muskavitch MA, Artavanis-Tsakonas S (1990) Molecular interactions between the protein products of the neurogenic loci *Notch* and *Delta*, two EGF-homologous genes in *Drosophila*. Cell 61:523–534.

Fleming RJ (1998) Structural conservation of Notch receptors and ligands. Semin Cell Dev Biol 9:599–607.

Fleming RJ, Scottgale TN, Diederich RJ, Artavanis-Tsakonas S (1990) The gene *Serrate* encodes a putative EGF-like transmembrane protein essential for proper ectodermal development in *Drosophila melanogaster*. Genes Dev 4:2188–2201.

Fleming RJ, Gu Y, Hukriede NA (1997) Serrate-mediated activation of Notch is specifically blocked by the product of the gene *fringe* in the dorsal compartment of the *Drosophila* wing imaginal disc. Development 124:2973–2981.

Fortini ME, Artavanis-Tsakonas S (1994) The suppressor of hairless protein participates in notch receptor signaling. Cell 79:273–282.

Fortini ME, Rebay I, Caron LA, Artavanis-Tsakonas S (1993) An activated Notch receptor blocks cell-fate commitment in the developing *Drosophila* eye. Nature 365: 555–557.

Fritzsch B (2003) Development of inner ear afferent connections: forming primary neurons and connecting them to the developing sensory epithelia. Brain Res Bull 60: 423–433.

Furukawa T, Maruyama S, Kawaichi M, Honjo T (1992) The *Drosophila* homolog of the immunoglobulin recombination signal-binding protein regulates peripheral nervous system development. Cell 69:1191–1197.

Furukawa T, Mukherjee S, Bao Z-Z, Morrow EM, Cepko C (2000) *Rax, Hes1*, and *notch1* promote the formation of Müller glia by postnatal retinal progenitor cells. Neuron 26: 383–394.

Gaiano N, Nye JS, Fishell G (2000) Radial glial identity is promoted by *Notch1* signaling in the murine forebrain. Neuron 26:395–404.

Gallahan D, Callahan R (1997) The mouse mammary tumor associated gene *INT3* is a unique member of the NOTCH gene family (NOTCH4). Oncogene 14:1883–1890.

Gray GE, Mann RS, Mitsiadis E, Henrique D, Carcangiu ML, Banks A, Leiman J, Ward D, Ish-Horowitz D, Artavanis-Tsakonas S. (1999) Human ligands of the Notch receptor. Am J Pathol 154:785–794.

Greenwald I (1985) *lin-12*, a nematode homeotic gene, is homologous to a set of mammalian proteins that includes epidermal growth factor. Cell 43:583–590.

Gridley T (1997) Notch signaling in vertebrate development and disease. Mol Cell Neurosci 9:103–108.

Gu Y, Hukriede NA, Fleming RJ (1995) *Serrate* expression can functionally replace *Delta* activity during neuroblast segregation in the *Drosophila* embryo. Development 121: 855–865.

Guillemot F, Joyner AL (1993) Dynamic expression of the murine *Achaete-Scute* homologue *Mash-1* in the developing nervous system. Mech Dev 42:171–185.

Gupta BP, Rodrigues V (1997) *Atonal* is a proneural gene for a subset of olfactory sense organs in *Drosophila*. Genes Cells 2:225–233.

Haddon C, Jiang YJ, Smithers L, Lewis J (1998) Delta–Notch signaling and the patterning of sensory cell differentiation in the zebrafish ear: evidence from the *mind bomb* mutant. Development 125:4637–4644.

Hartenstein AY, Rugendorff A, Tepass U, Hartenstein V (1992) The function of the neurogenic genes during epithelial development in the *Drosophila* embryo. Development 116:1203–1220.

Hartley DA, Xu TA, Artavanis-Tsakonas S (1987) The embryonic expression of the *Notch* locus of *Drosophila melanogaster* and the implications of point mutations in the extracellular EGF-like domain of the predicted protein. EMBO J 6:3407–3417.

Heitzler P, Simpson P (1993) Altered epidermal growth factor-like sequences provide evidence for a role of Notch as a receptor in cell fate decisions. Development 117: 1113–1123.

Heitzler P, Bourouis M, Ruel L, Carteret C, Simpson P (1996) Genes of the *Enhancer of split* and *achaete–scute* complexes are required for a regulatory loop between *Notch* and *Delta* during lateral signalling in *Drosophila*. Development 122:161–171.

Henrique D, Adam J, Myat A, Chitnis A, Lewis J, Ish-Horowicz D (1995) Expression of a *Delta* homologue in prospective neurons in the chick. Nature 375:787–790.

Hicks C, Johnston SH, diSibio G, Collazo A, Vogt TF, Weinmaster G (2000) Fringe differentially modulates Jagged1 and Delta1 signalling through Notch1 and Notch2. Nat Cell Biol 2:515–520.

Hsieh JJ, Henkel T, Salmon P, Robey E, Peterson MG, Hayward SD (1996) Truncated mammalian Notch1 activates *CBF1/RBPJk*-repressed genes by a mechanism resembling that of Epstein-Barr virus EBNA2. Mol Cell Biol 16:952–959.

Hsieh JJ, Nofziger DE, Weinmaster G, Hayward SD (1997) Epstein-Barr virus immortalization: Notch2 interacts with CBF1 and blocks differentiation. J Virol 71:1938–1945.

Ishibashi M, Sasai Y, Nakanishi S, Kageyama R (1993) Molecular characterization of HES-2, a mammalian helix-loop-helix factor structurally related to *Drosophila* hairy and Enhancer of split. Eur J Biochem 215:645–652.

Ishibashi M, Ang SL, Shiota K, Nakanishi S, Kageyama R, Guillemot F (1995) Targeted disruption of mammalian *hairy* and *enhancer of split homolog 1* (*HES1*) leads to upregulation of neural helix–loop–helix factors, premature neurogenesis, and severe neural tube defects. Genes Dev 9:3136–3148.

Itoh M, Kim CH, Palardy G, Oda T, Jiang YJ, Maust D, Yeo SY, Lorick K, Wright GJ, Ariza-McNaughton L, Weissman AM, Lewis J, Chandrasekharappa SC, Chitnis AB (2003) Mind bomb is a ubiquitin ligase that is essential for efficient activation of Notch signaling by Delta. Dev Cell 4:67–82.

Jacobsen TL, Brennan K, Arias AM, Muskavitch MA (1998) Cis-interactions between Delta and Notch modulate neurogenic signalling in *Drosophila*. Development 125: 4531–4540.

Jarman AP, Grau Y, Jan LY, Jan YN (1993) *Atonal* is a proneural gene that directs chordotonal organ formation in the *Drosophila* peripheral nervous system. Cell 73:1307–1321.

Jarman AP, Grell EH, Ackerman L, Jan LY, Jan YN (1994) *Atonal* is the proneural gene for *Drosophila* photoreceptors. Nature 369:398–400.

Jarriault S, Brou C, Logeat F, Schroeter EH, Kopan R, Israel A (1995) Signaling downstream of activated mammalian Notch. Nature 377:288–289.

Jehn BM, Bielke W, Pear WS, Osborne BA (1999) Cutting edge:protective effects of notch-1 on TCR-induced apoptosis. J Immunol 162:635–638.

Jen WC, Wettstein D, Turner D, Chitnis A, Kintner C (1997) The Notch ligand, X-Delta-2, mediates segmentation of the paraxial mesoderm in *Xenopus* embryos. Development. 124:1169–1178.

Jensen J, Pedersen EE, Galante P, Hald J, Heller RS, Ishibashi M, Kageyama R, Guillemot F, Serup P, Madsen OD (2000) Control of endodermal endocrine development by Hes-1. Nat Genet 24:36–44.

Jiang R, Lan Y, Chapman HD, Shawber C, Norton CR, Serreze DV, Weinmaster G, Gridley T (1998) Defects in limb, craniofacial and thymic development in *Jagged2* mutant mice. Genes Development 12:1046–1057.

Jimenez G, Ish-Horowicz D (1997) A chimeric enhancer-of-split transcriptional activator drives neural development and *acheate–scute* expression. Mol Cell Biol 17:4355–4362.

Johnston SH, Rauskolb C, Wilson R, Prabhakaran B, Irvine KD, Vogt TF (1997) A family of mammalian *Fringe* genes implicated in boundary determination and the Notch pathway. Development 124:2245–2254.

Joutel A, Tournier-Lasserve E (1998) Notch signalling pathway and human diseases. Semin Cell Dev Biol 9:619–625.

Joutel A, Corpechot C, Ducros A, Vahedi K, Chabriat H, Mouton P, Alamowitch S, Domenga V, Cecillion M, Marechal E, Maciazek J, Vayssiere C, Cruaud C, Cabanis EA, Ruchoux MM, Weissenbach J, Bach JF, Bousser MG, Tournier-Lasserve E (1997) *Notch3* mutations in cerebral autosomal dominant arteriopathy with subcortical infarcts and leukoencephalopathy (CADASIL), a mendelian condition causing stroke and vascular dementia. Ann NY Acad Sci 826:213–217.

Kageyama R, Nakanishi S (1997) Helix-loop-helix factors in growth and differentiation of the vertebrate nervous system. Curr Opin Genet Dev 7:659–665.

Kaufman MH (1992) The Atlas of Mouse Development. London: Academic Press, pp. 123–370.

Kawamoto K, Ishimoto S, Minoda R, Brough DE, Raphael Y J (2003) *Math1* gene transfer generates new cochlear hair cells in mature guinea pigs in vivo. Neurosci 23: 4395–4400.

Kelley MW, Talreja DR, Corwin JT (1995) Replacement of hair cells after laser microbeam irradiation in cultured organs of corti from embryonic and neonatal mice. J Neurosci 15:3013–3026.

Kidd S, Kelley MR, Young MW (1986) Sequence of the notch locus of *Drosophila melanogaster*: relationship of the encoded protein to mammalian clotting and growth factors. Mol Cell Biol 6:3094–3108.

Kiernan AE, Ahituv N, Fuchs H, Balling R, Avraham KB, Steel KP, Hrabe deAngelis M (2001) The Notch ligand Jagged1 is required for inner ear sensory development. Proc Natl Acad Sci USA 98:3873–3878.

Kiyota T, Jono H, Kuriyama S, Hasegawa K, Miyatani S, Kinoshita T (2001) X-Serrate-1 is involved in primary neurogenesis in *Xenopus laevis* in a complementary manner with X-Delta-1. Dev Genes Evol 211:367–376.

Klueg KM, Muskavitch MA (1999) Ligand-receptor interactions and trans-endocytosis of Delta, Serrate and Notch: members of the Notch signalling pathway in *Drosophila*. J Cell Sci 112:3289–3297.

Knust E, Dietrich U, Tepass U, Bremer KA, Weigel D, Vassin H, Campos-Ortega JA (1987) EGF homologous sequences encoded in the genome of *Drosophila melanogaster*, and their relation to neurogenic genes. EMBO J 6:761–766.

Kopan R, Nye JS, Weintraub H (1994) The intracellular domain of mouse Notch a constitutively activated repressor of myogenesis directed at the basic helix-loop-helix region of MyoD. Development 120:2385–2396.

Kopczynski CC, Alton AK, Fechtel K, Kooh PJ, Muskavitch MA. (1988) *delta,* a *Drosophila* neurogenic gene, is transcriptionally complex and encodes a protein related to blood coagulation factors and epidermal growth factor of vertebrates. Genes Dev 2: 1723–1735.

Kunisch M, Haenlin M, Campos-Ortega JA (1994) Lateral inhibition mediated by the *Drosophila* neurogenic gene *delta* is enhanced by proneural proteins. Proc Natl Acad Sci USA 91:10139–10143.

Lan Y, Jiang R, Shawber C, Weinmaster G, Gridley T (1997) The *Jagged2* gene maps to chromosome 12 and is a candidate for the lgl and sm mutations. Mamm Genome 8:875–876.

Lanford PJ, Lan Y, Jiang R, Lindsell C, Weinmaster G, Gridley T, Kelley MW (1999) Notch signaling pathway mediates hair cell development in mammalian cochlea. Nat Gen 21:289–292.

Lanford PJ, Shailam R, Norton CR, Gridley T, Kelley MW (2000) Expression of *Math1* and *HES5* in the cochleae of wildtype and *Jag2* mutant mice. J Assoc Res Otolaryngol 1:161–171.

Lardelli M, Lendahl U (1993) *Motch A* and *motch B*—two mouse Notch homologues coexpressed in a wide variety of tissues. Exp Cell Res. 204:364–372.

Lardelli M, Dahlstrand J, Lendahl U (1994) The novel Notch homologue mouse *Notch 3* lacks specific epidermal growth factor-repeats and is expressed in proliferating neuroepithelium. Mech Dev 46:123–136.

Lecourtois M, Schweisguth F (1997) Role of suppressor of hairless in the delta-activated Notch signaling pathway. Perspect Dev Neurobiol 4:305–311.

Lehmann R, Jiménez F, Dietrich U, Campo-Ortega JA (1983) On the phenotype and development of mutants of early neurogenesis in *Drosophila melanogaster*. Roux Arch Dev Biol 192:62–74.

Lewis AK, Franz GD, Carpenter DA, de Sauvage FJ, Gao WQ (1998) Distinct expression patterns of Notch family receptors and ligands during development of the mammalian inner ear. Mech Dev 78:159–163.

Lewis J (1991) Rules for the production of sensory cells. Ciba Found Symp 160:25–39.

Li L, Krantz ID, Deng Y, Genin A, Banta AB, Collins CC, Qi M, Trask BJ, Kuo WL, Cochran J, Costa T, Pierpont ME, Rand EB, Piccoli DA, Hood L, Spinner NB (1997) Alagille syndrome is caused by mutations in human *Jagged1*, which encodes a ligand for Notch1. Nat Genet 16:243–251.

Lim DJ, Anniko M (1985) Developmental morphology of the mouse inner ear: a scanning electron microscopic observation. Acta Otolaryngol Suppl 422:1–69.

Lim DJ, Rueda J (1992) Structural development of the cochlea. In: Romand R (ed), Development of Auditory and Vestibular Systems 2. New York: Elsevier, pp. 33–58.

Lindsell CE, Shawber CJ, Boulter J, Weinmaster G (1995) Jagged: a mammalian ligand that activates Notch1. Cell 80:909–917.

Lindsell C, Boulter J, diSibio G, Gossler A, Weinmaster G (1996) Expression patterns of *Jagged*, *Delta1*, *Notch1*, *Notch2* and *Notch3* genes identify ligand-receptor pairs that may function in neural development. Mol Cell Neurosci 8:14–27.

Logeat F, Bessia C, Brou C, LeBail O, Jarriault S, Seidah NG, Israel A (1998) The Notch1 receptor is cleaved constitutively by a furin-like convertase. Proc Natl Acad Sci USA 95:8108–8112.

Luo B, Aster CJ, Hasserjian RP, Kuo F, Sklar J (1997) Isolation and functional analysis of a cDNA for human *Jagged2*, a gene encoding a ligand for the N1 receptor. Mol Cell Biol 17:6057–6067.

Lutolf S, Radtke F, Aguet M, Suter U, Taylor V (2002) *Notch1* is required for neuronal and glial differentiation in the cerebellum. Development 129:373–385.

Ma Q, Chen Z, del Barco Barrantes I, de la Pompa JL, Anderson DJ (1998) *neurogenin1* is essential for the determination of neuronal precursors for proximal cranial sensory ganglia. Neuron 20:469–482.

Mango SE, Thorpe CJ, Martin PR, Chamberlain SH, Bowerman B (1994) Two maternal genes, *apx-1* and *pie-1*, are required to distinguish the fates of equivalent blastomeres in the early *Caenorhabditis elegans* embryo. Development 120: 2305–2315.

Mbiene JP, Sans A (1986) Differentiation and maturation of the sensory hair bundles in the fetal and postnatal vestibular receptors of the mouse: a scanning electron microscopy study. J Comp Neurol 254:271–278.

Miele L, Osborne BJ (1999) Arbiter of differentiation and death: Notch signaling meets apoptosis. Cell Physiol 181:393–409.

Milner LA, Bigas A (1999) Notch as a mediator of cell fate determination in hematopoiesis: evidence and speculation. Blood 93:2431–2448.

Modollel J (1997) Patterning of the adult peripheral nervous system of *Drosophila*. Perspect Dev Neurobiol 4:285–296.

Mohr, OL (1919) Character changes caused by mutation of an entire region of a chromosome in *Drosophila melanogaster*. Genetics 4:275–282.

Moloney DJ, Panin VM, Johnston SH, Chen J, Shao L, Wilson R, Wang Y, Stanley P, Irvine KD, Haltiwanger RS, Vogt TF (2000) Fringe is a glycosyltransferase that modifies Notch. Nature 406:369–375.

Morrison A, Hodgetts C, Gossler A, Lewis J (1999) Expression of *Delta1* and *Serrate1* (*Jag1*) in the mouse inner ear. Mech Dev 84:169–172.

Morrison SJ, Perez SE, Qiao Z, Verdi JM, Hicks C, Weinmaster G, Anderson DJ (2000) Transient Notch activation initiates an irreversible switch from neurogenesis to gliogenesis by neural crest stem cells. Cell 101:499–510.

Morsli H, Choo D, Ryan A, Johnson R, Wu DK (1998) Development of the mouse inner ear and origin of its sensory organs. J Neurosci 18:3327–3335.

Myat A, Henrique D, Ish-Horowicz D, Lewis J (1996) A chick homologue of *Serrate* and its relationship with *Notch* and *Delta* homologues during central neurogenesis. Dev Biol 174:233–247.

Nakao K, Campos-Ortega JA (1996) Persistent expression of genes of the *enhancer of split* complex suppresses neural development in *Drosophila*. Neuron 16:275–286.

Nishimura S, Isaka F, Ishibashi M, Tomita K, Tsuda H, Nakanishi S, Kageyama R (1998) Structure, chromosomal locus and promoter of mouse *HES2* gene, a homolog of *Drosophila hairy* and *enhancer of split*. Genomics 49:69–75.

Oda T, Elkahloun AG, Pike BL, Okajima K, Krantz ID, Genin A, Piccoli DA, Meltzer PS, Spinner NB, Collins FS, Chandrasekharappa SC (1997) Mutations in the human *Jagged1* gene are responsible for Alagille syndrome. Nat Genet 16:235–242.

Ohtsuka T, Ishibashi M, Gradwohl G, Nakanishi S, Guillemot F, Kageyama R (1999) *HES1* and *HES5* as Notch effectors in mammalian neuronal differentiation. EMBO J 18:2196–2207.

Panin VM, Papayannopoulos V, Wilson R, Irvine KD (1997) Fringe modulates Notch-ligand interactions. Nature 387:908–912.

Piccoli DA, Spinner NB (2001) Alagille syndrome and the *Jagged1* gene. Semin Liver Dis 21:525–534.

Posakony JW (1999) News and views: birds on a wire and tiling the inner ear. Nat Gen 21:253–254.

Poulson DF (1937) Chromosomal deficiencies and embryonic development of *Drosophila melanogaster*. Proc Natl Acad Sci USA 23:133–137.

Poulson DF (1940) The effects of certain X-chromosome deficiencies on the embryonic development of *Drosophila melanogaster*. J Exp Zool 83:271–325.

Presson JC, Lanford PJ, Popper AN (1996) Hair cell precursors are ultrastructurally indistinguishable from mature support cells in the ear of a postembryonic fish. Hear Res 100:10–20.

Rebay I, Fleming RJ, Fehon RG, Cherbas L, Cherbas P, Artavanis-Tsakonas S (1991) Specific EGF repeats of Notch mediate interactions with Delta and Serrate: implications for Notch as a multifunctional receptor. Cell 67:687–699.

Robey E, Chang D, Itano A, Cado D, Alexander H, Lans D, Weinmaster G, Salmon P

(1996) An activated form of Notch influences the choice between CD4 and CD8 T cell lineages. Cell 87:483–492.

Ruben RJ (1967) Development of the inner ear of the mouse: a radioautographic study of terminal mitosis. Acta Otolaryngol (Suppl) 220:1–44.

Sasai Y, Kageyama R, Tagawa Y, Shigemoto R, Nakanishi S (1992) Two mammalian helix-loop-helix factors structurally related to *Drosophila* hairy and enhancer of split. Genes Dev 6:2620–2634.

Schweisguth F, Posakony JW (1992) *Suppressor of Hairless*, the *Drosophila* homolog of the mouse recombination signal-binding protein gene, controls sensory organ cell fates. Cell 69:1199–1212.

Shailam R, Lanford PJ, Dolinsky CM, Norton CR, Gridley T, Kelley MW (1999) Expression of proneural and neurogenic genes in the embryonic mammalian vestibular system. J Neurocytol 28:809–819.

Shawber C, Boulter J, Lindsell CE, Weinmaster G. (1996) *Jagged2*: a serrate-like gene expressed during rat embryogenesis. Dev Biol 180:370–376.

Sherwood DR, McClay DR (1997) Identification and localization of a sea urchin Notch homologue: insights into vegetal plate regionalization and Notch receptor regulation. Development 124:3363–3374.

Shindo K, Kawashima N, Sakamoto K, Yamaguchi A, Umezawa A, Takagi M, Katsube K, Suda H (2003) Osteogenic differentiation of the mesenchymal progenitor cells, Kusa is suppressed by Notch signaling. Exp Cell Res 290:370–380.

Simpson P (1990) Notch and the choice of cell fate in *Drosophila* neuroepithelium. Trends Genet 6:343–345.

Smith P (2001) Epstein-Barr virus complementary strand transcripts (CSTs/BARTs) and cancer. Semin Cancer Biol 11:469–476.

Stifani S, Blaumueller CM, Redhead NJ, Hill RE, Artavanis-Tsakonas S (1992) Human homologs of a Drosophila *enhancer of split* gene product define a novel family of nuclear proteins. Nat Genet 2:119–127.

Sugaya K, Fukagawa T, Matsumoto K, Mita K, Takahashi E, Ando A, Inoko H, Ikemura T (1994) Three genes in the human MHC class III region near the junction with the class II: gene for receptor of advanced glycosylation end products, *PBX2* homeobox gene and a notch homolog, human counterpart of mouse mammary tumor gene *int-3*. Genomics 23:408–419.

Swiatek PJ, Lindsell CE, del Amo FF, Weinmaster G, Gridley T (1994) *Notch1* is essential for postimplantation development in mice. Genes Dev 8:707–719.

Tax FE, Yeargers JJ, Thomas JH (1994) Sequence of *C. elegans* lag-2 reveals a cell-signalling domain shared with Delta and Serrate of *Drosophila*. Nature 368:150–154.

Tomita K, Ishibashi M, Nakahara K, Ang SL, Nakanishi S, Guillemot F, Kageyama R (1996) Mammalian *hairy* and *enhancer of split homolog 1* regulates differentiation of retinal neurons and is essential for eye morphogenesis. Neuron 16:723–734.

Tournier-Lasserve E, Joutel A, Melki J, Weissenbach J, Lathrop GM, Chabriat H, Mas JL, Cabanis EA, Baudrimont M, Maciazek J, et al. (1993) Cerebral autosomal dominant arteriopathy with subcortical infarcts and leukoencephalopathy maps to chromosome 19q12. Nat Genet 3:256–259.

Tsai H, Hardisty RE, Rhodes C, Kiernan AE, Roby P, Tymowska-Lalanne Z, Mburu P, Rastan S, Hunter AJ, Brown SD, Steel KP (2001) The mouse *slalom* mutant demonstrates a role for *Jagged1* in neuroepithelial patterning in the organ of Corti. Hum Mol Genet 10:507–512.

Uyttendaele H, Marazzi G, Wu G, Yan Q, Sassoon D, Kitajewski J (1996) *Notch4/int-3*, a mammary proto-oncogene, is an endothelial cell-specific mammalian *Notch* gene. Development 122:2251–2259.

Valsecchi C, Ghezzi C, Ballabio A, Rugarli EI (1997) *JAGGED2*: a putative Notch ligand expressed in the apical ectodermal ridge and in sites of epithelial-mesenchymal interactions. Mech Dev 69:203–207.

Weinmaster G (1997) The ins and outs of Notch signaling. Mol Cell Neurosci 9:91–102.

Weinmaster G (2000) Notch signal transduction: a real rip and more. Curr Opin Genet Dev 10:363–369.

Weinmaster G, Roberts VJ, Lemke G (1991) A homolog of *Drosophila Notch* expressed during mammalian development. Development 113:199–205.

Weinmaster G, Roberts VJ, Lemke G (1992) *Notch2*: a second mammalian *Notch* gene. Development 116:931–941.

Wharton KA, Johansen KM, Xu T, Artavanis-Tsakonas S (1985) Nucleotide sequence from the neurogenic locus *notch* implies a gene product that shares homology with proteins containing EGF-like repeats. Cell 43:567–581.

Williams R, Lendahl U, Lardelli M (1995) Complementary and combinatorial patterns of *Notch* gene family expression during early mouse development. Mech Dev 53:357–368.

Woods C, Montcouquiol, M., Kelley, MW (2004) *Math1* regulates development of the sensory epithelium in the mammalian cochlea. Nat Neurosci 7:1310–1318.

Xue Y, Gao X, Lindsell CE, Norton CR, Chang B, Hicks C, Gendron-Maguire M, Rand EB, Weinmaster G, Gridley T (1999) Embryonic lethality and vascular defects in mice lacking the Notch ligand *Jagged1*. Hum Mol Genet 8:723–730.

Yoneya T, Tahara T, Nagao K, Yamada Y, Yamamoto T, Osawa M, Miyatani S, Nishikawa M (2001) Molecular cloning of *delta-4*, a new mouse and human Notch ligand. J Biochem (Tokyo) 129:27–34.

Zhang N, Martin GV, Kelley MW, Gridley T (2000) A mutation in the *Lunatic fringe* gene suppresses the effects of a *Jagged2* mutation on inner hair cell development in the cochlea. Curr Biol 10:659–662.

Zheng JL, Gao WQ (2000) Overexpression of *Math1* induces robust production of extra hair cells in postnatal rat inner ears. Nat Neurosci 3:580–586.

Zheng JL, Shou J, Guillemot F, Kageyama R, Gao WQ (2000) *Hes1* is a negative regulator of inner ear hair cell differentiation. Development 127:4551–4560.

Zhong W, Feder JN, Jiang MM, Jan LY, Jan YN (1996) Asymmetric localization of a mammalian numb homolog during mouse cortical neurogenesis. Neuron 17:43–53.

Zhou S, Hayward SD (2001) Nuclear localization of CBF1 is regulated by interactions with the SMRT corepressor complex. Mol Cell Biol 21:6222–6232.

Zhou S, Fujimuro M, Hsieh JJ, Chen L, Miyamoto A, Weinmaster G, Hayward SD (2000) SKIP, a CBF1-associated protein, interacts with the ankyrin repeat domain of NotchIC to facilitate NotchIC function. Mol Cell Biol 20:2400–2410.

Zine A, Aubert A, Qiu J, Therianos S, Guillemot F, Kageyama R, de Ribaupierre F (2001) *Hes1* and *Hes5* activities are required for the normal development of the hair cells in the mammalian inner ear. J Neurosci 21:4712–4720.

6

The Differentiation of Hair Cells

JANE E. BRYANT, ANDREW FORGE, AND GUY P. RICHARDSON

1. Introduction

Hair cells are found in the sensory epithelia of auditory and vestibular organs of all vertebrates, and in the neuromasts of the lateral line systems of fish and amphibia. They detect mechanical displacements and transduce these stimuli into electrical signals. These electrical signals are transmitted to the contacting afferent nerves fibers and thereby relayed to the brain. All hair cells are mechanotransducers and have a number of common structural features such as the hair bundle, cuticular plate, and specialized intercellular junctions. At the same time, however, these cells respond to different types of stimuli and this is manifested in their morphological diversity. In addition, outer hair cells in the mammalian cochlea can act as reverse transducers, and this capacity is based on a unique basolateral membrane specialization that supports the electromotile activity of this particular type of hair cell.

1.1 The Common Structural Features of Hair Cells

The hair bundle is a collection of rigid, erect projections present on the apical surface of the hair cell. Deflections of the hair bundle modulate a potassium current flowing through the hair cell and thereby generate an intracellular receptor potential. The hair bundles are formed of rows of stereocilia, which are modified microvilli that are each packed with a paracrystalline array of F-actin filaments (Flock et al. 1982; Sobin and Flock 1983). Stereocilia increase in height in one particular direction across the bundle. In all hair cells except in the mature hair cells of the mammalian cochlea, there is also a single kinocilium, a true cilium formed of microtubules, located behind the row of longest stereocilia. In the mammalian cochlea the kinocilium is present during embryonic development, but is lost during the latter stages of cochlear maturation, leaving only the associated basal body in the apical cytoplasm of the cell.

The hair bundle is an asymmetrical, morphologically polarized structure, and the position of the kinocilium (or its basal body) and the row of longest stereocilia defines its orientation or polarity. Deflections of the hair bundle along its axis of mechanosensitivity, toward and away from the longest row of stereocilia, regulate the opening and closing of nonselective cation channels (Hudspeth 1989). These transduction channels are thought to be located at the tips of the stereocilia and gated by the fine tip-link that is seen connecting the top of one stereocilium to the shaft of the adjacent longer stereocilium. These tip links lie along the axis of sensitivity (Pickles et al. 1984; Markin and Hudspeth 1995). In addition to the tip link, the stereocilia in an individual hair bundle are connected by a variety of other extracellular crosslinks (Goodyear and Richardson 1992; Goodyear et al. 2005). The orientation of hair bundles in a particular sensory epithelium is not random and is related to that of its neighbors. Thus the hair cells in each sensory organ exhibit planar polarity.

The stereocilia are supported on the cuticular plate, a meshwork of actin filaments (Hirokawa and Tilney 1982) and actin-associated proteins. The cuticular plate is positioned in the apical cytoplasm of the hair cell lying immediately beneath, and attached to the underside of, the apical plasma membrane. Actin filaments descend from the center of each stereocilium into the cuticular plate, forming a rootlet that is crosslinked into the actin meshwork (Hirokawa and Tilney 1982). The basal body of the kinocilium lies within an eccentrically positioned hole within the cuticular plate. The presence of numerous intracellular vesicles and vesicles fusing with apical plasma membrane in the region of the basal body suggest it is a site of exo- or endocytotic activity (Kachar et al. 1997).

Each hair cell in a sensory organ is separated from its neighbors by surrounding supporting cells so that adjacent hair cells do not contact one another. The cell bodies of the supporting cells also intervene between the base of the hair cell and the basal lamina that underlies the sensory epithelium. Mature hair cells are therefore characterized by a lack of contact with the basement membrane. Hair cells are also functionally isolated from the surrounding supporting cells that are themselves all interconnected via a system of intercellular gap junctions. There are no gap junction plaques associated with plasma membranes of hair cells (Forge et al. 2003). Hair and supporting cells are, however, linked to one another via a specialized tight-adherens junctional complex located at the apical-most region of their basolateral membranes. This junctional complex is particularly extensive in the outer hair cell region of the mature organ of Corti.

2. Morphological Variety

While all hair cells can be defined by the features described above, a variety of hair-cell types are recognized in the different vertebrate classes and in different auditory and vestibular sensory organs. This variability encompasses cell shape, innervation pattern, hair-bundle morphology and the composition of the lateral plasma membrane.

2.1 Different Types of Hair Cell

In the vestibular sensory epithelia in both birds and mammals there are two types of hair cell (Bergstrom and Engstrom 1973; Goldberg 1991). Type 1 hair cells are pear shaped and the basolateral portion of the cell is entirely enclosed by a single, calyxlike afferent nerve ending. Type 2 hair cells are cylindrical and both afferent and efferent nerve endings synapse at their base. The vestibular sensory epithelia are essentially two-dimensional epithelial sheets with a differential distribution of type 1 and type 2 hair cells. Type 1 hair cells predominate across the striola, a region within the epithelium where hair-bundle orientation changes by 180° so that all hair bundles on one side are oriented in exactly the opposite direction from all those on the other. Type 2 hair cells predominate in the extrastriolar regions.

Two hair-cell types are also distinguished in the avian auditory epithelium, the basilar papilla (Smith et al. 1985). Tall hair cells are located on the inner side of the curved ribbonlike epithelium (the neural edge along which the nerves enter the papilla). They have a cylindrically shaped cell body with a long axis much greater than its width and are exclusively innervated by afferent fibers. Short hair cells, with a cell body whose width is much greater than its length, lie on the outer side of the papilla and synapse with efferent nerves.

A distinct separation of two hair-cell types in auditory organs is most clearly pronounced in the mammalian cochlea (Lim 1980; Slepecky 1996); spiraling along the organ of Corti there is a single row of flask-shaped inner hair cells (IHCs) with exclusively afferent innervation, and three (sometimes four) rows of cylindrical outer hair cells (OHCs) that predominantly receive efferent innervation. The cell bodies of the IHCs are angled toward the center of the spiral and those of OHCs toward the outside, creating a wide separation between the two hair cell types within the corpus of the organ of Corti formed by the intervening supporting cells.

In line with the differing innervation patterns, the number and type of ion channels in the lateral plasma membrane of hair cells also vary with hair-cell type, thereby influencing the characteristics of the neural signal that the cell can convey (Eatock and Rüsch 1997; Ashmore 2002). In addition to particular ion channels, the hair cells of the mammalian cochlea possess further unique specializations of their lateral plasma membrane. Freeze–fracture electron microscopy reveals distinct plaque regions just below the junctional complex on the lateral membrane of IHCs. Within these plaques, particles are organized in square arrays (Forge 1987). The nature and function of the plaques are not yet known.

OHCs possess a high density of a unique protein, prestin (Belyantseva et al. 2000; J. Zheng et al. 2000; Dallos and Fakler 2002). On deflection of their hair bundles, OHCs undergo fast (up to auditory frequencies), reversible length changes that are driven by the electrical changes induced by the transduction currents (Ashmore 2002; Dallos and Fakler 2002). This active response of OHCs to stimulation is thought to underlie the amplification that enhances cochlear sensitivity and frequency selectivity. Prestin is the motor protein that drives this motile activity (Oliver et al. 2001; Dallos and Fakler 2002; Liberman et al. 2002).

Freeze–fracture shows that the lateral plasma membrane of the OHC contains a uniquely high density of closely packed, large intramembrane particles covering the entire membrane from the junctional region at the apex to the synaptic area at the cell's base (Gulley and Reese 1977; Forge 1991; Kalinec et al. 1992). These particles are presumed to represent the sites where prestin is incorporated into the membrane.

OHCs are further differentiated in other ways. They possess a specialized smooth endoplasmic reticulum system that is organized parallel to and lies just inside the lateral plasma membrane (Lim 1980; Forge et al. 1993; Slepecky 1996). A Ca^{2+}-ATPase has been localized to this membrane system (Schulte 1993). Between this lateral (or subsurface) cisternal system and the lateral plasma membrane is an organized cytoskeletal framework, the cortical lattice (Holley and Ashmore 1990; Holley et al. 1992; Kalinec et al. 1992). The cortical lattice is comprised of actin filaments that are crosslinked by fodrin and organized in a springlike spiral around the cell body (Holley and Ashmore 1990; Ylikoski et al. 1990). The lattice may confer rigidity to the cell and exert a shape-restoring force when the motor is activated, and is connected to proteins in the plasma membrane by a series of regularly arranged micropillars that cross the subplasmalemal space (Flock et al. 1986; Forge 1991).

In addition to these morphological characteristics that distinguish different hair-cell types there are often more subtle variations within individual populations. In particular, hair cells in the auditory epithelia show systematic variations along the length of the organ that correspond to the place at which sounds of different frequency produce maximal stimulation (the tonotopic organization). For example, OHCs of the organ of Corti increase in length systematically from the basal end of the coil, where high-frequency sounds are detected, to the apical coil where low frequencies produce maximal stimulation (Lim 1980; Dannhof et al. 1991; Pujol et al. 1998). At any one point along the cochlear spiral there is also a radial gradient of increasing OHC length from the first (innermost) row to the outermost row, suggesting that the length of each individual outer hair cell is uniquely specified in relation to its position. Tonotopic variation is also apparent in several other hair-cell features including the number and properties of different calcium channels in the lateral membrane of some auditory hair cells (Rosenblatt et al. 1997; Jagger and Ashmore 1998; Pantelias et al. 2001; Ramanathan and Fuchs 2002; Duncan and Fuchs 2003) and hair-bundle morphology.

2.2 Variations in Hair-Bundle Morphology

Although all vertebrate hair bundles conform to the same common structure, there is considerable variability in their morphology in the different sensory organs and in the different types of hair cell. As illustrated in Figure 6.1, this

A
B
C
D
E
F
G
H
I
J
K

generally concerns the shape of the bundle, the length of the kinocilium, and the number and length of the stereocilia. In the neuromast organs of the lateral line (Fig. 6.1A, B), the rows of stereocilia are arranged in approximately straight lines lying at right angles to the axis of sensitivity, and the kinocilium (which is about 10 μm in length) extends several micrometers above the stereocilia (the longest of which are about 1 μm) that are relatively few in number.

In the inner ear of the newt, there are distinct hair-bundle types in the different sensory patches as well as within the same sensory patch. In the cristae (Fig. 6.1C), the bundle is rounded in shape and the kinocilium and stereocilia at one edge of the bundle are both very long (approximately 60 μm and 35 to 40 μm, respectively), although the kinocilium extends greatly beyond the longest stereocilia. Both the kinocilium and the stereocilia gradually decrease in length toward the center of the organ. In the utricular macula of the newt, the hair cells across the striola (Fig. 6.1D) have stereocilia in a distinct staircase pattern and a kinocilium of roughly the same proportions as the longest stereocilia. In contrast, the bundles on hair cells in the extrastriolar region (Fig. 6.1E) display a kinocilium that is much longer than the stereocilia, and the stereocilia are much shorter than those of the striolar hair cells. The apical surface of the extrastriolar hair cells is also much smaller than that of striolar hair cells.

At least four types of hair-bundle morphology have been described in individual maculae of the vestibular system of mammals (Bagger-Sjoback and Takumida 1988): tall and short hair bundles, each of which may have either tight or loose organization. In tight bundles the stereocilia in each row across the

FIGURE 6.1. Variation in hair-bundle morphology.

(**A, B**) Neuromast of the zebrafish lateral line system. (**A**), Kinocilia extend several micrometers above the apical surface of the epithelium. (**B**) Individual hair bundle viewed from the side, showing relatively short stereocilia in linear rows of increasing height. Scale bar = (A) 2 μm, (B) 1 μm.

(**C–E**) Newt inner ear. (**C**) Crista. The kinocilia and longest row of stereocilia are extremely long but both decrease in height gradually from the periphery toward the center. Scale bar = 10 μm. (**D**) Striolar and (**E**) extrastriolar hair bundles in utricular macula. Extrastriolar hair bundles are shorter than those of the striolar region, the staircase-like arrangement is less pronounced, and the kinocilium extends some distance above the longest stereocilia. Scale bars = 2 μm.

(**F, G**) Utricular macula of mouse. A variety of hair-bundle morphologies is evident. In some hair bundles (**F**), the stereocilia of individual transverse rows are of equal height. In others (**G**), the stereocilia in an individual transverse row vary in height. Scale bar = 2 μm.

(**H, I**) Basilar papilla of chicken. Hair bundles on cell at proximal end (**H**) and distal end (**I**). The bundle at the proximal end has shorter stereocilia and is wider than the bundle from the proximal end. Scale bars = 2 μm.

(**J, K**) Organ of Corti of the guinea pig. (**J**) Outer hair cell; (**K**) inner hair cell viewed from the inner (neural) toward the outer side of the organ of Corti. Scale bars = 2 μm.

bundle (transverse to the axis of sensitivity) are of equal height but the height of each row increases along the line of polarity, as in a typical hair bundle (Fig. 6.1F). In loose bundles, stereocilia in each transverse row increase in height across the bundle creating steps of increasing stereocilial height parallel to the line of polarity (Fig. 6.1G). In the avian basilar papilla the height and number of stereocilia in the hair bundle and the apical surface area of the hair cells vary systematically along the length and across the width of the papilla such that these parameters are different for every hair cell (Tilney et al. 1987, 1988a, b). The shortest and widest hair bundles are at the proximal (basal, high-frequency) end of the papilla (Fig. 6.1H), the longest and narrowest are at the distal (apical, low-frequency) end (Fig. 6.1I), with hair-bundle height ranging from approximately 1.65 μm at the proximal end to approximately 5.5 μm at the distal end (Tilney et al. 1987). A similar phenomenon is observed in the mammalian organ of Corti. On outer hair cells, the bundle of stereocilia forms a W-shape, with the base of the W, where the basal body of the kinocilium is located, directed toward the outer side of the epithelium (Fig. 6.1J).

In most rodent species, there are usually only three rows of stereocilia but the number of rows is often greater in the hair bundles of the human cochlea. The bundles of inner hair cells (Fig. 6.1K) exhibit an open U-shape or straight-line configuration, and the stereocilia of IHCs are noticeably thicker than those of the OHCs. The stereocilia of the OHCs, but not IHCs, like those of the avian basilar papilla, vary in height systematically with position (Lim 1980; Wright et al. 1987). The height of the longest stereocilia increases systematically from the basal (high-frequency) end to the apex (low-frequency region) (from approximately 2 μm to approximately 6 μm). A radial gradient of increasing stereocilial height is seen across the organ of Corti from the first (innermost) to the third row of OHCs. Thus, as with the basilar papilla, the characteristics of the hair bundle of each OHC are unique, indicating that each individual OHC is specified for its position in the organ of Corti.

As yet it is not known how this structural and functional diversity is generated during development. However, many aspects of hair-cell differentiation have been described in detail and the molecular mechanisms underlying some of these processes are beginning to be unraveled. This chapter reviews what is known about the initial events associated with hair-cell differentiation and focuses on the development of the four most striking structural features of hair cells. These are the hair bundle, the cuticular plate, the intercellular junctions, and the basolateral plasma membrane of the highly specialized OHCs. The development of innervation patterns and of the synaptic specialization of hair cells is covered elsewhere in this volume and is not discussed here.

3. The Earliest Signs of Hair-Cell Differentiation

All the factors necessary for the initiation and completion of hair-cell differentiation appear to reside within the presumptive sensory epithelium. In explants of the avian basilar papilla maintained in organotypic culture, hair cells differentiate in the absence of exogenous growth factors (Warchol and Corwin 1996). Innervation is also unnecessary for hair-cell differentiation; when developing sensory epithelia of chicks are explanted, prior to the entry of the relevant axons, to nonneuronal areas, hair cells develop normally (Corwin and Cotanche 1989; Swanson et al. 1990). Apparently normal synaptic specializations in hair cells are also acquired in the absence of ingrowing nerves (Hirokawa 1977).

The presumptive sensory epithelium in which hair and supporting cells differentiate consists initially of an apparently homogeneous population of relatively unspecialized columnar epithelial cells. These cells are all interconnected by gap junctions (Ginzberg and Gilula 1979), appear to be attached to the basement membrane, and express cytokeratins (Kuijpers et al. 1991). Once a hair cell becomes specified these features are lost. The stage at which gap junctional communication is lost in relation to the expression of genes that commit a cell to a hair cell fate is not yet known, but it is likely that down-regulation of connexin genes is one of the earliest events in hair-cell differentiation. This probably allows hair cells and supporting cells to differentiate along separate pathways, and studies of chick embryos suggest that the removal of gap junctions occurs just before the morphological characteristics of hair cells become apparent (Ginzberg and Gilula 1979). The loss of contact between the differentiating hair cell and the extracellular matrix may also be of significance. Signaling between extracellular matrix molecules and cell-surface receptors such as integrins is known to play a crucial role in controlling cell-cycle progression and cell differentiation, and it is possible that the loss of cytokeratin expression by differentiating hair cells is a consequence of a change in cell–substratum interaction. Undifferentiated progenitor cells in the presumptive sensory epithelia round up and detach from the basal lamina as they enter terminal mitosis and exit from the cell cycle, before cell fate is determined (Katayama and Corwin 1993). However, it has been recently shown that many of the cells expressing Math1, a potential molecular marker for hair-cell differentiation, appear to have contact with the underlying extracellular matrix (Chen et al. 2002).

Examination of hair-cell regeneration in the avian basilar papilla has also shown that proteins specific for hair cells in the sensory epithelia, such as myosin VIIa and Tuj1, are present in cells that have basement membrane contact (Stone and Rubel 2000b). Likewise, hair cells with immature hair bundles that retain attachment to the basement membrane have been reported to be present during the reappearance of hair cells observed in the vestibular organs of mammals following gentamicin-induced hair-cell loss (Li and Forge 1997). Loss of contact with the basement membrane may therefore not be necessary for hair-cell differentiation. *Pirouette* and *shaker2* mutant mouse strains, which are deaf and have balance dysfunction, do maintain basement membrane contact and concomitantly show abnormalities in the construction of the actin-based components in hair cells (Beyer et al. 2000). Thus, loss of basement membrane contact may be important for the proper progression of differentiation.

Once specified, a hair cell almost immediately begins to express a number of particular proteins that are, within the sensory epithelium, specific to hair cells. These include calcium binding proteins such as calmodulin and calretinin

(Zheng and Gao 1997), the microtubule protein Tuj1 (Stone and Rubel 2000b), and certain unconventional myosins, including VI and VIIa (Hasson et al. 1997; Self et al. 1999). Immunolabeling for these proteins provides a means for the identification of early differentiating hair cells and some of them have been suggested to be present in the body of the differentiating hair cell prior to hair-bundle formation (Hasson et al. 1997). Although these marker proteins may provide the first indications of hair cells, the elongation of the surface microvilli and the initiation of hair-bundle formation are, however, among the first obvious defining characteristics of hair cells identifiable in the developing sensory epithelia.

4. Hair-Bundle Differentiation

Despite the numerous differences in hair-bundle morphology, the processes involved in the formation of the many different types of hair bundle appear to be broadly similar across organ types and species boundaries. The appearance and maturation of the hair bundle were first described in detail for the chick basilar papilla by Tilney and colleagues (Tilney and DeRosier 1986), and certain aspects of this description are likely to be applicable to the formation of hair bundles in both the auditory and vestibular organs of mammals. Molecules responsible for the process of hair-bundle development are being currently discovered at a rapid rate owing to the analysis of mouse and zebrafish mutants and the study of comparable processes in invertebrate systems.

4.1 Timing and Progression of Hair-Bundle Emergence

4.1.1 Appearance of Hair Bundles in the Vestibular System

Hair bundles are first observed in the vestibular organs of the developing inner ear. A monoclonal antibody to the hair-cell antigen (HCA) identifies hair bundles in the cristae of the chick inner ear at embryonic day 4 (E4) (Bartolami et al. 1991), and hair bundles can be tentatively identified in the saccule and utricle of the mouse inner ear at E12.5 by scanning electron microscopy (SEM), although they are difficult to distinguish from the microvilli on the surrounding supporting cells at this stage (Denman-Johnson and Forge 1999). Hair bundles can be definitively identified by SEM in both the cristae and the maculae of the mouse inner ear at E13.5 (Mbiene and Sans 1986; Denman-Johnson and Forge 1999), and antibodies to cadherin 23, harmonin b, and the mammalian HCA (Ptprq) first recognize hair bundles in the vestibule at this stage of development (Boeda et al. 2002; Goodyear et al. 2003).

4.1.2 Appearance of Hair Bundles in the Auditory System

In the basilar papilla of the chick, hair bundles are first observed in the distal, low-frequency end (equivalent to the apex of the mammalian cochlea) at E6.5, based on studies using either SEM or antibodies to the HCA (Cotanche and Sulik 1983; 1984; Bartolami et al. 1991; Goodyear and Richardson 1997). At this early stage of development, the characteristic staircase architecture is absent and the hair bundle can be discerned from the surrounding supporting-cell microvilli only by the relative thickness of the emerging stereocilia and their concentric packing around a central kinocilium (Cotanche 1987). Transmission electron microscope (TEM) studies have described the presence of hair bundles in the mouse cochlea at E15 (Anniko 1983), and antibodies against cadherin 23 first detect hair bundles in the cochlea at the same stage (Boeda et al. 2002). According to SEM studies, hair bundles first appear in the rat cochlea at E18 (Zine and Romand 1996). At this stage there is a single central kinocilium surrounded by short microvilli of uniform height (Zine and Romand 1996). In the postnatal cochlea of the golden hamster, the microvilli of the emerging hair bundle are distinguishable from the surrounding supporting-cell microvilli only by their short, fat, clustered appearance (Kaltenbach et al. 1994).

4.1.3 Gradients of Hair-Bundle Emergence and Development in Various Organs

In the basilar papilla of the chick, hair-bundle emergence spreads in a distal to proximal direction between E6.5 and E10 (Tilney et al. 1986; Goodyear and Richardson 1997), although some isolated bundles are seen in the proximal region prior to the complete proximal-ward expansion of this distal patch (Goodyear and Richardson 1997). In contrast, the first hair cells to be born in this organ withdraw from the cell cycle between E4 an E5 and are located in a strip extending along the entire superior edge of the basilar papilla. This strip of postmitotic hair cells enlarges across and at both ends of the organ as terminal mitosis proceeds until it incorporates the inferior edge at E8 (Katayama and Corwin 1989). Hair-bundle emergence therefore does not simply initiate as soon as hair cells reach a certain age; if this were so hair cells along the superior edge would be the first to develop hair bundles.

Hair bundles initially appear in the basal end of the mammalian cochlea and appear more apically with time (Kaltenbach et al. 1994; Zine and Romand 1996), in a direction opposite to the wave of hair-cell terminal mitosis that progresses in an apical to basal direction (Ruben 1967; Marowitz and Shugar 1976). As in the bird, hair-bundle formation in the organ of Corti does not initiate in the hair cells that are born first. The appearance of bundles in the mammalian cochlea occurs in a basal to apical direction in contrast to the distal (apical) to proximal (basal) direction observed in birds.

In the developing cristae of the mouse, hair bundles at the apex of the crest are usually more mature than those around its base, as might be expected from the pattern described for hair-cell birth dates in these organs (Sans and Chat 1982), although very immature bundles are seen interspersed throughout both regions at any given stage. Similarly, hair bundles in the maculae are more developmentally advanced in the central striolar portion than in the surrounding extrastriolar region, although immature bundles are seen alongside more mature ones in both regions, suggesting that two or more waves of differentiation may occur at different times in the developing vestibular organs (Denman-Johnson and Forge 1999).

4.2 Development of the Chick Auditory Hair Bundle

Hair bundles vary in terms of stereocilial width, length, and number in the avian basilar papilla and these parameters correlate precisely with their location along the cochlear duct. The development of hair-bundle morphology in the chick follows a sequence that can be separated into three distinct phases (Fig. 6.2).

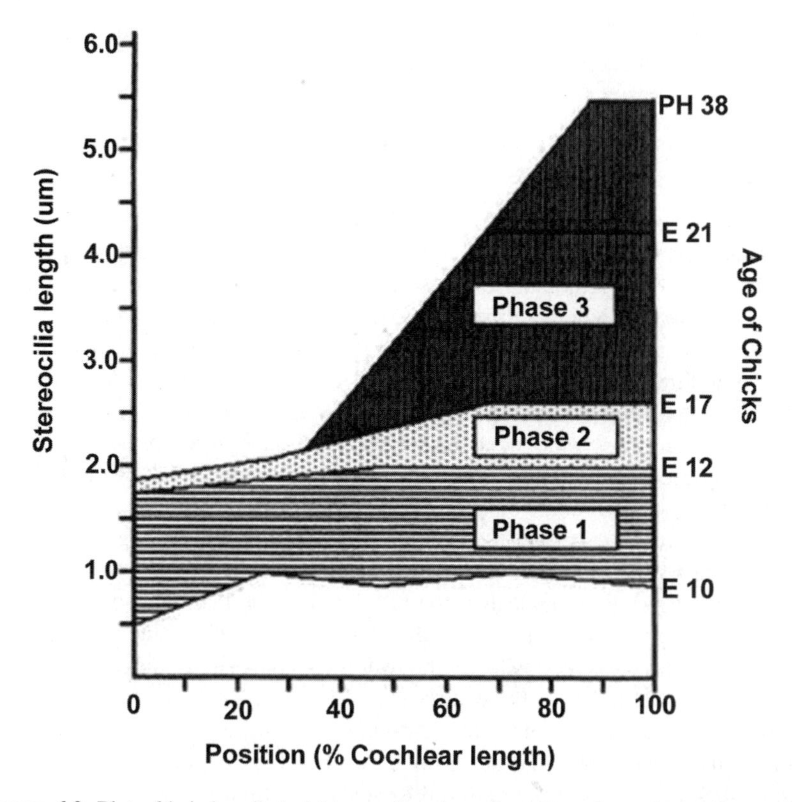

FIGURE 6.2. Plot of hair-bundle height as a function of position along the basilar papilla (cochlea; 0% = proximal, 100% = distal) at different stages of development. The graph illustrates the three different phases of hair-bundle growth observed in the avian papilla. During phase 1 growth is initiated along the full length of the organ, simultaneously generating the stereocilial staircase. Elongation practically ceases during phase 2 as the stereocilia increase in width. In phase 3 the stereocilia that are not at their full height continue their elongation. This is largely seen in the longer, distal hair bundles (> 30%). From Cotanche (1987).

The first phase of hair-bundle development involves changes that create a template for the subsequent fine tuning of hair-bundle shape and size during the subsequent steps. Prior to E9, the kinocilium is located in the center of the hair cell's apical surface and is surrounded by numerous very short microvilli of uniform height. After E9, the kinocilium is found at the periphery of the bundle; this shift in location to the edge of the apical surface specifies the future morphological and functional polarity of the hair bundle (Cotanche and Sulik 1984; Tilney and DeRosier 1986; Tilney et al. 1992). By E10, immature hair bundles, each with an eccentrically located kinocilium, are present along the entire length and breadth of the basilar papilla (Tilney et al. 1992). Following the migration of the kinocilium to the periphery, the microvilli in the rows nearest the kinocilium in each bundle commence their growth, followed successively by those located in rows progressively further and further away. The onset of stereociliary elongation occurs simultaneously in hair bundles along the whole of the organ (Tilney et al. 1986), and delay between the growth-onset time in the different rows creates the staircase pattern of ranked stereocilia in each bundle. The signal for the onset of elongation is unknown, although it has been suggested that it may be Ca^{2+} entry through the transducer channel mediated by tip-link stretch, which in turn is caused by stereociliary elongation and would therefore propagate across the bundle (Tilney et al. 1988). However, studies with FM1-43, a permeant fluorescent blocker of the hair cells' mechanotransducer channel (Gale et al. 2001), indicate that transduction may begin only after the stereocilial staircase has formed in both the chick basilar papilla and the mouse utricle (Geleoc and Holt 2003; (Si et al. 2003). Between E8 and E11, as the hair bundle orientates and elongates, the actin cytoskeleton of the stereocilia becomes increasingly cross-bridged by the actin-bundling proteins fimbrin (Tilney and DeRosier 1986) and espin (Bartles et al. 1996). These proteins provide the hair bundle with the rigidity necessary for effective mechanotransduction (Pack and Slepecky 1995; Zine et al. 1995; Bartles et al. 1996). The tallest row of stereocilia appears to straighten between E10 and E12 (Tilney et al. 1992) as nascent stereocilia are added to the ends of each of the rows generating a semicircular profile that more closely resembles the shape of the adult hair bundle. Extra stereocilia are continually added to the expanding hair-cell surface during the first stage of hair-bundle development, such that by E12 each hair cell has 1.5 to 2.0 times the number of stereocilia than that present on mature hair cells (Tilney et al. 1992).

The second phase of hair-bundle development takes place between E13 and E16. Stereocilia pause in their elongation during this stage and begin to widen owing to an increase in the number of actin filaments per stereocilium (Tilney and DeRosier 1986). At the same time, the actin core of each stereocilium lengthens basally in order to form the rootlets that will anchor the stereocilium in the developing cuticular plate. This is thought to be due to the addition of actin monomers to the proximal, nonpreferred, pointed ends of the few filaments that pass through the constricting taper developing at the stereociliary base (Tilney and DeRosier 1986; DeRosier and Tilney 1989). Bundle maturation continues with the resorption of excess stereocilia at the foot of the stereociliary staircase. This not only creates the mature rectangular-shaped hair bundle but also produces the reduced final number of stereocilia per hair cell (Tilney et al. 1992).

The third and final phase of bundle development is initiated at E17 and continues for some weeks after hatching. Stereocilia, particularly in the distal end of the cochlear duct, begin to elongate again. This occurs synchronously in all rows, but stops first in the shortest row and last in the longest thus exaggerating the difference in height ranking between the rows. Further stereociliary elongation occurring during the third and final phase of development creates a basilar papilla with proximal bundles that have numerous short stereocilia and distal bundles that have fewer but taller stereocilia (100 to 140 stereocilia per bundle in the proximal end and 40 to 50 stereocilia per bundle in the distal end [Cotanche and Sulik 1984]).

It has been suggested that stereocilia elongation during the first and third stages of hair-bundle development in the chick basilar papilla occur via the addition of actin monomers to the distal barbed end of the actin filaments in the stereocilia (Tilney et al. 1988, 1992). Although there is as yet no direct evidence for this, transfection of GFP-actin into hair cells of the postnatal rat cochlea (P5 to P12) indicates that new actin is continuously incorporated at the tip of the stereocilium indicating that stereocilia grow at their tips (Schneider et al. 2002).

4.3 Development of the Mammalian Hair Bundles

Information on the development of hair bundles in the mammalian cochlea is relatively sparse compared to data available for the avian auditory organ. As mentioned previously, the mammalian hair bundle develops following the pattern that is broadly, but not exactly, similar to that described above for the chick basilar papilla. In mammals, as in the chick, the longest hair bundles are present in the apex of the cochlea. In the chick basilar papilla, a constant growth rate and variation in the duration of growth, with shorter bundles ending their elongation first, creates differences in bundle heights. In contrast, studies on the postnatal hamster show that the growth rate is not constant along the organ of Corti, and that the taller apical hair bundles grow faster than their shorter basal counterparts. Growth of the bundles initiates in the basal coil (unlike in the basilar papilla where hair-bundle growth begins simultaneously along the entire length of the organ), and the elongation and widening of stereocilia occurs concurrently (Kaltenbach et al. 1994). In all mammalian cochleae studied so far, hair-bundle development in the apical coil lags appreciably behind that in the basal coil (mouse: Nishida et al. 1998; rat: Lenoir et al. 1987; Zine and Romand 1996; human: Lavigne-Rebillard and Pujol 1986). Whether the other trends described in the hamster are applicable to mammalian auditory development in general is not known.

In the rat, movement of the kinocilium to the bundle periphery occurs at approximately E18 and stereocilial growth first begins at E20 in the basal-coil IHCs. Complete development of all hair bundles in the rat cochlea was observed at postnatal day 6 (P6) (Zine and Romand 1996). A time lag between the development of hair bundles on IHCs (fully mature at P1) and those on OHCs (mature at between P5 and P7) was also observed in the mouse cochlea, with bundles in the first row of OHCs maturing more quickly than those in the second row, which in turn developed faster than the bundles of OHCs in the third row (furthest from the IHCs and limbus) (Anniko 1983). The loss of the kinocilium by P14 in the mouse gives the stereocilial bundles their adult morphology (Lim and Rueda 1990). The relative maturity of the IHC bundle in comparison with that of the OHC bundle in the developing cochlea has also been observed in the human fetus (Lavigne-Rebillard and Pujol 1986).

Two studies have focused on the morphological characterization of hair-bundle development in the mouse vestibular system (Mbiene and Sans 1986; Denman-Johnson and Forge 1999). Once the nascent stereocilia have emerged around a central kinocilium of similar height, the subsequent maturation is extremely rapid. Hair bundles are first seen at E13.5 in the mouse cristae. By E14.5, they have begun to develop the height differences characteristic of the stereociliary staircase (Mbiene and Sans 1986). In the maculae, a variety of hair-bundle morphologies are already visible by E13.5, 1 day after hair bundles can first be potentially identified. These include a single cilium (Fig. 6.3A, a), a small bundle of microvilli with a central kinocilium of nearly equal height (Fig. 6.3A,b), an unranked bundle with a peripheral kinocilium (Fig. 6.3B), and finally a bundle with some height-ranked stereocilia and a peripheral kinocilium (Fig. 6.3C). By E15.5 hair bundles displaying distinctive staircase morphology with numerous interstereocilial links can be seen, in addition to more immature bundles similar to those encountered at E13.5 (Denman-Johnson and Forge 1999). The rapidity with which the hair bundles mature from relatively undifferentiated structures into the final mature architecture is a noticeable feature of hair-bundle development in the murine vestibular system.

4.4 Development of Hair-Bundle Links

There are four morphologically distinct link types present on certain types of hair bundles such as those in the auditory papilla of the bird: tip-links, horizontal top connectors, shaft connectors, and ankle links (Goodyear and Richardson 1992; Goodyear et al. 2005). These links are morphologically and immunologically distinct, and can also be operationally distinguished on the basis of their relative sensitivities to the calcium chelator BAPTA (bis-aminophenoxyethane-tetraacetic acid) and the protease subtilisin (Goodyear and Richardson 1999, 2003). The tip links are thought to gate the hair cell's mechanotransducer channel, and the other link types are assumed to hold the stereocilia together as a coherent unit. Immunofluorescence studies with antibodies to antigens associated with three of these link types—tip-links, ankle links, and shaft connectors—

FIGURE 6.3. Scanning electron micrographs illustrating the initial stages of hair-bundle differentiation observed in the mouse vestibule at E13.5. The hair bundle emerges as a single cilium surrounded by numerous, tiny stereocilia (**A**, a). The bundle develops to produce stereocilia of equal height located around a centrally located, taller kinocilium (**A**, b). The kinocilium (*arrow*) relocates to the periphery of the hair bundle (**B**). The stereocilia closest to the kinocilium (*arrow*) elongate (**C**). Scale bars = 1 μm. From Denman-Johnson and Forge (1999).

suggest that the proteins that form these links are expressed at high levels during the very early stages of hair-bundle development. The HCA, a component of shaft connectors, is expressed at E6.5 in papillae, concomitant with the emergence of hair bundles (Bartolami et al. 1991), and the tip-link antigen (TLA) and ankle-link antigen (ALA) are expressed shortly after, by E7 (Goodyear and Richardson 1999, 2003). The TLA and ALA appear to be uniformly distributed over the surface of these immature hair bundles and the distribution of each antigen becomes progressively restricted as development proceeds, with the ALA becoming restricted to the bundle base and the TLA becoming localized to the tip of the hair bundle (Goodyear and Richardson 1999, 2003). Tip-links have been reported in the chick papilla at E9 (Pickles et al. 1991), in the mouse macula at E15.5 (Denman-Johnson and Forge 1999), on rat inner hair cells at birth (P0) (Zine and Romand 1996), on gerbil outer hair cells at P2 (Souter et al. 1995), and on cultured mouse cochlear outer hair cells at the equivalent of P3 (Furness et al. 1989). In chick and mouse, tip links appear to emerge from arrays of multiple fine horizontal filaments that radiate from the tip of each stereocilium in all directions to make contact with their adjacent neighboring stereocilia (Pickles et al. 1991, Forge et al. 1997) (see Fig. 6.4). A pruning process appears to eliminate all but one of these links from these arrays as the stereocilia begin to elongate, leaving a single tip-link running along the hair-

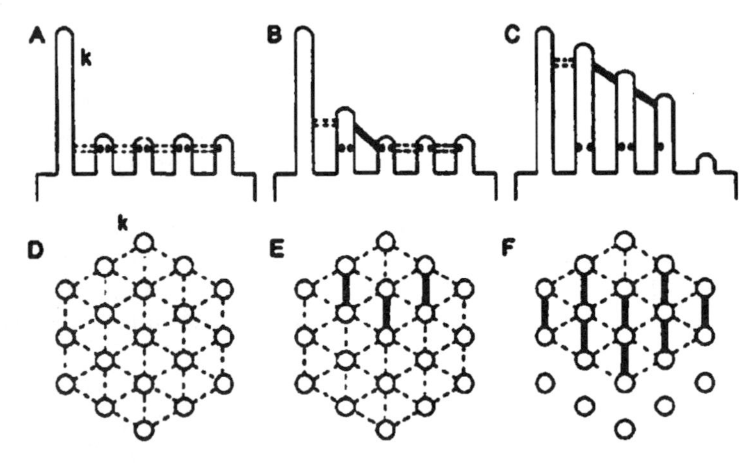

FIGURE 6.4. Tip-links emerge from arrays of multiple horizontal filaments that radiate from the tip of each stereocilium in all directions to make contact with neighboring stereocilia. As the stereocilia begin to elongate the extraneous links are lost, leaving a single tip link running along the hair bundle's axis of mechanosensitivity and projecting upward to the side of adjacent, taller stereocilium. From Pickles et al. (1991).

bundle's axis of mechanosensitivity and projecting upwards at an angle from the tip of each stereocilium to the side of an adjacent stereocilium. Cadherin 23, the ALA, and the TLA are all concentrated at the tips of developing hair bundles (Goodyear and Richardson 1999, 2003; Boeda et al. 2002) and may therefore be components of these spoke like arrays from which tip links emerge. The morphological development of the other link types has yet to be described in detail.

4.5 Development of Planar Polarity

As outlined previously, hair cells are polarized epithelial cells that have an apical membrane specialized for the detection of mechanical stimuli, and a basolateral membrane specialized for a number of tasks including the shaping of the hair-cell's receptor potential, the release of neurotransmitter onto the contacting afferent nerve fiber, and other more specialized functions such as the somatic electromotility exhibited by cochlear OHCs. The hair bundle itself is also a morphologically and functionally polarized structure, designed such that stimuli along one axis, that running parallel to the hair bundles axis of mirror symmetry, opens or closes mechanotransducer channels. For the different sensory organs to function as they do, the hair bundles must also have a coordinated orientation with respect to one another and with the epithelium as a whole. This level of organization is referred to as planar polarity. In its simplest terms planar polarity means that the hair bundles in the organ all have their axis of mechanosensitivity aligned in the same direction. However, each type of sensory organ in the inner ear has an individual and precise pattern of planar polarity. In the mammalian and avian cochleae, the kinocilium is always located closest to the abneural edge of the organ so that the bundle faces away from the neural edge. In the maculae of the vestibule this pattern is complicated by the fact that hair bundles have different polarities on either side of the epithelium. Polarity reverses around a line known as the striola that roughly bisects each macula into two halves (see Section 2.1). The hair bundles lying either side of the striola are either orientated toward this strip, as in the utricle, or away from it, as in the saccule. The initial movement of the central kinocilium to the edge of the developing hair bundle is partially responsible for establishing the planar polarity.

In the avian basilar papilla, the kinocilium moves to the periphery of the hair bundle by E9. At this stage, hair bundles exhibit a broad but unimodal distribution of orientations centered around 200° (as judged by the position of their kinocilium, where 0° denotes a bundle facing the neural, superior edge of the papilla and 180° denotes a bundle facing the abneural, inferior edge of the papilla), with individual bundle orientations ranging from 60° to 340° (Fig. 6.5). By posthatch day 3 all bundles on the organ are tightly orientated toward the 200° point, implying that the hair bundle, or the entire hair cell, must rotate after the initial migration of the kinocilium to one point around the cell surface (Cotanche and Corwin 1991). Although this rotation was not observed directly, SEM analysis at successive time points indicate that papillar bundles acquire correct planar polarity with proximal bundles reorienting fairly rapidly (by E11, Fig. 5) and distal bundles much more slowly (by posthatch day 3) (Cotanche and Corwin 1991). It was suggested that the differential growth of layers of matrix within the overlying tectorial membrane could provide a traction force that the kinocilium uses to align the hair bundles in the developing basilar papilla (Cotanche and Corwin 1991). However, differences in sites of otoconial membrane production that could account for the opposing hair-bundle polarities seen in the maculae of the vestibular organs of the mouse have not been observed (Denman-Johnson and Forge 1999).

Although it was originally suggested that the movement of the kinocilium from the center to the periphery of the cell surface in the mammalian cochlea was precise from the outset and dictated the final planar polarity of the hair cell (Kaltenbach et al. 1994; Zine and Romand 1996), recent work (Dabdoub et al. 2003) has shown otherwise. The position of the kinocilium initiates the formation of the hair bundle's morphological asymmetry (by being at one edge of the bundle where the tallest stereocilia grow to form the staircase), but the hair bundle is not, at first, precisely oriented with respect to the cochlea as a whole. At E17, hair-bundle orientation in the base of the cochlea deviates from 0° (facing the abneural edge) by as much as 15° for IHCs and first-row OHCs, and by up to 30° for second and third row OHCs. Variation in hair-bundle orientation diminishes over time so that by P10 the hair bundles of basal IHCs deviate by 2°, those of first row OHCs by 3°, and those of second and third row OHCs

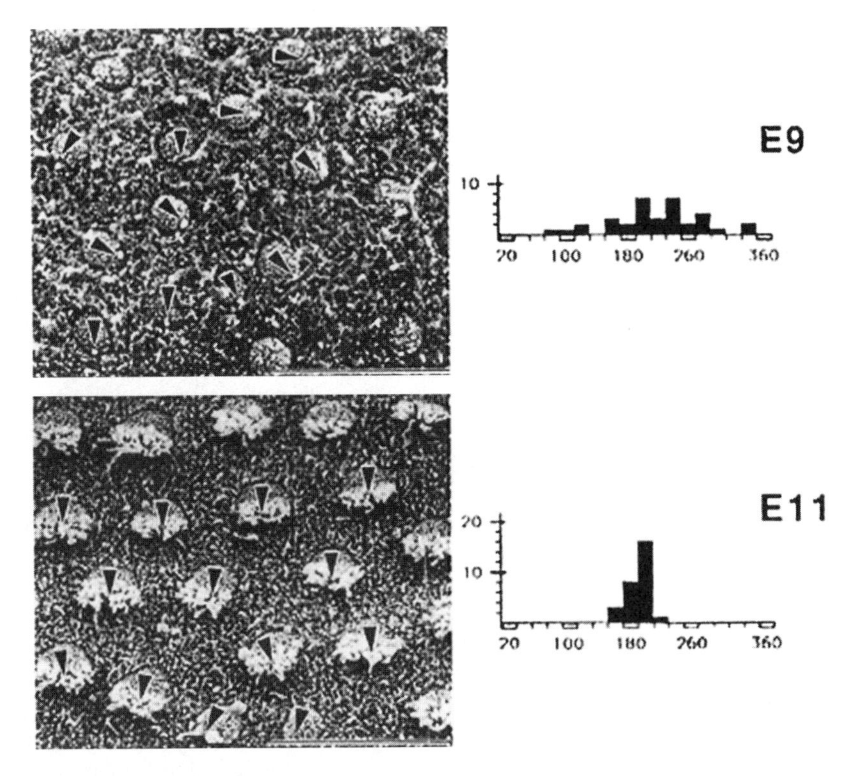

FIGURE 6.5. Scanning electron micrographs (*right*) and bar charts (*left*) illustrating how hair bundles reorient during development of the avian basilar papilla. Hair bundles in the proximal end of the papilla show a broad distribution of polarities around a mean orientation of 240° at E9. By E11 these hair bundles reorient so that their polarity is more precisely defined with respect to their neighbors and with respect to the organ as a whole. Scale bars = 10 μm. From Cotanche and Corwin (1991).

by between 2° and 5°. The time delay observed in reorientation for the hair bundles of the second- and third-row OHCs compared to those of IHCs is accompanied by a basal-to-apical gradient of maturation in planar polarity. Hair bundles in the apex exhibit greater deviation than their counterparts in the basal coil at all stages (Dabdoub et al. 2003).

In a comprehensive study of utricular and saccular development in the mammalian inner ear, it was also found that hair bundles are planar polarized to a certain degree from the very outset. At E13.5, the orientation of the hair bundles on neighboring cells is similar and varies systematically across the epithelium but not in as precise a manner as that observed in the adult. A very sharp reversal of hair-bundle polarities with respect to the striola becomes apparent by E15.5 (Fig. 6.6) (Denman-Johnson and Forge 1999). Together these studies

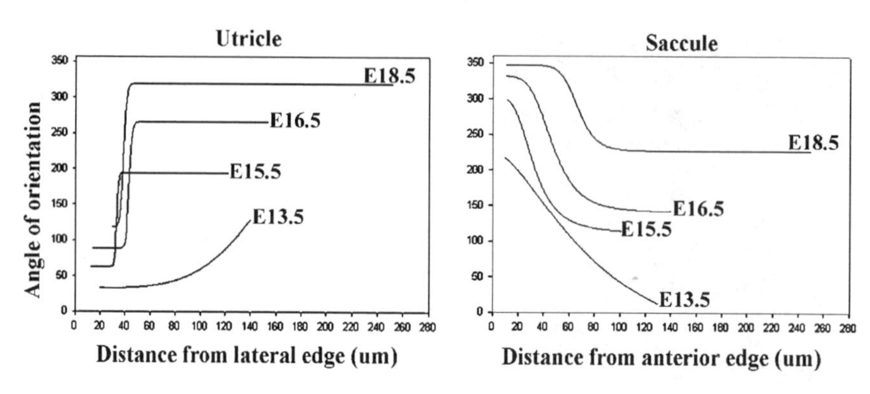

FIGURE 6.6. Plots of hair-bundle orientation angles as a function of distance from the lateral and anterior edges of the utricle and saccule respectively. At E13.5 hair-bundle orientation already varies systematically across the epithelium. By E15.5 a very abrupt reversal in hair-bundle polarity becomes apparent with respect to the striola in both the utricle and the saccule. From Denman-Johnson and Forge (1999).

in birds and mammals show that reorientation of the hair bundle after the initial movement of the kinocilium to the periphery of the cell surface is a common step in the development of the hair bundle in both birds and mammals.

The question of how cells within sensory epithelia communicate to coordinate polarities may be answered by looking at sensory bristles in the fruit fly *Drosophila melanogaster*. The bristle of the shaft cell, like the hair bundle, is part of a mechanotransducer complex and is comprised of densely packed actin filaments. Studies of the bristle-shaft cell have isolated a group of genes that, when mutated, disrupt planar polarity. In addition to this, the protein products of these genes are characterized by asymmetric localizations and are either transmembrane or membrane associated, suggesting the signal is delivered from one pole of the cell at a unidirectional cell–cell junction. These protein products are called Frizzled, Dishevelled, Flamingo, Dachsous, Strabismus (also known as Van Gogh), and Prickle (Adler 2002; Lewis and Davies 2002; Strutt et al. 2002).

Frizzled is a seven-pass transmembrane receptor for the Wnt family of signaling molecules. Dishevelled is an intracellular protein that contains a PDZ domain, a domain thought to act as a binding scaffold for large protein complexes (see Section 4.6.4). Flamingo is a cadherin family member with an ectodomain containing cadherin, epidermal growth factor (EGF), and laminin AG repeats, a seven-pass transmembrane region, and an intracellular domain with a putative G-protein binding motif. Dachsous is also a member of the cadherin superfamily, but has a single transmembrane domain and 27 extracellular cadherin repeats. Strabismus has four transmembrane regions and a PDZ binding interface, suggesting that it may bind to Dishevelled. Prickle has three LIM protein interaction domains. It is thought that activated Frizzled recruits Dishevelled from the cytoplasm to the membrane in order to transmit a polarization signal from the distal edge of a bristle shaft cell to the proximal edge of the next cell. Signal transmission could take place if the products of bristle building genes were activated at the distal side of the junction and inhibited at the proximal side, and at least four of the proteins described above are located at the distal/proximal boundary of adjacent bristle cells. Dishevelled and Frizzled are located on the distal side of this junction while Prickle and Strabismus are found on the proximal side (see Lewis and Davies 2002 for review).

Three mouse mutants have been recently described that have planar polarity defects in the organ of Corti. Heterozygous *spin cycle* and *crash* mutant mice display head shaking behavior and have misoriented hair bundles in the cochlea. This planar polarity defect is restricted to the OHCs and is evident from as early as E16.5. These mutations are due to point mutations in the *Celsr1* gene, a mammalian homolog of the *Drosophila* planar polarity gene *flamingo* (Curtin et al. 2003). The *looptail* mouse has a mutation in *Vangl2*, the mammalian homolog of *Strabismus* (Kibar et al. 2001; Murdoch et al. 2003). In homozygous *looptail* mice, the hair bundles of both IHCs and OHCs are misoriented (Montcouquiol et al. 2003). This indicates that *Drosophila* planar cell polarity genes are conserved in the vertebrate inner ear. The specific roles played by *Vangl2* and *Celsr1* in the cochlea, however, remain to be characterized.

In addition to these planar polarity cassette genes, soluble wingless type signaling molecules (Wnts) may play a role in orienting the hair bundles in the inner ear (Dabdoub et al. 2003). During normal development in the mouse cochlea, hair bundles gradually acquire their final precise orientation over a period of 12 days from E17 to P10, and the hair bundles of the second- and third-row OHCs do not reorient with the same efficacy as those of the IHCs and the first row OHCs (see above). *Wnt7a* is expressed by the developing pillar cells that lie between the IHCs and OHCs, and may be a morphogen that controls hair-bundle orientation. The absence or overexpression of Wnt7a protein, however, resulted in OHC planar polarity being disrupted in a complex manner (e.g., the deviation of bundles increased in the base but not the apex), one that was not entirely consistent with the pillar cells acting as a simple line source for a polarity morphogen (Dabdoub et al. 2003). Wnt proteins may activate the asymmetrically localized receptor protein Frizzled, which in turn would recruit downstream proteins such as Dishevelled promoting movement of the hair bundle to a particular orientation in one cell while simultaneously recruiting inhibitory proteins in neighboring cells to deflect the bundle from the incorrect orientation.

It remains to be determined how the orientation signals encoded by the planar cell polarity genes could be transduced in the inner ear. For the hair bundle to reorient, cytoskeletal rearrangements are required. Rho GTPases are part of the Ras superfamily of small G-proteins that alternate between a GTP-bound, active form and a GDP-bound, inactive form and are key regulators of actin assembly. RhoA lies downstream of *Dishevelled* in the noncanonical Wnt Jnk pathway and in the fly, disruption of the *RhoA* gene results in the misorientation of wing hairs (Strutt et al. 1997). The JNK cascade would be responsible for the recruitment of further downstream effectors of tissue polarity (Heisenberg and Tada 2002). Rho GTPases appear to have many functions in the inner ear and their roles are discussed in a more general sense later in this chapter (see Section 4.6.14).

4.6 Molecules Required for Hair-Bundle Development

A number of molecules are now known to be required for the development and maintenance of hair bundle structure. These include the cytoskeletal proteins that form the rigid core of the stereocilium, the molecular motors that are associated with the actin core and its ensheathing plasma membrane, the proteins of the stereociliary membrane, and the proteins that may act as scaffolds for various multiprotein complexes within the hair bundle. Potential interactions between these various components of the hair bundle are summarised diagrammatically in Figure 6.7.

4.6.1 Cytoskeletal Core Protein—Actin

The stereocilia of the hair bundle are comprised of hexagonally packed actin filaments. The growth in stereocilial length that occurs during development is assumed to occur by addition of actin monomers to the preferred (barbed) end of the filament at the tip of the stereocilium (Tilney and DeRosier 1986). Although this has yet to be demonstrated directly for the very early stages of hair-bundle development, there is now conclusive evidence that there is a constant turnover of actin filaments in the hair bundles of the rat cochlea between P5 and P10 and that this occurs via the addition of new monomers at the distal tip of the stereocilium (Schneider et al. 2002; Rzadzinska et al. 2004). The growth in stereocilial width observed during development is achieved by the incorporation of additional actin filaments into the packed array. In the chick basilar papilla at E8, there are only 25 to 50 actin filaments per stereocilium. These are not tightly packed and considerable space can be seen between the individual filaments. By E9 this space reduces and the actin core becomes more tightly packed until, by E10.5, the filaments are hexagonally arranged, as in the adult. The number of actin filaments per stereocilium increases over the next 6 days, and approximately 250 actin filaments are finally found in each stereocilium (Tilney and DeRosier 1986). In the *tasmanian-devil* mouse mutant the hair bundles exhibit thin stereocilia from P0 onward (Erven et al. 2002). The mutation responsible for the abnormality of these stereocilia has not yet been identified, but the absence of normal width expansion suggests that actin filaments are not accumulating or bundling normally within the stereocilia.

4.6.2 Cytoskeletal Core Protein—Fimbrin

Fimbrin has been characterized as an actin filament bundling protein of the vestibule and the cochlea via immunohistochemistry (guinea pig: Flock et al. 1982; mouse: Slepecky and Chamberlain 1985; chick: Tilney et al. 1989; gerbil

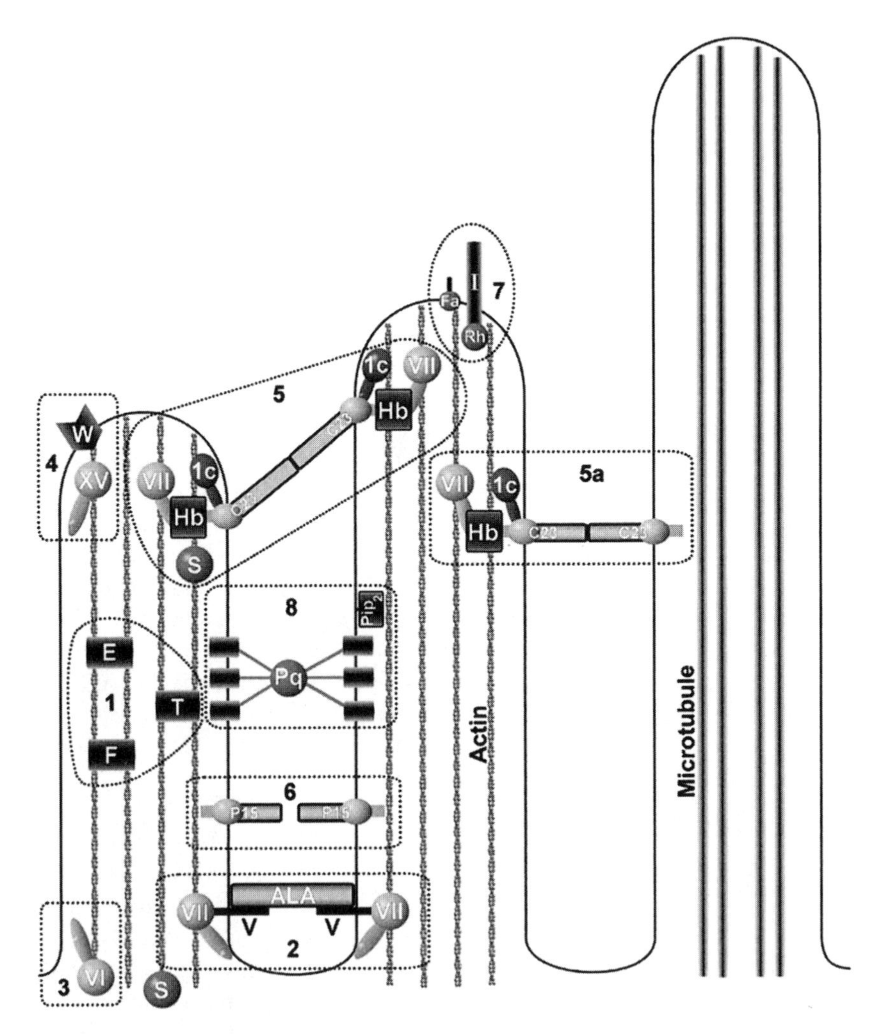

FIGURE 6.7. Diagram of possible molecular interactions responsible for the development and maintenance of the hair bundle. (*1*) Espin (E), fimbrin (F), and T-plastin (T) are responsible for actin bundling at different times during development and maturation. (*2*) Myosin VIIa (VII), vezatin (V), and the ankle link antigen (ALA) may interact to form the ankle link. (*3*) Myosin VI (VI) may anchor the membrane at the base of each stereocilium to the cuticular plate. (*4*) Whirlin (w) and myosin XV (XV) may interact at the tip of the stereocilium to regulate and maintain stereocilial length. (*5*, *5a*) Cadherin 23 (C23), harmonin b (Hb), and myosin VIIa may interact to form transient lateral links, kinocilial links, and even the tip-link complex. (*6*) Protocadherin 15 (P15) may be part of an as yet uncharacterized stereociliary connector protein complex. (*7*) Integrins (I) recruit FAK (focal adhesion kinase) to the stereocilial surface for cytoskeletal maintenance and also trigger downstream signaling molecules such as Rho-GTPases. (*8*) Ptprq (Pq) is a component of the shaft connectors.

and guinea pig: Pack and Slepecky 1995) and immunoblotting of stereocilial preparations (Tilney et al. 1989). The fimbrin family includes the plastin isoforms referred to as I-plastin, T-plastin, and L-plastin. The fimbrin molecule has two structural and functional domains, a 12 kDa N-terminal regulatory domain (headpiece) and a 58-kDa C-terminal actin-crosslinking fragment (core), which has two repeated actin binding domains (ABDs). The headpiece controls actin crosslinking activity via its calcium binding domains (Hanein et al. 1997). The classic fimbrin found in hair-cell stereocilia is the chicken homolog of human I-plastin (Lin et al. 1994), referred to here as fimbrin.

Fimbrin and actin are first observed in the basal-coil IHCs of the rat cochlea at E18 (Zine and Romand 1996), coincident with when hair bundles are first observed using SEM (Zine et al. 1995), suggesting fimbrin plays a role as an actin crosslinker during early stages of hair-bundle development (Fig. 6.7, box 1). Expression of fimbrin persists in the adult cochlea (Zine et al. 1995). Recent work has shown that a second fimbrin isoform, T-plastin (Fig. 6.7, box 1), is expressed transiently in the developing hair bundles of the rat cochlea (Daudet and Lebart 2002). T-plastin is found at differing intensities throughout the rat organ of Corti at P0, the earliest age studied. The transient T-plastin expression in these cells suggests that this protein plays a role in the development of hair bundles rather than being required for their maintenance in the adult. The basal-to-apical sequence of disappearance of the protein is similar to the known basal-to-apical wave of hair-bundle maturation. By P60, T-plastin immunoreactivity is only seen in supporting-cell microvilli (Daudet and Lebart 2002).

4.6.3 Cytoskeletal Core Protein—Espin

The actin bundling protein espin is found in the hair-cell stereocilia (Fig. 6.7, box 1) and to a lesser extent in the cell body and the cuticular plate. In the chick, espin is first detected in the stereocilia of the vestibular organs at E8, and in the auditory hair bundles at E8.5 (Li et al. 2004). In auditory hair cells, espin expression therefore begins just before the onset of stereocilial elongation. Espin has been characterized as a high-affinity, Ca^{2+}-insensitive, actin-bundling protein (Bartles et al. 1996), and recent studies have indicated that it may play a role in maintaining the fixed, steady-state length of each stereocilium. In a transfected kidney cell-line expressing variable amounts of GFP-espin, the length of microvilli was directly proportional to the amount of espin expressed (Loomis et al. 2003). In the adult rat cochlea, espin levels vary along an apical-to-basal gradient that mimics the increase in stereocilial (and therefore actin filament) length that occurs along the cochlea (Loomis et al. 2003). Espin's function can therefore be likened to the rate-determining step in a chemical reaction; the amount of espin contained within a stereocilium reflects the length at which its actin core can be maintained. While the stereocilium continues to treadmill at the correct length, sufficient espin will be produced to support it along its length, but any attempt to become longer will not be supported by crosslinking and the hair bundle will therefore collapse. In *jerker* mice that are deficient for espin, however, disorganized hair bundles are not observed until P11 (L. Zheng et al. 2000). There may therefore be some redundancy in the function of fimbrin and espin.

4.6.4 Molecular Motor—Myosin VIIA

Myosins are motor proteins that produce force or move along actin filaments using the energy derived from the hydrolysis of ATP. In the mature mammalian auditory hair bundle, the unconventional myosin, myosin VIIa, is uniformly distributed along the stereocilia; it is also present in the cuticular plate, pericuticular necklace (the vesicle-rich zone at the apex of the hair cell) and cell bodies of cochlear inner and outer hair cells (Hasson et al. 1997). In frog saccular hair cells, myosin VIIa is concentrated in a band around the base of the hair bundle, in the region where the ankle links are located. Its distribution in the hair bundle suggests that myosin VIIa may anchor the connectors between stereocilia that maintain cohesion of the hair bundle. Myosin VIIa is expressed from the earliest stages of hair-cell differentiation (Self et al. 1998). In shaker1 mice that lack functional myosin VIIa, disorganized hair bundles are observed as early as E18 (Self et al. 1998), possibly because the integrity of the hair bundle cannot be maintained if the interstereociliary links are not anchored to the actin core of the stereocilium (Steel and Brown 1996). Vezatin, a ubiquitous component of the adherens junction, interacts with myosin VIIa and localizes to the base of the hair bundle in the mouse and chick inner ear (Fig. 6.7, box 2). In the hair bundles of the chick, vezatin colocalizes with the ankle link antigen, a protein antigen associated with the ankle links (Küssel-Andermann et al. 2000). In the mammalian cochlea, the expression of vezatin in the hair bundles and the presence of the ankle links are transient. By P30, vezatin is almost undetectable at the base of the hair bundle, suggesting that its interaction with myosin VIIa ensures the maintenance of bundle shape during development and is not required following maturation.

In addition to tensioning links (Kros et al. 2002), myosin VIIa may act as a molecular transporter during hair-cell development. The ototoxic aminoglycoside antibiotics most likely enter hair cells via the transduction channel (Gale et al. 2001) and the accumulation of aminoglycosides in sensory hair cells does not occur in myosin VIIA mutant mice (Richardson et al. 1997). The transducer channel may therefore be transported to the hair cell's apical membrane and up the stereocilial actin filaments to its functionally optimal position by myosin VIIa (Richardson et al. 1997, 1999).

4.6.5 Molecular Motor—Myosin VI

Myosin VI is unusual as it moves toward the nonpreferred, minus end of actin filaments in the opposite direction to the all other myosins (Wells et al. 1999). Immunofluorescence studies indicate that myosin VI is found along the entire length of frog saccular hair bundles and also at high concentrations in the rootlets, cuticular plate, and pericuticular necklace (Hasson et al. 1997). Mammalian cochlear and vestibular stereocilia do not express high levels of myosin VI; instead it is concentrated at the cuticular plate and to a lesser extent in the soma of the hair cell (Avraham et al. 1995; Hasson et al. 1997). Myosin VI can be detected in the mouse organ of Corti from E13.5 (Xiang et al. 1998). The production of giant stereocilia in *Snell's waltzer* mutant mice suggests that myosin VI may anchor the membrane at the base of each stereocilium to the cuticular plate (Fig. 6.7, box 3), and tension the stereociliary rootlets, thus ensuring the maintenance of hair-bundle integrity (Self et al. 1999; Cramer 2000). Studies in polarised epithelial cell lines have indicated that myosin VI binds to nascent endocytotic vesicles to transport them in the negative, nonpreferred direction along actin filaments away from the periphery of the cell (Buss et al. 2002; Aschenbrenner et al. 2003). Endocytotic vesicles are present in high numbers at the apical pole of the developing hair cell (Forge and Richardson 1993), so myosin VI may also provide the force for membrane remodeling during development of the hair cell's apical surface.

4.6.6 Molecular Motor—Myosin XV

Myosin XVa is the product of the *shaker2* locus and *Shaker2* mice have extremely short stereocilia. Myosin XVa (Fig. 6.7, box 4) is located at the extreme tips of the stereocilia in the cochlear and vestibular hair bundles of the mouse, rat, and guinea pig, lying under the apical membrane and overlapping with the barbed ends of the actin filaments (Belyantseva et al. 2003a, b; Rzadzinska et al. 2004). In *shaker2* mice that have mutations in the motor or tail domains of myosin XVa, the protein is mislocalized to the base of the stereocilium, indicating both the motor and tail are required for its correct placement at the tip (Belyantseva et al. 2003a). Myosin XVa is also found in the cuticular plate and cell body of hair cells in the mammalian vestibular and auditory organs (Anderson et al. 2000). In the mouse, myosin XVa is first seen at E14.5 in the vestibular organs. It is seen from E18.5 in the basal turn of the cochlea, and expression spreads in an apical direction in a manner that corresponds with the basal-to-apical gradient of hair-bundle development seen in mammals (Belyantseva et al. 2003a).

Actin turnover rates within the hair bundle depend on the length of each individual stereocilium, with longer stereocilia within the bundle treadmilling their actin cores faster than shorter ones (Rzadzinska et al. 2004). Correspondingly, hair bundles in the apex of the cochlea experience greater actin turnover than those in the basal turn of the cochlea. Myosin XVa immunoreactivity in the tip of the stereocilium depends on its position in the hair bundle. The tallest stereocilia at the back of the bundle have a greater concentration of this protein at their tips than the shorter stereocilia in the rows in front of them, indicating that the level of myosin XVa in the tip is proportional to the actin incorporation rate, which is in turn proportional to the length of the stereocilium within the hair bundle. Myosin XVa may therefore also regulate stereociliary growth during development, a supposition supported by the presence of abnormally short stereocilia in *shaker2* mice with mutations in myosin XVa.

4.6.7 Cell Membrane Protein—Cadherin 23

Cadherins mediate cell–cell adhesion in a Ca^{2+}-dependent manner. Defects in cadherin 23 are responsible for USH1D syndromic deafness in humans (Bork et al. 2001) and inner ear disorders in the mutant mouse strain *waltzer* (Di Palma et al. 2001). Cadherin 23 has 27 cadherin repeats in its ectodomain, a single-pass transmembrane region, and an intracellular domain that, unlike that of classical cadherins, is not predicted to interact with β-catenin. Recent studies (Siemens et al. 2004; Sollner et al. 2004) have provided evidence that cadherin 23 may be a component of the both the tip link and the kinocilial link, and that it interacts with myosin Ic (Fig. 6.7, box 5). Immunofluorescence studies using an antibody to the ectodomain of cadherin 23 have indicated that it is initially located over the entire surface of the hair bundle as it emerges, and that it becomes concentrated at its apical tip during subsequent development (Boeda et al. 2002). The hair bundles of *waltzer* mice are disorganized and fragmented into several smaller clusters of stereocilia. They resemble those seen in shaker1 mice with mutations in myosin VIIA, and it has been suggested that cadherin 23 may play a cohesive role in maintaining hair-bundle integrity during development (Boeda et al. 2002).

4.6.8 Cell Membrane Protein—Pcdh15

Protocadherin 15 (Pcdh15) is also a member of the cadherin superfamily of cell–cell adhesion molecules. It has eleven cadherin repeats, a single-pass transmembrane region, and a proline rich intracellular domain (Angst et al. 2001). Defects in *Pcdh15* are responsible for USH1F syndromic deafness in humans (Ahmed et al. 2001) and underlie the symptoms observed in the *Ames waltzer* mutant mouse strain (Alagramam et al. 2001). In the cochleae of wild type mice at P16 to P21, Pcdh15 is distributed along the entire shaft of the stereocilium (Fig. 6.7, box 6). In the utricle and ampullae at P15, the longer stereocilia exhibited brighter staining than their shorter counterparts in the bundle, a trend that was indicative of the organ as a whole (Ahmed et al. 2003). In *Ames waltzer* mice by P16 (the earliest stage studied), the amount of actin present in the stereocilia and cuticular plate of IHCs and OHCs is abnormally low, and most hair bundles cannot be detected using phalloidin staining for F-actin; those bundles that can be seen are tall but disorganized (Raphael et al. 2001). The entire auditory epithelium in these animals is absent by P50 (Alagramam et al. 2001). The distribution of protocadherin 15 suggests that it may be a component of some as yet undefined lateral link (Fig. 6.7), and the collapse of the hair bundle in the absence of protocadherin 15 (as observed in *Ames waltzer* mice at P16) supports this possibility.

4.6.9 Cell Membrane Protein—Integrins

Integrins are heterodimeric cell-surface receptors for extracellular matrix molecules such as fibronectin and laminin. On ligand binding, integrins form signaling complexes that regulate actin accumulation. In the cochlea and vestibule, *Itga8* mRNA is located in the apical cytoplasm of hair cells. By E16 the α8 protein is present in the apex and hair bundle of all hair cells and this expression persists into maturation (Littlewood-Evans and Muller 2000). The expression pattern suggests Itga8, presumably in conjunction with the β1 subunit, may interact with extracellular matrix associated with the apical surface of the sensory epithelium, possibly with the tectorial membrane of the cochlea and the otoconial membrane of the maculae. Transgenic mice homozygous for a null mutation in the *Itga8* gene encoding the α8 integrin subunit have defects in hair-bundle development, but only in the utricle and then only in a proportion of the hair cells. In the stereocilia of the utricle this signaling complex appears to consist of fibronectin (the α8β1 ligand) and focal adhesion kinase (FAK), which is recruited to the apex of the hair cell by wild type receptors (Fig. 6.7, box 7). In homozygous *Itga8* null mutant mice FAK and fibronectin do not accumulate in the utricular stereocilia by birth, and a proportion of the hair bundles become disorganized and collapse. As integrins are known regulators of actin dynamics, the α8β1 receptor may be responsible for maintaining the hair-cell cytoskeleton and ensuring the ECM forms properly (Littlewood-Evans and Muller 2000). However, the role of α8β1 is unique to the utricle, and alternative mechanisms and molecules must carry out these roles in the other sensory organs of the inner ear.

4.6.10 Cell Membrane Protein—Ptprq

Ptprq is a receptor-like inositol lipid phosphatase with an ectodomain containing 18 fibronectin type III repeats, a single-pass transmembrane domain, and an intracellular domain that has catalytic activity against inositol phospholipids (Wright et al. 1998; Oganesian et al. 2003). It is a component of the hair-bundle shaft connectors (Fig. 6.7, box 8) and is most likely to be the HCA (Goodyear et al. 2003). Although expressed very early during hair-bundle development in the vestibular organs of the mouse and chick inner ear, and from the onset of hair-bundle emergence in the development of the chick basilar papilla, the HCA/Ptprq is first expressed in the mouse cochlea at E17.5 in IHCs and at E18.5 in OHCs. This is at least 2 days after the hair bundles have appeared, indicating it is dispensable for the very early stages of cochlear hair-bundle development. In transgenic mice that do not express detectable levels of Ptprq in hair cells, although shaft connectors are completely absent, vestibular hair bundles appear to develop and mature normally (Goodyear et al. 2003). However, in the cochlea, the hair bundles begin to degenerate and collapse just after birth, leading eventually to the complete loss of hair cells from the basal turn of the cochlea. As an inositol lipid phosphatase, Ptprq may regulate PIP2 levels in the stereocilia membrane and thereby control the polymerization of actin filaments in the maturing hair bundle. In the vestibular organs, another phosphatase may compensate for the loss of Ptprq.

4.6.11 PDZ-Domain Protein—Harmonin b

Harmonin is the central core of the *Usher syndrome type 1* (*USH1*) gene cassette thought to be responsible for hair-bundle organization during the development of the inner ear. Usher syndrome is an autosomal recessive disease that can be subcategorized depending on the severity of the patient phenotype. USH1 is the most acute of these diseases and manifests itself in the human sufferer as profound, congenital deafness, vestibular dysfunction, and prepubertal retinitis pigmentosa (Otterstedde et al. 2001). Five genes corresponding to five of the seven defined loci for USH1 (*USH1A-G*) have now been isolated and these include the genes encoding three of the proteins that have already been described above; myosin VIIa (*USH1B*), cadherin 23 (*USH1D*), and protocadherin 15 (*USH1F*). Harmonin b is the product of the *USH1C* locus, and sans is the product of the *USH1G* locus. Defects in each of these five genes in the corresponding mouse mutants result in splayed hair bundles, deafness, and circling behavior. There are multiple harmonin isoforms and these can be divided into three subclasses: harmonin a, b, and c. Harmonin a and c have a broad expression profile in the mouse, whereas harmonin b (the longest harmonin isoform) is found only in the inner ear, suggesting a structure-specific function in these epithelia (Verpy et al. 2000). The harmonin b isoform is present at the first emergence of the hair bundles in both the cochlea and the vestibule, and is located along the entire length of each stereocilium before its expression becomes restricted to the tips of the stereocilia (Boeda et al. 2002). Harmonin b expression is absent in the hair bundles of mice at P30 and older, suggesting that this protein has a major role in development of the hair bundles (Boeda et al. 2002). In HeLa cell cultures, GFP-harmonin constructs and immunofluorescence indicated that harmonin-b functions as an actin-bundling protein that, like espin, is independent of Ca^{2+} for its action and is preferentially located at the barbed, preferred end of actin filaments (Boeda et al. 2002).

Harmonin b contains three PDZ (*p*ostsynaptic density, *d*iscs large, *z*ona occludens) domains that are classically thought to only bind to C-terminally located PDZ binding interfaces. Harmonin b interacts with myosin VIIa and cadherin 23 (Fig. 6.7, box 5) (Boeda et al. 2002; Siemens et al. 2002). The interaction between harmonin b and myosin VIIa occurs between the PDZ1 domain of harmonin b and the C-terminal tail of myosin VIIa. In the absence of myosin VIIa, harmonin b is found only in the cuticular plate of the stereocilial bundle and not in the stereocilia, suggesting that the localization of harmonin b depends on myosin VIIa. The spatial and temporal distribution of cadherin 23 in stereocilia, as determined with an antibody to the ectodomain of cadherin 23, and that of harmonin b are similar but the distribution of cadherin 23 does not depend on myosin VIIa. Harmonin b may link cadherin 23 to the actin cytoskeleton of the stereocilia once transported to this location by myosin VIIa (Boeda et al. 2002; Siemens et al. 2002).

4.6.12 PDZ-Domain Protein—Whirlin

There are two alternate splice isoforms of whirlin; the long form has a proline-rich domain and three PDZ domains—PDZ1, 2, and 3—while the short form does not contain PDZ domains 1 or 2 (Mburu et al. 2003). Both isoforms are found in the inner ear. Defects in *whirlin* are responsible for congenital deafness (DFNB31) and cause inner-ear defects in the mutant mouse strain *whirler* (Mburu et al. 2003). The hair bundles of cochlear hair cells in homozygote *whirler* mice have height ranked stereocilia and display normal orientation. The bundles are, however, consistently shorter than those found in age-matched wild types. In addition, the hair bundles of OHCs have a U-shaped morphology in contrast to their normal W-shaped bundle; this abnormality is ascribed to the presence of unusually short and ungraded stereocilia at the periphery of the bundle. *Whirler* mutant mice also display cochlear degeneration from P60 onwards (Holme et al. 2002). The shortened-bundle phenotype is due to a deletion in the *whirlin* gene that causes a frameshift at amino acid 433, resulting in a premature termination that results in the loss of one of the three whirlin PDZ domains (Mburu et al. 2003).

RT-PCR indicates whirlin is expressed from E12.5 in the mouse inner ear and the protein localizes to the stereocilia of cochlear and vestibular stereocilia. In transfected HeLa cells, whirlin is found at the barbed end of growing actin filaments (Mburu et al. 2003). Two other mutant mice strains, in addition to the *whirler* mouse, exhibit stunted hair bundles. These are the jerker (J. Zheng et al. 2000) and shaker2 (Belyantseva et al. 2003a) mice that are due to mutations in *espin* and *Myo15a,* respectively. The shared mouse mutant phenotypes raise the intriguing possibility that whirlin can bind with espin and/or myosin XVa to create a protein complex that controls the maintenance of hair-bundle length (Fig. 6.7, box 4). Belyantseva et al. (2003b) have suggested that myosin XVa and whirlin could interact via a PDZ binding motif in the C-terminal tail of this unconventional myosin. A potential site for espin–whirlin interactions has yet to be found.

4.6.13 PDZ-Binding Domain Protein—Sans

Sans is a protein with three ankyrin (Ank) repeats at its N-terminal end, a sterile alpha motif (SAM) domain, and a PDZ binding domain at its C-terminus. Although Ank repeats are thought to provide a site for protein–protein interactions and interaction with the cytoskeleton, their function in the Sans protein is as yet unknown. Cotransfection experiments have provided evidence for an interaction between Sans and harmonin b (Fig. 6.7, box 5) (Weil et al. 2003). The SAM domain normally mediates dimerization and proteins with a similar structure have been shown to allow the crosslinking of several proteins at the same time, suggesting Sans may act as a scaffolding or anchoring protein in the hair bundle (Kikkawa et al. 2003).

4.6.14 Rho-GTPases

The role of Rho-GTPases as downstream effectors of polarity genes has been discussed above (see Section 4.5). Rho-GTPases are also downstream targets for the integrins (Fig. 6.7, box 7) that may, as mentioned above (see Section 4.6.9), mediate communication between the extracellular matrix (ECM) and the growing bundle. Rho-GTPases exert their effects via further downstream molecules such as the profilins and their ligands, the formin homology (FH) proteins. In humans, mutations in the gene encoding the FH protein diaphanous-1 (DIA1) underlie DFNA1, a dominant, autosomal deafness syndrome that begins at 10 years of age (Lynch et al. 1997). DFNA1 was found to be due to a guanine to thymidine substitution in the penultimate exon of *DFNA1*. This single nucleotide substitution leads to a frameshift in the corresponding mRNA, resulting in a truncated protein product (Lynch et al. 1997). In the *Drosophila* bristle shaft cell, diaphanous recruits profilin to the membrane for the initiation of actin polymerization (Evangelista et al. 1997). The DFNA1 deafness phenotype therefore raises the intriguing possibility that Rho-GTPases, activated by interactions between integrin receptors and ECM molecules (Muller and Littlewood-Evans 2001), activate diaphanous which then binds to profilin and initiates actin polymerization in the hair bundle.

5. Development of the Cuticular Plate

Myosins VI and VIIa, which localize to the mature cuticular plate, have been identified in the apical cytoplasm of early differentiating hair cells (Hasson et al. 1997). Labeling with a monoclonal antibody that specifically recognizes an unidentified component of the mature cuticular plate is detected at E14 to E15 in the mouse cochlea. The antigen is located below the apical membrane during the initial stages of hair-cell differentiation, and is detected prior to detectable phalloidin labeling of stereocilial actin filaments (Nishida et al. 1998). These findings suggest that an incipient cuticular plate may be formed very early during hair-cell differentiation. The cytoskeletal network of the cuticular plate is not identifiable at the ultrastructural level when stereocilia first emerge, although parallel microfilaments, continuous with those in the immature stereocilium, descend into the cell (Tilney et al. 1992; Denman-Johnson and Forge 1999). In thin sections, the apical cytoplasm of the immature hair cell initially appears relatively undifferentiated, with the exception of microtubules that run parallel to the cell surface (Mbiene et al. 1988; Troutt et al. 1994; Souter et al. 1995; Denman-Johnson and Forge 1999). These persist immediately below the cuticular plate as it becomes more pronounced and thus may play a role in its formation. Coated and uncoated vesicles are also present in the apical cytoplasm beneath the developing stereocilia, and numerous coated pits open to the apical membrane (Forge and Richardson 1993; Souter et al. 1997; Denman-Johnson and Forge 1999), suggesting that there is extensive endocytotic activity and membrane turnover at the apical membrane both of auditory (Forge and Richardson 1993; Souter et al. 1995) and vestibular (Denman-Johnson and Forge 1999) hair cells during their development. These vesicles are infrequently observed associated with the apical membrane of the mature hair cell, suggesting that such activity must become restricted with continuing differentiation. The disappearance of coated pits coincides with the further maturation of the cuticular plate, a process that involves an increase in its thickness via the continual addition and crosslinking of the microfilaments. This basalward extension of the cuticular plate occurs in parallel with the development of the junctional complex around the neck of the cell (Souter et al. 1995) to which the cuticular plate becomes attached (Hirokawa and Tilney 1982; Forge et al. 1991). The formation of the tightly crosslinked microfilament assembly may impede vesicular traffic across the apical cytoplasm and may therefore account for the developmental decrease in vesicular traffic to the apical membrane. The cuticular plate becomes anchored to the apical membrane during development (Hirokawa and Tilney 1982). This may act to stabilize the membrane and in addition impede membrane trafficking, and may also be important in stereocilial maintenance. In *Snell's waltzer* mutant mice that are deficient for myosin VI, the stereocilia fuse over a postnatal period that coincides temporally with maturation of the cuticular plate. These membrane fusions are found between adjacent stereocilia and originate at the proximal end of the stereocilia (Self et al. 1999), suggesting a loss of apical plasma membrane stability (see above, Section 4.6.2). The filaments that are seen to crosslink the cuticular plate to the underside of the apical membrane may therefore contain myosin VI.

6. Development of the Intercellular Junctional Complexes

In the presumptive sensory epithelium, the junctions between cells at the lumenal side seen using freeze–fracture are of relatively simple morphology, consisting of two to three strands parallel to the apical surface (Ginzberg and Gilula 1979; Bagger-Sjoback and Anniko 1984; Souter et al. 1995). These represent the tight (occluding) junction that maintains the fluid separation between the lumen of the otocyst and the intercellular spaces of the epithelium. An adherens-type component of distinct morphology is not revealed by freeze–fracture at this stage, but an adherens junction can be identified just below the occluding junction in thin sections (Ginzberg and Gilula 1979). As differentiation proceeds, the junction associated with a hair cell becomes more complex, and manifests itself as strands that run perpendicular to the cell surface (Ginzberg and Gilula 1979; Anniko 1983; Bagger-Sjoback and Anniko 1984; Souter et al. 1995).

During development of hair cells in the organ of Corti, the two parts of the junction that can be recognized morphologically (the apical band of parallel elements and the basal network of strands) mature at different rates. The apical region, reaches a mature configuration of five to eight parallel strands around the time when the endocochlear potential (EP—the positive DC potential that is recorded in the endolymph fluid bathing the apical surface of the cell) is first generated, namely at P8 in gerbils (Souter et al. 1995) and P6 in mice. The development of this part of the junction is probably necessary to allow the EP, which provides a driving force for current flow into the hair cells, to develop fully. Tight junction development also plays a role in establishing polarity in epithelial cells (Gumbiner 1990), separating the apical and lateral membranes and allowing targeting of specific proteins to each plasma membrane domain. Freeze–fracture studies of the immature cochleae of mice and gerbils (Forge and Richardson 1993; Souter et al. 1995), however, indicate that particular structural characteristics of the apical plasma membrane of mature hair cell are present before maturation of the tight junction, indicating that cell polarity is established early in hair-cell development.

The basal part of the junction continues to develop after the apical region of parallel strands has reached maturity. Perpendicular strands increase in number and degree of branching to eventually form a network composed of branched, linear elements (Souter et al. 1995). This basal region of the junctional complex develops in basalward extension and concomitantly with development of the cuticular plate on the hair-cell side of the junction and the deposition of membrane-associated cytoskeletal elements on the supporting-cell side. Thus, this junctional region is clearly of an adherens type, providing a site for membrane anchoring to cytoskeletal assemblies. It is unusual because most adherens junctions do not show a distinctive morphology in freeze–fracture replicas.

The strands that represent the morphological appearance of tight junctions in freeze–fracture replicas are composed of a variety of proteins including occludin (Furuse et al. 1993) and claudins (Tsukita and Furuse 2000a,b). Claudins are a family of proteins and different members are differentially expressed in various tissues and confer ion-selective permeability across the tight junction (Colegio et al. 2002). Several accessory proteins including ZO-1, ZO-2, ZO-3, and cingulin are also associated with tight junctions and interact with claudins (Itoh et al. 1999; D'Atri and Citi 2002). ZO-1—which is ubiquitous to most tight junctions—is present in the hair-cell junctions from early stages of development in the mammalian cochlea (M. Souter, unpublished observations). The time course of morphological maturation of the tight junctions of mammalian cochlear hair cells coincides with the onset and increase of claudin 14 expression by hair cells (Ben-Yosef et al. 2003), indicating that this protein is added to the junctions as the number of apical parallel strands increases. A defect in the gene for claudin 14 has been shown to be the cause of profound congenital deafness (Wilcox et al. 2001), a deafness disorder that may result from a disturbance of the permeability barrier at the apical surface of the sensory epithelium (Ben-Yosef et al. 2003). However, the morphology of the tight junction complex around cochlear hair cells in claudin 14 null mice is normal and there is no vestibular defect suggesting the presence of additional claudins and accompanying functional redundancy in the sensory epithelia. The mature organ of Corti has been shown to contain claudins, 1, 2, 3, 9, 10, 12, and 18 in addition to claudin 14, as well as occludin (Kitajiri et al. 2004) but there is little knowledge of the expression pattern of these proteins during development nor whether the sensory epithelia in different species contain similar complements of claudin family members.

The protein(s) that forms the complex network of strands in the adherens component of the junction is also not known. Epithelial cadherin (E-cadherin) is a common constituent of epithelial adherens junctions. The intracellular domain of E-cadherin links to the actin cytoskeleton through α- and β-catenins. E-cadherin facilitates assembly of intercellular junctions (Gumbiner et al. 1988), but cadherin-based cell adhesion may also be involved in the intercellular signalling that regulates cell division and differentiation (Takeichi 1995). However, while E-cadherin is expressed at cell junctions in the organ of Corti of mice and rats (Whitlon 1993; Simonneau et al. 2003), and the avian equivalent L-CAM (L-cell adhension molecule) is expressed in the developing basilar papilla (Richardson et al. 1987; Raphael et al. 1988), E-cadherin is absent from hair cell-supporting cell junctions in the mature organ of Corti of mice (Whitlon 1993) and gerbils (M. Souter, unpublished observations). Likewise E-cadherin is reportedly absent from the hair-cell junctions of the mammalian vestibular organs, although other members of the cadherin family appear to be present as the junction is labeled with a pan-cadherin antibody (Hackett et al. 2002). In the mouse organ of Corti, the down-regulation of E-cadherin appears to occur during the early postnatal period (Whitlon 1993) over the same time course as the complex network of junctional strands is developing. This suggests that, unusually, the adherens component of the junctional complex does not contain E-cadherin. The period over which E-cadherin disappears also corresponds temporally to the formation of large extracellular spaces around the hair cells (owing to a change in supporting cell morphology) and the acquisition of the characteristic specialization of the OHC lateral wall. However, whether the modulation of cadherin expression, a known differentiation signal in other systems (Takeichi 1995), is directly related to the initiation of these phases of development in the organ of Corti is not known.

7. Generation of the Different Types of Hair Cell Found in the Inner Ear

The various specializations of different hair-cell types appear to be acquired for the most part at later stages of development. However, understanding the factors that influence hair-cell specialization currently presents something of a puzzle: what is it that makes an infant hair cell of essentially common morphology mature into a particular hair cell type? Hair cells of different types in a particular sensory epithelium are usually differentially distributed suggesting positional information may play a role. In vestibular macular organs type 1 and type 2 hair cells are differentially distributed in relation to the striola, and in cristae type 1 hair cells predominate at the crest of the saddle-shaped sensory epithelium, with type 2 hair cells predominating at the flanking skirts. In the organ of Corti, IHCs and OHCs develop in anatomically distinct regions of the Kolicker's organ referred to as the greater and lesser epithelial ridge, respectively. These regions are separated by cells that eventually form the inner and outer pillar cells of the mature organ of Corti. It is thought that this separation may have a role in differential differentiation. Likewise, the systematic variations seen within a hair-cell population in an individual sensory epithelium suggest that there are positional cues that influence hair-cell differentiation. Interestingly, the regenerated hair cells in the mature avian basilar papilla that appear after hair-cell loss in a restricted region of the sensory epithelium, for example after noise damage, acquire the correct characteristics for their position (Cotanche et al. 1994; Cotanche 1999; Smolders 1999). This indicates that a memory for hair-cell organization is retained in the mature epithelium. However, the nature and action of any positional cues that regulate hair-cell specialization are not yet known.

Investigations of vestibular hair-cell differentiation during development in mammals (Sans and Chat 1982; Rüsch et al. 1998) and during regeneration of the avian vestibular organs (Rubel et al. 1991) have both suggested that, prior to the obvious distinction of two hair-cell types, there is a single cell type that morphologically resembles the mature type 2 hair cell. From the later appearance of morphologically distinct type 1 hair cells during regeneration in the avian utricle it was suggested that type 1 hair cells differentiate from the type 2s in a serial progression of development. However, earlier studies of spatiotemporal patterns of hair cell birth had suggested that hair cells in regions where type 1 cells predominate in the mature sensory epithelium (across the striola of the maculae and at the crest of the cristae) are born first, while hair cells in those regions where type 2 cells predominate are born later, indicating that type 1 and type 2 hair cells may be specified separately. An investigation of the acquisition of ion-channel phenotypes over the period of early postnatal development in mice (Rüsch et al. 1998) in which, as in the cochlea, the maturation of the vestibular organs continues after birth, reported the presence of immature hair cells that morphologically resembled type 2 hair cells but had an ion channel complement unlike either mature type 1 or mature type 2 hair cells. Cells appeared to acquire ion channel characteristics to become either mature type 1-like or mature type-2 like but without obvious morphological changes again suggesting separate, but parallel, differentiation of the two hair-cell types. The acquisition of the physiological phenotypes and subsequently of the distinct morphological features of type 1 cells also proceeded in cultured explants devoid of innervation indicating that hair-cell shape, and other morphological and physiological properties, are all determined by the hair cell autonomously. In addition to genes that specify a hair cell, there is also likely to be a set of genes that regulate hair-cell type.

8. Differentiation of the Lateral Membrane in Mammalian Cochlear Hair Cells

Cochlear hair cells acquire their particular specializations in a progressive manner that occurs postnatally in altricial animals. In early neonatal mice and gerbils, during the period when the OHCs are closely surrounded by adjacent supporting cells, the lateral membranes of both IHCs and OHCs appear relatively unspecialized (Forge and Richardson 1993; Souter et al. 1995) and little different from the equivalent regions of mature vestibular hair cells and avian auditory hair cells. In gerbils at about P8, the supporting cells begin to separate from the hair cells along their lateral borders eventually thinning to create the large extracellular spaces that are present around each OHC in the mature organ of Corti. At this time, large particles begin to be incorporated into the lateral membranes of OHCs, and increase in number and density over the next few days so that they become closely packed over the entire lateral membrane, as in mature hair cells, by P16. Over an equivalent period in the rat cochlea, immunohistochemical labeling shows the motor protein prestin becomes incorporated into the membrane and to increase in concentration (Belyantseva et al. 2000). The appearance and subsequent progression of development of these particles and the increase in prestin incorporation coincide temporally with the first emergence of the electromotile response and its subsequent development in isolated OHCs (He et al. 1994; Belyantseva et al. 2000). With further development additional proteins become incorporated into the lateral plasma membrane. The glucose transporter Glut-5 can first be detected in hair cells of the rat at P15 (Belyantseva et al. 2000) at about the onset of auditory function. At around the time that the mid regions of supporting cells separate from the lateral membrane of the outer hair cell, a single layer of endoplasmatic cisternal elements becomes organized parallel to the inner aspect of the OHC's lateral membrane. Ca^{2+}-ATPase becomes localized to this region of the cell some days later (Schulte 1993). Elements of the cortical cytoskeleton that first appear as the cisternal membranes become organized after the initiation of electromotility (Souter et al. 1995). The pillar structures that link the plasma membrane proteins to the cortical lattice are seen as the cisternae develop, but they are initially quite widely separated. They increase in number and become more closely spaced in line with further development of the cisternae and the lateral plasma membrane. The appearance of actin in the lateral wall occurs in a similar temporal pattern (Weaver et al. 1994). It thus appears that there is a sequence of development of the structures in the lateral wall of the OHC. The motor proteins are incorporated into the lateral plasma membrane first and electromotility can be initiated. This is followed by the formation of the cortical lattice and its links to the plasma membrane in parallel with the organization of the cisternal membranes, but these may undergo further maturation. Functionally, the increasing development of the links between the plasma membrane and the cortical lattice may provide the necessary structural coupling to allow the motor to influence cochlear mechanics. This is suggested by the initiation and growth of otoacoustic emissions, which first appear and grow over a time scale that corresponds to the organization of the cytoskeletal elements in the lateral wall (Souter et al. 1995, 1997).

The structural specializations of the lateral membranes of IHCs appear later than those of the OHC. Organizations of particles in rows (the functional correlate of which is unknown) appear on the lateral membrane of the IHC just prior to the acquisition of the plaque regions of particles in square array. These appear at about the time when OHCs have almost reached maturity. The time of appearance of the plaques coincides with final maturation of the EP (M. Souter and A. Forge, unpublished observations). This potential is uniquely high in the mammalian cochlea, creating a large potential difference across the apical membrane of the hair cells and providing the driving force for the transduction current. The appearance of plaques at the apical end of the IHC's lateral membrane at the time when the EP reaches its mature level suggests that they have some role in maintaining homeostasis in IHCs during the rapid changes in ion flux through the cell.

9. Conclusion

From the work presented here it is clear that the development of the hair cell's basic morphology is well described. Our understanding of the molecular mechanisms that underlie this developmental process is also improving rapidly. This is largely due to the analysis of mouse mutants, the cloning of human deafness genes, and from the study of homologous systems in *Drosophila*. In light of this review it would seem, however, that many puzzles still remain to be solved. For example, the process that creates an apical to basal gradient in hair-cell morphology remains unknown, as does the mechanism responsible for generating such a wide variety of different hair cell types.

References

Adler PN (2002) Planar signaling and morphogenesis in *Drosophila*. Dev Cell 2:525–535.

Ahmed ZM, Riazuddin S, Bernstein SL, Ahmed Z, Khan S, Griffith AJ, Morell RJ, Friedman TB, Riazuddin S, Wilcox ER (2001) Mutations of the protocadherin gene *PCDH15* cause Usher syndrome type 1F. Am J Hum Genet 69:25–34.

Ahmed ZM, Riazuddin S, Ahmad J, Bernstein SL, Guo Y, Sabar MF, Sieving P, Riazuddin S, Griffith AJ, Friedman TB, Belyantseva IA, Wilcox ER (2003) PCDH15 is expressed in the neurosensory epithelium of the eye and ear and mutant alleles are responsible for both *USH1F* and *DFNB23*. Hum Mol Genet 12:3215–3223.

Alagramam KN, Murcia CL, Kwon HY, Pawlowski KS, Wright CG, Woychik RP (2001) The mouse Ames waltzer hearing-loss mutant is caused by mutation of Pcdh15, a novel protocadherin gene. Nat Genet 27:99–102.

Anderson DW, Probst FJ, Belyantseva IA, Fridell RA, Beyer L, Martin DM, Wu D,

Kachar B, Friedman TB, Raphael Y, Camper SA (2000) The motor and tail regions of myosin XV are critical for normal structure and function of auditory and vestibular hair cells. Hum Mol Genet 9:1729–1738.

Angst BD, Marcozzi C, Magee AI (2001) The cadherin superfamily. J Cell Sci 114: 625–626.

Anniko M (1983) Postnatal maturation of cochlear sensory hairs in the mouse. Anat Embryol 166:355–368.

Aschenbrenner L, Lee T, Hasson T (2003) Myo6 facilitates the translocation of endocytic vesicles from cell peripheries. Mol Biol Cell 14:2728–2743.

Ashmore J (2002) Biophysics of the cochlea—biomechanics and ion channelopathies. Br Med Bull 63:59–72.

Avraham KB, Hasson T, Steel KP, Kingsley DM, Russell LB, Mooseker MS, Copeland NG, Jenkins NA (1995) The mouse *Snell's waltzer* deafness gene encodes an unconventional myosin required for structural integrity of inner ear hair cells. Nat Genet 11:369–375.

Bagger-Sjoback D, Anniko M (1984) Development of intercellular junctions in the vestibular end-organ. A freeze-fracture study in the mouse. Ann Otol Rhinol Laryngol 93:89–95.

Bagger-Sjoback D, Takumida M (1988) Geometrical array of the vestibular sensory hair bundle. Acta Otolaryngol 106:393–403.

Bartles JR, Zheng L, Li A, Wierda A, Chen B (1996) Small espin: a third actin-bundling protein and potential forked protein ortholog in brush border microvilli. J Cell Biol 143:107–119.

Bartolami S, Goodyear R, Richardson G (1991) Appearance and distribution of the 275 kD hair-cell antigen during development of the avian inner ear. J Comp Neurol 314: 777–788.

Belyantseva IA, Adler HJ, Curi R, Frolenkov GI, Kachar B (2000) Expression and localization of prestin and the sugar transporter GLUT-5 during development of electromotility in cochlear outer hair cells. J Neurosci 20:RC116.

Belyantseva IA, Boger ET, Friedman TB (2003a) Myosin XVa localizes to the tips of inner ear sensory cell stereocilia and is essential for staircase formation of the hair bundle. Proc Natl Acad Sci USA 100:13958–13963.

Belyantseva IA, Labay V, Boger ET, Griffith AJ, Friedman TB (2003b) Stereocilia: the long and the short of it. Trends Mol Med 9:458–461.

Ben-Yosef T, Belyantseva IA, Saunders TL, Hughes ED, Kawamoto K, Van Itallie CM, Beyer LA, Halsey K, Gardner DJ, Wilcox ER, Rasmussen J, Anderson JM, Dolan DF, Forge A, Raphael Y, Camper SA, Friedman TB (2003) Claudin 14 knockout mice, a model for autosomal recessive deafness *DFNB29*, are deaf due to cochlear hair cell degeneration. Hum Mol Genet 12:2049–2061.

Bergstrom B, Engstrom H (1973) The vestibular sensory cells and their innervation. Int J Equilib Res 3:27–32.

Beyer LA, Odeh H, Probst FJ, Lambert EH, Dolan DF, Camper SA, Kohrman DC, Raphael Y (2000) Hair cells in the inner ear of the *pirouette* and *shaker 2* mutant mice. J Neurocytol 29:227–240.

Boeda B, El-Amraoui A, Bahloul A, Goodyear R, Daviet L, Blanchard S, Perfettini I, Fath KR, Shorte S, Reiners J, Houdusse A, Legrain P, Wolfrum U, Richardson G, Petit C (2002) Myosin VIIa, harmonin and cadherin 23, three Usher I gene products that cooperate to shape the sensory hair cell bundle. EMBO J 21:6689–6699.

Bork JM, Peters LM, Riazuddin S, Bernstein SL, Ahmed ZM, Ness SL, Polomeno R, Ramesh A, Schloss M, Srisailpathy CR, Wayne S, Bellman S, Desmukh D, Ahmed Z, Khan SN, Kaloustian VM, Li XC, Lalwani A, Riazuddin S, Bitner-Glindzicz M, Nance WE, Liu XZ, Wistow G, Smith RJ, Griffith AJ, Wilcox ER, Friedman TB, Morell RJ (2001) Usher syndrome 1D and nonsyndromic autosomal recessive deafness *DFNB12* are caused by allelic mutations of the novel cadherin-like gene *CDH23*. Am J Hum Genet 68:26–37.

Buss F, Luzio JP, Kendrick-Jones J (2002) Myosin VI, an actin motor for membrane traffic and cell migration. Traffic 3:851–858.

Chen P, Johnson JE, Zoghbi HY, Segil N (2002) The role of *Math1* in inner ear development: uncoupling the establishment of the sensory primordium from hair cell fate determination. Development 129:2495–2505.

Colegio OR, Van Itallie CM, McCrea HJ, Rahner C, Anderson JM (2002) Claudins create charge-selective channels in the paracellular pathway between epithelial cells. Am J Physiol Cell Physiol 283:C142–147.

Corwin JT, Cotanche DA (1989) Development of location-specific hair cell stereocilia in de-enervated embryonic ears. J Comp Neurol 288:529–537.

Cotanche DA (1987) Development of hair cell stereocilia in the avian cochlea. Hear Res 28:35–44.

Cotanche DA (1999) Structural recovery from sound and aminoglycoside damage in the avian cochlea. Audiol Neurootol 4:271–285.

Cotanche DA, Corwin JT (1991) Stereociliary bundles reorient during hair cell development and regeneration in the chick cochlea. Hear Res 52:379–402.

Cotanche DA, Sulik KK (1983) Early differentiation of hair cells in the embryonic chick basilar papilla. Arch Otorhinolaryngol 237:191–195.

Cotanche DA, Sulik KK (1984) The development of stereociliary bundles in the cochlear duct of chick embryos. Brain Res 318:181–193.

Cotanche DA, Lee KH, Stone JS, Picard DA (1994) Hair cell regeneration in the bird cochlea following noise damage or ototoxic drug damage. Anat Embryol (Berl) 189:1–18.

Cramer LP (2000) Myosin VI: roles for a minus end-directed actin motor in cells. J Cell Biol 150:F121–126.

Curtin JA, Quint E, Tsipouri V, Arkell RM, Cattanach B, Copp AJ, Henderson DJ, Spurr N, Stanier P, Fisher EM, Nolan PM, Steel KP, Brown SD, Gray IC, Murdoch JN (2003) Mutation of *Celsr1* disrupts planar polarity of inner ear hair cells and causes severe neural tube defects in the mouse. Curr Biol 13:1129–1133.

Dabdoub A, Donohue MJ, Brennan A, Wolf V, Montcouquiol M, Sassoon DA, Hseih JC, Rubin JS, Salinas PC, Kelley MW (2003) *Wnt* signaling mediates reorientation of outer hair cell stereociliary bundles in the mammalian cochlea. Development 130: 2375–2384.

Dallos P, Fakler B (2002) Prestin, a new type of motor protein. Nat Rev Mol Cell Biol 3:104–111.

Dannhof BJ, Roth B, Bruns V (1991) Length of hair cells as a measure of frequency representation in the mammalian inner ear? Naturwissenschaften 78:570–573.

D'Atri F, Citi S (2002) Molecular complexity of vertebrate tight junctions (review). Mol Membr Biol 19:103–112.

Daudet N, Lebart MC (2002) Transient expression of the t-isoform of plastins/fimbrin in the stereocilia of developing auditory hair cells. Cell Motil Cytoskeleton 53:326–336.

Denman-Johnson K, Forge A (1999) Establishment of hair bundle polarity and orientation in the developing vestibular system of the mouse. J Neurocytol 28:821–835.

DeRosier DJ and Tilney LG (1989) The structure of the cuticular plate, an in vivo actin gel. J Cell Biol 109:2853–2867.

Di Palma F, Holme RH, Bryda EC, Belyantseva IA, Pellegrino R, Kachar B, Steel KP, Noben-Trauth K (2001) Mutations in *Cdh23*, encoding a new type of cadherin, cause stereocilia disorganization in *waltzer*, the mouse model for *Usher syndrome type 1D*. Nat Genet 27:103–107.

Duncan RK, Fuchs PA (2003) Variation in large-conductance, calcium-activated potassium channels from hair cells along the chicken basilar papilla. J Physiol 547:357–371.

Eatock RA, Rusch A (1997) Developmental changes in the physiology of hair cells. Semin Cell Dev Biol 8:265–275.

Erven A, Skynner MJ, Okumura K, Takebayashi S, Brown SD, Steel KP, Allen ND (2002) A novel stereocilia defect in sensory hair cells of the deaf mouse mutant *Tasmanian devil*. Eur J Neurosci 16:1433–1441.

Evangelista M, Blundell K, Longtine MS, Chow CJ, Adames N, Pringle JR, Peter M, Boone C (1997) Bni1p, a yeast formin linking cdc42p and the actin cytoskeleton during polarized morphogenesis. Science 276:118–122.

Flock A, Bretscher A, Weber K (1982) Immunohistochemical localization of several cytoskeletal proteins in inner ear sensory and supporting cells. Hear Res 7:75–89.

Flock A, Flock B, Ulfendahl M (1986) Mechanisms of movement in outer hair cells and a possible structural basis. Arch Otorhinolaryngol 243:83–90.

Forge A (1987) Specialisations of the lateral membrane of inner hair cells. Hear Res 31: 99–109.

Forge A (1991) Structural features of the lateral walls in mammalian cochlear outer hair cells. Cell Tissue Res 265:473–483.

Forge A, Richardson G (1993) Freeze fracture analysis of apical membranes in cochlear cultures: differences between basal and apical-coil outer hair cells and effects of neomycin. J Neurocytol 22:854–867.

Forge A, Davies S, Zajic G (1991) Assessment of ultrastructure in isolated cochlear hair cells using a procedure for rapid freezing before freeze-fracture and deep-etching. J Neurocytol 20:471–484.

Forge A, Zajic G, Li L, Nevill G, Schacht J (1993) Structural variability of the subsurface cisternae in intact, isolated outer hair cells shown by fluorescent labelling of intracellular membranes and freeze-fracture. Hear Res 64:175–183.

Forge A, Souter M, Denman-Johnson K (1997) Structural development of sensory cells in the ear. Semin Cell Dev Biol 8:225–237.

Forge A, Becker D, Casalotti S, Edwards J, Marziano N, Nevill G (2003a) Gap junctions in the inner ear: comparison of distribution patterns in different vertebrates and assessement of connexin composition in mammals. J Comp Neurol 467:207–231.

Furness DN, Richardson GP, Russell IJ (1989) Stereociliary bundle morphology in organotypic cultures of the mouse cochlea. Hear Res 38:95–109.

Furuse M, Hirase T, Itoh M, Nagafuchi A, Yonemura S, Tsukita S (1993) Occludin: a novel integral membrane protein localizing at tight junctions. J Cell Biol 123:1777–1788.

Gale JE, Marcotti W, Kennedy HJ, Kros CJ, Richardson GP (2001) FM1-43 dye behaves as a permeant blocker of the hair-cell mechanotransducer channel. J Neurosci 21: 7013–7025.

Geleoc GS, Holt JR (2003) Developmental acquisition of sensory transduction in hair cells of the mouse inner ear. Nat Neurosci 6:1019–1020.

Ginzberg RD, Gilula NB (1979) Modulation of cell junctions during differentiation of the chicken otocyst sensory epithelium. Dev Biol 68:110–129.

Goldberg JM (1991) The vestibular end organs: morphological and physiological diversity of afferents. Curr Opin Neurobiol 1:229–235.

Goodyear RJ, Richardson GP (1992) Distribution of the 275 kD hair cell antigen and cell surface specialisations on auditory and vestibular hair bundles in the chicken inner ear. J Comp Neurol 325:243–256.

Goodyear R, Richardson G (1997) Pattern formation in the basilar papilla: evidence for cell rearrangement. J Neurosci 17:6289–6301.

Goodyear R, Richardson G (1999) The ankle-link antigen: an epitope sensitive to calcium chelation associated with the hair-cell surface and the calycal processes of photoreceptors. J Neurosci 19:3761–3772.

Goodyear RJ, Richardson GP (2003) A novel antigen sensitive to calcium chelation that is associated with the tip links and kinocilial links of sensory hair bundles. J Neurosci 23:4878–4887.

Goodyear RJ, Legan PK, Wright MB, Marcotti W, Oganesian A, Coats SA, Booth CJ, Kros CJ, Seifert RA, Bowen-Pope DF, Richardson GP (2003) A receptor-like inositol lipid phosphatase is required for the maturation of developing cochlear hair bundles. J Neurosci 23:9208–9219.

Goodyear RJ, Marcotti W, Kros CJ, Richardson GP (2005) Development and properties of stereociliary link types in hair cells of the mouse cochlea. J Comp Neurol 485:75–85.

Gulley RL, Reese TS (1977) Regional specialization of the hair cell plasmalemma in the organ of corti. Anat Rec 189:109–123.

Gumbiner B (1990) Generation and maintenance of epithelial cell polarity. Curr Opin Cell Biol 2:881–887.

Gumbiner B, Stevenson B, Grimaldi A (1988) The role of the cell adhesion molecule uvomorulin in the formation and maintenance of the epithelial junctional complex. J Cell Biol 107:1575–1587.

Hackett L, Davies D, Helyer R, Kennedy H, Kros C, Lawlor P, Rivolta MN, Holley M (2002) E-cadherin and the differentiation of mammalian vestibular hair cells. Exp Cell Res 278:19–30.

Hanein D, Matsudaira P, DeRosier DJ (1997) Evidence for a conformational change in actin induced by fimbrin (N375) binding. J Cell Biol 139:387–396.

Hasson T, Heintzelman MB, Santos-Sacchi J, Corey DP, Mooseker MS (1995) Expression in cochlea and retina of myosin VIIa, the gene product defective in *Usher syndrome type 1B*. Proc Natl Acad Sci USA 92:9815–9819.

Hasson T, Gillespie PG, Garcia JA, MacDonald RB, Zhao Y, Yee AG, Mooseker MS, Corey DP (1997) Unconventional myosins in inner-ear sensory epithelia. J Cell Biol 137:1287–1307.

He DZ, Evans BN, Dallos P (1994) First appearance and development of electromotility in neonatal gerbil outer hair cells. Hear Res 78:77–90.

Heisenberg CP, Tada M (2002) Wnt signalling: a moving picture emerges from *van gogh*. Curr Biol 12:R126–128.

Hirokawa N (1977) Disappearance of afferent and efferent nerve terminals in the inner ear of the chick embryo after chronic treatment with beta-bungarotoxin. J Cell Biol 73:27–46.

Hirokawa N, Tilney LG (1982) Interactions between actin filaments and between actin filaments and membranes in quick-frozen and deeply etched hair cells of the chick ear. J Cell Biol 95:249–261.

Holley MC, Ashmore JF (1990) Spectrin, actin and the structure of the cortical lattice in mammalian cochlear outer hair cells. J Cell Sci 96:283–291.

Holley MC, Kalinec F, Kachar B (1992) Structure of the cortical cytoskeleton in mammalian outer hair cells. J Cell Sci 102:569–580.

Holme RH, Kiernan BW, Brown SD, Steel KP (2002) Elongation of hair cell stereocilia is defective in the mouse mutant whirler. J Comp Neurol 450:94–102.

Hudspeth AJ (1989) How the ear's works work. Nature 341:397–404.

Itoh M, Furuse M, Morita K, Kubota K, Saitou M, Tsukita S (1999) Direct binding of three tight junction-associated MAGUKs, ZO-1, ZO-2, and ZO-3, with the COOH termini of claudins. J Cell Biol 147:1351–1363.

Jagger DJ, Ashmore JF (1998) A potassium current in guinea-pig outer hair cells activated by ion channel blocker DCDPC. NeuroReport 9:3887–3891.

Kachar B, Battaglia A, Fex J (1997) Compartmentalized vesicular traffic around the hair cell cuticular plate. Hear Res 107:102–112.

Kalinec F, Holley MC, Iwasa KH, Lim DJ, Kachar B (1992) A membrane-based force generation mechanism in auditory sensory cells. Proc Natl Acad Sci USA 89:8671–8675.

Kaltenbach JA, Falzarano PR, Simpson TH (1994) Postnatal development of the hamster cochlea. II. Growth and differentiation of stereocilia bundles. J Comp Neurol 350: 187–198.

Katayama A, Corwin JT (1989) Cell production in the chicken cochlea. J Comp Neurol 281:129–135.

Katayama A, Corwin JT (1993) Cochlear cytogenesis visualized through pulse labeling of chick embryos in culture. J Comp Neurol 333:28–40.

Kibar Z, Vogan KJ, Groulx N, Justice MJ, Underhill DA, Gros P (2001) *Ltap*, a mammalian homolog of *Drosophila Strabismus/Van Gogh*, is altered in the mouse neural tube mutant Loop-tail. Nat Genet 28:251–255.

Kikkawa Y, Shitara H, Wakana S, Kohara Y, Takada T, Okamoto M, Taya C, Kamiya K, Yoshikawa Y, Tokano H, Kitamura K, Shimizu K, Wakabayashi Y, Shiroishi T, Kominami R, Yonekawa H (2003) Mutations in a new scaffold protein Sans cause deafness in Jackson shaker mice. Hum Mol Genet 12:453–461.

Kitajiri SI, Furuse M, Morita K, Saishin-Kiuchi Y, Kido H, Ito J, Tsukita S (2004) Expression patterns of claudins, tight junction adhesion molecules, in the inner ear. Hear Res 187:25–34.

Kros CJ, Marcotti W, van Netten SM, Self TJ, Libby RT, Brown SD, Richardson GP, Steel KP (2002) Reduced climbing and increased slipping adaptation in cochlear hair cells of mice with *Myo7a* mutations. Nat Neurosci 5:41–47.

Kuijpers W, Tonnaer EL, Peters TA, Ramaekers FC (1991) Expression of intermediate filament proteins in the mature inner ear of the rat and guinea pig. Hear Res 52(1): 133–46.

Küssel-Andermann P, El-Amraoui A, Safieddine S, Nouaille S, Perfettini I, Lecuit M, Cossart P, Wolfrum U, Petit C (2000) Vezatin, a novel transmembrane protein, bridges myosin VIIA to the cadherins–catenins complex. EMBO J 19:6020–6029.

Lavigne-Rebillard M, Pujol R (1986) Development of the auditory hair cell surface in human fetuses. A scanning electron microscopy study. Anat Embryol 174:369–377.

Lenoir M, Puel JL, Pujol R (1987) Stereocilia and tectorial membrane development in the rat cochlea. A SEM study. Anat Embryol 175:477–487.

Lewis J, Davies A (2002) Planar cell polarity in the inner ear: how do hair cells acquire their oriented structure? J Neurobiol 53:190–201.

Li H, Liu H, Balt S, Mann S, Corrales CE, Heller S (2004) Correlation of expression of the actin filament-bundling protein espin with stereociliary bundle formation in the developing inner ear. J Comp Neurol 468:125–134.

Li L, Forge A (1997) Morphological evidence for supporting cell to hair cell conversion in the mammalian utricular macula. Int J Dev Neurosci 15:433–446.

Liberman MC, Gao J, He DZ, Wu X, Jia S, Zuo J (2002) Prestin is required for electromotility of the outer hair cell and for the cochlear amplifier. Nature 419:300–304.

Lim DJ (1980) Cochlear anatomy related to cochlear micromechanics. A review. J Acoust Soc Am 67:1686–1695.

Lim DJ, Rueda J (1990) Distribution of glycoconjugates during cochlea development. A histochemical study. Acta Otolaryngol 110:224–1233.

Lin CS, Shen W, Chen ZP, Tu YH, Matsudaira P (1994) Identification of I-plastin, a human fimbrin isoform expressed in intestine and kidney. Mol Cell Biol 14:2457–2467.

Littlewood-Evans A, Muller U (2000) Stereocilia defects in the sensory hair cells of the inner ear in mice deficient in integrin alpha8beta1. Nat Genet 24:424–428.

Liu XZ, Walsh J, Tamagawa Y, Kitamura K, Nishizawa M, Steel KP, Brown SD (1997) Autosomal dominant non-syndromic deafness caused by a mutation in the *myosin VIIA* gene. Nat Genet 17:268–269.

Loomis PA, Zheng L, Sekerkova G, Changyaleket B, Mugnaini E, Bartles JR (2003) Espin cross-links cause the elongation of microvillus-type parallel actin bundles in vivo. J Cell Biol 163:1045–1055.

Lynch ED, Lee MK, Morrow JE, Welcsh PL, Leon PE, King MC (1997) Nonsyndromic deafness *DFNA1* associated with mutation of a human homolog of the *Drosophila* gene *diaphanous*. Science 278:1315–1318.

Markin VS, Hudspeth AJ (1995) Gating-spring models of mechanoelectrical transduction by hair cells of the internal ear. Annu Rev Biophys Biomol Struct 24:59–83.

Marowitz WF, Shugar JM (1976) Single mitotic center for rodent cochlear duct. Ann Otol Rhinol Laryngol 85:225–233.

Mbiene JP, Sans A (1986) Differentiation and maturation of the sensory hair bundles in the fetal and postnatal vestibular receptors of the mouse: a scanning electron microscopy study. J Comp Neurol 254:271–278.

Mbiene JP, Favre D, Sans A (1988) Early innervation and differentiation of hair cells in the vestibular epithelia of mouse embryos: SEM and TEM study. Anat Embryol (Berl) 177:331–340.

Mburu P, Mustapha M, Varela A, Weil D, El-Amraoui A, Holme RH, Rump A, Hardisty RE, Blanchard S, Coimbra RS, Perfettini I, Parkinson N, Mallon AM, Glenister P, Rogers MJ, Paige AJ, Moir L, Clay J, Rosenthal A, Liu XZ, Blanco G, Steel KP, Petit C, Brown SD (2003) Defects in whirlin, a PDZ domain molecule involved in stereocilia elongation, cause deafness in the *whirler* mouse and families with *DFNB31*. Nat Genet 34:421–428.

Montcouquiol M, Rachel RA, Lanford PJ, Copeland NG, Jenkins NA, Kelley MW (2003) Identification of *Vangl2* and *Scrb1* as planar polarity genes in mammals. Nature 423: 173–177.

Muller U, Littlewood-Evans A (2001) Mechanisms that regulate mechanosensory hair cell differentiation. Trends Cell Biol 11:334–342.

Murdoch JN, Henderson DJ, Doudney K, Gaston-Massuet C, Phillips HM, Paternotte C, Arkell R, Stanier P, Copp AJ (2003) Disruption of scribble (*Scrb1*) causes severe neural tube defects in the *circletail* mouse. Hum Mol Genet 12:87–98.

Nishida Y, Rivolta MN, Holley MC (1998) Timed markers for the differentiation of the cuticular plate and stereocilia in hair cells from the mouse inner ear. J Comp Neurol 395:18–28.

Oganesian A, Poot M, Daum G, Coats SA, Wright MB, Seifert RA, Bowen-Pope DF (2003) Protein tyrosine phosphatase RQ is a phosphatidylinositol phosphatase that can regulate cell survival and proliferation. Proc Natl Acad Sci USA 100:7563–7568.

Oliver D, He DZ, Klocker N, Ludwig J, Schulte U, Waldegger S, Ruppersberg JP, Dallos P, Fakler B (2001) Intracellular anions as the voltage sensor of prestin, the outer hair cell motor protein. Science 292:2340–2343.

Otterstedde CR, Spandau U, Blankenagel A, Kimberling WJ, Reisser C (2001) A new clinical classification for Usher's syndrome based on a new subtype of *Usher's syndrome type I.* Laryngoscope 111:84–86.

Pack AK, Slepecky NB (1995) Cytoskeletal and calcium-binding proteins in the mammalian organ of Corti: cell type-specific proteins displaying longitudinal and radial gradients. Hear Res 91:119–135.

Pantelias AA, Monsivais P, Rubel EW (2001) Tonotopic map of potassium currents in chick auditory hair cells using an intact basilar papilla. Hear Res 156:81–94.

Pickles JO, Comis SD, Osborne MP (1984) Cross-links between stereocilia in the guinea pig organ of Corti, and their possible relation to sensory transduction. Hear Res 15: 103–112.

Pickles JO, von Perger M, Rouse GW, Brix J (1991) The development of links between stereocilia in hair cells of the chick basilar papilla. Hear Res 54:153–63.

Pujol R, Lavigne-Rebillard M, Lenoir M (1998) Development of the sensory and neural structures in the mammalian cochlea. In: Rubel EW, Popper AN, Fay RR (eds), Development of the Auditory System. New York: Springer-Verlag, pp. 146–193.

Ramanathan K, Fuchs PA (2002) Modeling hair cell tuning by expression gradients of potassium channel beta subunits. Biophys J 82:64–75.

Raphael Y, Volk T, Crossin KL, Edelman GM, Geiger B (1988) The modulation of cell adhesion molecule expression and intercellular junction formation in the developing avian inner ear. Dev Biol 128:222–235.

Raphael Y, Kobayashi KN, Dootz GA, Beyer LA, Dolan DF, Burmeister M (2001) Severe vestibular and auditory impairment in three alleles of *Ames waltzer* (*av*) mice. Hear Res 151:237–249.

Richardson GP, Crossin KL, Chuong CM, Edelman GM (1987) Expression of cell adhesion molecules during embryonic induction. III. Development of the otic placode. Dev Biol 119:217–230.

Richardson GP, Forge A, Kros CJ, Fleming J, Brown SD, Steel KP (1997) Myosin VIIA is required for aminoglycoside accumulation in cochlear hair cells. J Neurosci 17: 9506–9519.

Richardson GP, Forge A, Kros CJ, Marcotti W, Becker D, Williams DS, Thorpe J, Fleming J, Brown SD, Steel KP (1999) A missense mutation in *myosin VIIA* prevents aminoglycoside accumulation in early postnatal cochlear hair cells. Ann N Y Acad Sci 884:110–124.

Rosenblatt KP, Sun ZP, Heller S, Hudspeth AJ (1997) Distribution of Ca^{2+}-activated K^+ channel isoforms along the tonotopic gradient of the chicken's cochlea. Neuron 19: 1061–1075.

Rubel EW, Oesterle EC, Weisleder P (1991) Hair cell regeneration in the avian inner ear. Ciba Found Symp 160:77–96; discussion 96–102.

Ruben RJ (1967) Development of the inner ear of the mouse: a radioautographic study of terminal mitoses. Acta Otolaryngol Suppl 220:1–44.

Rüsch A, Lysakowski A, Eatock RA (1998) Postnatal development of type I and type II hair cells in the mouse utricle: acquisition of voltage-gated conductances and differentiated morphology. J Neurosci 18:7487–7501.

Rzadzinska AK, Schneider ME, Davies C, Riordan GP, Kachar B (2004) An actin molecular treadmill and myosins maintain stereocilia functional architecture and self-renewal. J Cell Biol 164:887–897.

Sans A, Chat M (1982) Analysis of temporal and spatial patterns of rat vestibular hair cell differentiation by tritiated thymidine radioautography. J Comp Neurol 206:1–8.

Scarfone E, Dememes D, Perrin D, Aunis D, Sans A (1988) Alpha-fodrin (brain spectrin) immunocytochemical localization in rat vestibular hair cells. Neurosci Lett 93:13–18.

Schneider ME, Belyantseva IA, Azevedo RB, Kachar B (2002) Rapid renewal of auditory hair bundles. Nature 418:837–838.

Schulte BA (1993) Immunohistochemical localization of intracellular Ca-ATPase in outer hair cells, neurons and fibrocytes in the adult and developing inner ear. Hear Res 65: 262–273.

Self T, Mahony M, Fleming J, Walsh J, Brown SD, Steel KP (1998) *Shaker-1* mutations reveal roles for myosin VIIA in both development and function of cochlear hair cells. Development 125:557–566.

Self T, Sobe T, Copeland NG, Jenkins NA, Avraham KB, Steel KP (1999) Role of myosin VI in the differentiation of cochlear hair cells. Dev Biol 214:331–341.

Si F, Brodie H, Gillespie PG, Vasquez AE, Yamoah EN (2003) Developmental assembly of transduction apparatus in chick basilar papilla. J Neurosci 23:10815–10826.

Siemens J, Kazmierczak P, Reynolds A, Sticker M, Littlewood-Evans A, Muller U (2002) The Usher syndrome proteins cadherin 23 and harmonin form a complex by means of PDZ-domain interactions. Proc Natl Acad Sci USA 99:14946–14951.

Siemens J, Lillo C, Dumont RA, Reynolds A, Williams DS, Gillespie PG, Muller U (2004) Cadherin 23 is a component of the tip link in hair-cell stereocilia. Nature 428: 950–955.

Simonneau L, Gallego M, Pujol R (2003) Comparative expression patterns of T-, N-, E-cadherins, beta-catenin, and polysialic acid neural cell adhesion molecule in rat cochlea during development: implications for the nature of Kolliker's organ. J Comp Neurol 459:113–126.

Slepecky N, Chamberlain SC (1985) Immunoelectron microscopic and immunofluorescent localization of cytoskeletal and muscle-like contractile proteins in inner ear sensory hair cells. Hear Res 20:245–260.

Slepecky NB (1996) Structure of the mammalian cochlea. In: Dallos P, Popper AN, Fay RR (eds), The Cochlea. New York: Springer-Verlag, pp. 44–129.

Smith CA, Konishi M, Schuff N (1985) Structure of the barn owl's (*Tyto alba*) inner ear. Hear Res 17:237–247.

Smolders JW (1999) Functional recovery in the avian ear after hair cell regeneration. Audiol Neurootol 4:286–302.

Sobin A, Flock A (1983) Immunohistochemical identification and localization of actin and fimbrin in vestibular hair cells in the normal guinea pig and in a strain of the waltzing guinea pig. Acta Otolaryngol 96:407–412.

Sollner C, Rauch GJ, Siemens J, Geisler R, Schuster SC, Muller U, Nicolson T (2004) Mutations in *cadherin 23* affect tip links in zebrafish sensory hair cells. Nature 428: 955–959.

Souter M, Nevill G, Forge A (1995) Postnatal development of membrane specializations of gerbil outer hair cells. Hear Res 91:43–62.

Souter M, Nevill G, Forge A (1997) Postnatal maturation of the organ of Corti in gerbils: morphology and physiological responses. J Comp Neurol 386:635–651.

Steel KP, Brown SD (1996) Genetics of deafness. Curr Opin Neurobiol 6:520–525.

Stone JS, Rubel EW (2000b) Temporal, spatial, and morphologic features of hair cell regeneration in the avian basilar papilla. J Comp Neurol 417:1–16.

Strutt DI, Weber U, Mlodzik M (1997) The role of RhoA in tissue polarity and Frizzled signalling. Nature 387:292–295.

Strutt D, Johnson R, Cooper K, Bray S (2002) Asymmetric localization of *frizzled* and the determination of notch-dependent cell fate in the *Drosophila* eye. Curr Biol 12: 813–824.

Swanson GJ, Howard M, Lewis J (1990) Epithelial autonomy in the development of the inner ear of a bird embryo. Dev Biol 137:243–257.

Takeichi M (1995) Morphogenetic roles of classic cadherins. Curr Opin Cell Biol 7: 619–627.

Tilney LG, DeRosier DJ (1986) Actin filaments, stereocilia, and hair cells of the bird cochlea. IV. How the actin filaments become organized in developing stereocilia and in the cuticular plate. Dev Biol 116:119–129.

Tilney LG, Tilney MS, Saunders JS, DeRosier DJ (1986) Actin filaments, stereocilia, and hair cells of the bird cochlea. III. The development and differentiation of hair cells and stereocilia. Dev Biol 116:100–118.

Tilney MS, Tilney LG, DeRosier DJ (1987) The distribution of hair cell bundle lengths and orientations suggests an unexpected pattern of hair cell stimulation in the chick cochlea. Hear Res 25:141–151.

Tilney LG, Tilney MS, Cotanche DA (1988a) New observations on the stereocilia of hair cells of the chick cochlea. Hear Res 37:71–82.

Tilney LG, Tilney MS, Cotanche DA (1988b) Actin filaments, stereocilia, and hair cells of the bird cochlea. V. How the staircase pattern of stereociliary lengths is generated. J Cell Biol 106:355–365.

Tilney MS, Tilney LG, Stephens RE, Merte C, Drenckhahn D, Cotanche DA, Bretscher A (1989) Preliminary biochemical characterization of the stereocilia and cuticular plate of hair cells of the chick cochlea. J Cell Biol 109:1711–1723.

Tilney LG, Tilney MS, DeRosier DJ (1992) Actin filaments, stereocilia, and hair cells: how cells count and measure. Annu Rev Cell Biol 8:257–274.

Troutt LL, van Heumen WR, Pickles JO (1994) The changing microtubule arrangements in developing hair cells of the chick cochlea. Hear Res 81:100–108.

Tsukita S, Furuse M (2000a) Pores in the wall: claudins constitute tight junction strands containing aqueous pores. J Cell Biol 149:13–16.

Tsukita S, Furuse M (2000b) The structure and function of claudins, cell adhesion molecules at tight junctions. Ann NY Acad Sci 915:129–135.

Verpy E, Leibovici M, Zwaenepoel I, Liu XZ, Gal A, Salem N, Mansour A, Blanchard S, Kobayashi I, Keats BJ, Slim R, Petit C (2000) A defect in harmonin, a PDZ domain-containing protein expressed in the inner ear sensory hair cells, underlies Usher syndrome type 1C. Nat Genet 26:51–55.

Warchol ME, Corwin JT (1996) Regenerative proliferation in organ cultures of the avian cochlea: identification of the initial progenitors and determination of the latency of the proliferative response. J Neurosci 16:5466–5477.

Weaver SP, Hoffpauir J, Schweitzer L (1994) Distribution of actin in developing outer hair cells in the gerbil. Hear Res 72:181–188.

Weil D, El-Amraoui A, Masmoudi S, Mustapha M, Kikkawa Y, Laine S, Delmaghani S, Adato A, Nadifi S, Zina ZB, Hamel C, Gal A, Ayadi H, Yonekawa H, Petit C (2003) Usher syndrome type I G (USH1G) is caused by mutations in the gene encoding SANS, a protein that associates with the USH1C protein, harmonin. Hum Mol Genet 12:463–471.

Wells AL, Lin AW, Chen LQ, Safer D, Cain SM, Hasson T, Carragher BO, Milligan RA, Sweeney HL (1999) Myosin VI is an actin-based motor that moves backwards. Nature 401:505–508.

Whitlon DS (1993) E-cadherin in the mature and developing organ of Corti of the mouse. J Neurocytol 22:1030–1038.

Wilcox ER, Burton QL, Naz S, Riazuddin S, Smith TN, Ploplis B, Belyantseva I, Ben-Yosef T, Liburd NA, Morell RJ, Kachar B, Wu DK, Griffith AJ, Friedman TB (2001) Mutations in the gene encoding tight junction claudin-14 cause autosomal recessive deafness *DFNB29*. Cell 104:165–172.

Wright A, Davis A, Bredberg G, Ulehlova L, Spencer H (1987) Hair cell distributions in the normal human cochlea. Acta Otolaryngol Suppl 444:1–48.

Wright MB, Hugo C, Seifert R, Disteche CM, Bowen-Pope DF (1998) Proliferating and migrating mesangial cells responding to injury express a novel receptor protein-tyrosine phosphatase in experimental mesangial proliferative glomerulonephritis. J Biol Chem 273:23929–23937.

Xiang M, Gao WQ, Hasson T, Shin JJ (1998) Requirement for Brn-3c in maturation and survival, but not in fate determination of inner ear hair cells. Development 125:3935–3946.

Ylikoski J, Pirvola U, Narvanen O, Virtanen I (1990) Nonerythroid spectrin (fodrin) is a prominent component of the cochlear hair cells. Hear Res 43:199–203.

Zheng J, Shen W, He DZ, Long KB, Madison LD, Dallos P (2000) Prestin is the motor protein of cochlear outer hair cells. Nature 405:149–155.

Zheng JL, Gao WQ (1997) Analysis of rat vestibular hair cell development and regeneration using calretinin as an early marker. J Neurosci 17:8270–8282.

Zheng L, Sekerkova G, Vranich K, Tilney LG, Mugnaini E, Bartles JR (2000) The *deaf jerker* mouse has a mutation in the gene encoding the espin actin-bundling proteins of hair cell stereocilia and lacks espins. Cell 102:377–385.

Zine A, Romand R (1996) Development of the auditory receptors of the rat: a SEM study. Brain Res 721:49–58.

Zine A, Hafidi A, Romand R (1995) Fimbrin expression in the developing rat cochlea. Hear Res 87:165–169.

7

Developmental Genes Associated with Human Hearing Loss

RONNA HERTZANO AND KAREN B. AVRAHAM

1. Introduction

Hearing loss due to genetic mutations affects approximately 60% of persons with a form of hearing impairment. In most of these cases, a reduction in the ability to hear is due to a mutation in a single gene. The genes known today to be involved in deafness encode a large variety of proteins, including transcription factors, ion channels, molecular motors, gap junctions, and proteins that form the extracellular matrix of the inner ear. Many of these genes are expressed in the inner ear during development and the mutations leading to deafness cause a portion of their damage during embryogenesis. In this chapter, the genes associated with hearing loss and the potential role of each during inner ear development are reviewed.

Human hearing impairment can be caused by environmental and/or genetic factors, including exposure to ototoxic drugs, rubella during pregnancy, trauma, excessive noise, or mutations in one of the 25,000 genes that define our genome. Even for hearing loss caused by environmental factors, modifying genes may influence the onset or severity of the hearing impairment.

Hearing loss may be isolated, in the form of nonsyndromic hearing loss (NSHL), often associated with vestibular dysfunction. NSHL accounts for 70% of genetic hearing loss. Alternatively, the hearing loss may be one of several abnormalities including diabetes, peripheral neuropathy, kidney disease, craniofacial abnormalities, dwarfism, or retinitis pigmentosa, to name a few, to define syndromic hearing loss (SHL). Approximately 30% of genetic hearing loss is syndromic in nature. More than 500 syndromes have been described with hearing impairment as one of the features (NCBI Online Mendelian Inheritance in Man, http://www.ncbi.nlm.nih.gov/Omim) (Petit 2001; Friedman and Griffith 2003).

Genetic hearing loss is most often monogenic, attributable to mutations in one gene per individual or family. To date, the chromosomal locations for almost 90 loci associated with human NSHL are known (Hereditary Hearing Loss Homepage, http://http://www.uia.ac.be/dnalab/hhh/). These loci are defined as *DFNA* for autosomal dominant inherited deafness, *DFNB* for autosomal recessive inherited deafness, or *DFN* for X-linked loci. Mitochondrial mutations are also responsible for a small portion, approximately 1%, of hereditary hearing loss. Recently, two new classes of loci, *DFNM* for modifiers, genes that influence the expression or function of other genes (Riazuddin et al. 2000), and *OTSC* for otosclerosis, have been added to the nomenclature. Otosclerosis is a common disorder of the otic capsule of the human temporal bone characterized by progressive conductive hearing impairment ranging up to 60 dB, which might develop into mixed or even sensorineural hearing loss (SNHL) (Menger and Tange 2003). Almost 40 of the *DFNA*, *DFNB*, and *DFN* loci have been cloned and mutations in these genes leading to different types of hearing loss defined (Hereditary Hearing Loss Homepage; Deafness Gene Mutation Database, http://hearing.harvard.edu/db/genelist.htm).

Onset of hearing loss varies for each form of deafness. Most often there is a correlation between recessive inheritance and early onset of hearing loss. Mutations in genes that lead to recessively inherited congenital hearing loss would be expected to lead to a defect during inner ear development. There are exceptions, however; three different myosin IIIA mutations lead to late-onset progressive hearing loss in recessively inherited DFNB30 (Walsh et al. 2002). Late-onset progressive hearing loss is usually associated with post-developmental changes, although these genes may have an important role in development of the inner ear. For example, late-onset DFNA15 deafness is due to mutations in the *POU4F3* gene, a transcription factor essential for hair-cell development and survival (Vahava et al. 1998; Xiang et al. 1998).

Most information regarding the role of human deafness genes in development has come from research with mouse models for deafness. As early as 1907, Robert M. Yerkes described the "dancing mouse" as a "structural variation or mutation that occasionally appears in *Mus musculus*, and causes those peculiarities of movement which are known as dancing" (Yerkes 1907). Today we know that many of these mouse mutants exhibit a circling or waltzing behavior caused by mutations in genes associated with both vestibular dysfunction and deafness (Ahituv and Avraham 2002; Anagnostopoulos 2002).

How do we define genes essential for development? The basis for developmental involvement may be whether the pathology is caused during embryonic development or it is sufficient that the gene is expressed during development, even if the damage is postnatal (which is often dependent on the type of mutation, rather than the developmental role of a gene). The development of the inner ear is achieved through a series of highly regulated and complex processes involving a multitude of intrinsic and extrinsic signaling cascades. Mutations in early developmental genes will cause severe malformations in the inner ear. Of the genes known to be involved in otic placode induction or patterning of the otocyst, only *POU3F4* and *EYA4* has been found to be involved in human hearing loss (de Kok et al. 1995; Wayne et al. 2001). However, hair cell and sensory epithelium development continues in the three weeks after birth, including establishment of mechanoelectrical transduction, development of electromotility, expression of basolateral channels, and synapse formation. All genes involved in these pathways may be candidates for human hearing loss. A selection of genes associated with NSHL will be described, grouped according to protein or functional classification. The present authors chose to summarize what is known about a significant number of genes associated with hearing loss, focusing primarily on NSHL. For a list of all genes associated with hearing loss, refer to the Hereditary Hearing Loss Homepage.

2. Myosins

Myosins were among the first group of proteins found to be associated with hereditary hearing loss. These motor molecules are divided into 17 classes based on analysis of their motor and tail domains and numbered in the order in which they were discovered (Myosin Homepage, http://www.mrc-lmb.cam.ac.uk/myosin/myosin.html). The myosins are comprised of a heavy chain with a conserved approximately 80 kDa catalytic domain (the head or motor domain), and most are followed by an α-helical light chain-binding region (the neck region) (Sellars 1999). All myosins contain an actin-binding domain and an ATP-binding domain in their head region, allowing them to move along actin filaments. Most myosins contain a C-terminal tail and in some cases, an N-terminal extension as well. The tail domains diverge from one another between myosin classes and are believed to confer the function of each different myosin. Following is a number of myosins that when mutated, were found to compromise both auditory and vestibular function, manifested clinically by various forms of hearing impairment.

2.1 MYO7A/DFNB3/DFNA11

Myosin VIIA mutations were first detected in patients with a syndromic form of HL, USH1B, associated with retinitis pigmentosa (RP) (Weil et al. 1995). The shaker 1 (*sh1*) mouse was instrumental in the elucidation of the human gene (Gibson et al. 1995). Two years later, mutations associated with NSHL, *DFNB2*, and *DFNA11* were discovered, although much rarer than those contributing to USH1B (Liu et al. 1997a, b). *DFNB2* is an autosomal recessive form of NSHL, identified in Chinese and Tunisian families. The mutation in the Tunisian family is a G1797A missense, changing a methionine to isoleucine at the end of exon 15, leading to a decrease in splicing efficiency. Affected DFNA11 family members from Japan have moderate progressive sensorineural hearing loss, with an in-frame nine-base-pair deletion that results in the loss of three amino acids (Tamagawa et al. 2002). This deletion occurs in the coiled-coiled region of the myosin VIIA tail and thus is predicted to prevent dimerization of myosin VII, leading to a dominant-negative effect. Most recently, an American family of English descent was described with a missense mutation in the motor domain (Street et al. 2004). This evolutionarily conserved glycine is converted to an arginine, and may cause a structural change in the protein by disrupting the hydrophobic pocket formed by amino acid side chains in the converter domain and relay loop of the myosin head. None of the patients examined with the recessive or dominant forms of hearing loss have RP.

The mouse model for DFNB3 is the shaker1 (*sh1*) mouse, discovered in the late 1920s. Ten recessive *sh1* alleles have been discovered over the years, having arisen spontaneously or generated by radiation. The *sh1* locus lies on mouse chromosome 7, in the region homologous to human chromosome 11q13.5 containing the *DFNB3/A11* and *USH1* loci. Indeed, the discovery of a myosin VIIa exon during the *sh1* cloning process lead to the identity of this gene's involvement in human hearing loss and RP (Gibson et al. 1995). However, the mice have no visual abnormalities, and while electroretinogram (ERG) amplitudes were lower, the thresholds were normal. However, no retinal degeneration has been seen in *sh1* mutant mice, despite these electrophysiological abnormalities (Libby and Steel 2001).

The myosin VIIA gene is expressed in the cochlear and vestibular neuroepithelia of otocyst derived from 7- and 8-week-old human embryos (Weil et al. 1996). In frog, mouse, rat, and guinea pig inner ears, myosin VIIa is expressed in both the cochlea and vestibular epithelium (Hasson et al. 1997). Myosin VIIa is expressed in the cell bodies, the cuticular plate (the region immediately under the stereocilia), and along the length of the stereocilia. Myosin VIIa is first seen in the epithelium of the otic vesicle at embryonic day 9 (E9), as well as in the statoacoustic ganglion (Sahly et al. 1997; Boeda et al. 2001). During development, myosin VIIa initiates expression at E13.5 in the otocyst and by E15.5 is strongly expressed in cochlear and vestibular sensory hair cells (Xiang et al. 1998) (Fig. 7.1A). In the zebrafish (*Danio rerio*), myosin VIIa is found in the otic vesicle 24 h post-fertilization (h.p.f.), expands to the sensory patches and is then detected in the neuromasts of the anterior and posterior lateral lines (Ernest et al. 2000) (Fig. 7.1B).

Several proteins have been shown to interact with myosin VIIA, revealed through the use of yeast two-hybrid screens, including MyRIP, a novel Rab effector (El-Amraoui et al. 2002); Keap1, a human homolog of the *Drosophila* ring canal protein, kelch (Velichkova et al. 2002); the type I alpha regulatory subunit (RI alpha) of protein kinase A (Kussel-Andermann et al. 2000a); and vezatin, a novel transmembrane protein that bridges myosin VIIA to the cadherin–catenins complex (Kussel-Andermann et al. 2000b). Myosin VIIA is implicated in the development of the hair bundle, revealed by a study of its expression with several interacting proteins, cadherin 23 and harmonin (Boeda et al. 2002). These interactions have provided evidence that myosin VIIa is essential for shaping the hair bundle. Harmonin, which is associated with mutations in *USH1C* (Bitner-Glindzicz et al. 2000; Verpy et al. 2000), directly interacts with cadherin 23. These two proteins are expressed together in the developing stereocilia, but disappear in the adult hair bundle.

In *sh1* mice lacking myosin VIIa, the resting tension that is maintained by the adaptation motor is absent, allowing channels to be open even in the absence

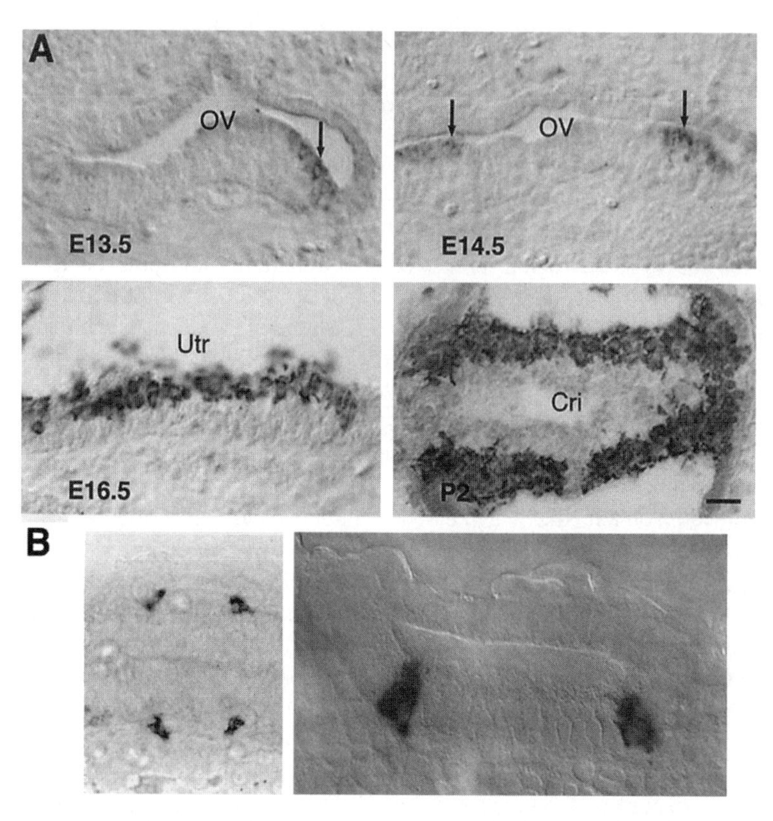

FIGURE 7.1. Myosin VIIa expression in (**A**) the developing mouse inner ear (adapted from Xiang et al. 1998) and from (**B**) zebrafish. In the wild type zebrafish, in situ hybridization reveals expression in the sensory cells of 24 hpf otic vesicles (adapted from Ernest et al. 2000). Cri, Crista; hpf, hours post-fertilization; OV, otic vesicle; Utr, utricle.

of stimuli, requiring unphysiologically large bundle deflections to open the transduction channels (Kros et al. 2002). *Myo7a* missense and nonsense mutations in mariner larvae lead to defects in mechanotransduction and inhibition of apical hair-cell endocytosis (Ernest et al. 2000). The mariner mutant is a circling zebrafish with inner ear hair bundle defects, manifested as splaying of the stereocilia.

2.2 MYO6/DFNB37/DFNA22

In humans, mutations in myosin VI (*MYO6*) underlie both syndromic and nonsyndromic hearing loss. The association of myosin VI (*Myo6*) with hereditary HL was first reported in *Snell's waltzer* mice (Avraham et al. 1995). In *sv/sv* mice lacking myosin VI protein, the stereocilia bundles become disorganized over time and the stereocilia fuse together, resulting in several giant stereocilia 20 days after birth (Self et al. 1999). In addition, studies of fibroblasts from *sv/sv* mice showed a reduction in both secretion and the size of the Golgi network (Warner et al. 2003). Mutations in *MYO6* were subsequently found to underlie NSHL DFNA22 and DFNB37 (Melchionda et al. 2001; Ahmed et al. 2003a) and SHL (Mohiddin et al. 2004).

Myosin VI functions as both an anchor and a transporter (Miller 2004). It was first discovered in *Drosophila* (Kellerman et al. 1992) but since then has been identified in many other organisms. Myosin VI, like other unconventional myosins, has a class-conserved head, neck and tail region (Mooseker and Cheney 1995). This motor has a number of defined features in the head region, including a an approximately 25-amino-acid insertion at the position of a surface loop and a conserved threonine residue. In the neck domain, myosin VI has a 53-amino-acid segment adjacent to its single IQ motif that differs from any known N-terminal junction of the neck domain of other myosins. The tail domain has a coiled-coil domain, like a number of other myosins, followed by a globular domain that is unique and highly conserved between the different organisms' myosin VI. The presence of a unique domain in the neck region of myosin VI led to the suspicion that myosin VI moves in the opposite direction along actin relative to other myosins, toward the "minus" end of actin filaments, which was validated using in vitro motility assays, where myosin VI was shown to move toward the pointed (minus) end of actin (Wells et al. 1999).

In mammalian cells, myosin VI is associated with the Golgi complex (Buss et al. 1998) and is involved in clathrin-coated vesicle formation, as well as in trafficking of uncoated nascent vesicles (Buss et al. 2001; Biemesderfer et al. 2002; Aschenbrenner et al. 2003). The mammalian myosin VI protein is expressed in a wide variety of tissues and cells (Hasson and Mooseker 1994; Avraham et al. 1995); however, in the inner ear, myosin VI is expressed solely in the sensory hair cells (Hasson et al. 1997; Fig. 7.2A) and is considered to be one of the earliest hair cell markers (Montcouquiol and Kelley 2003). In the hair cells, myosin VI is localized to the cuticular plate, pericuticular necklace and is throughout the cell body (Hasson et al. 1997). Known interacting proteins of myosin VI from other organ systems include GLUT1CBP (GIPC) (Bunn et al. 1999), DOC-2/DAB2 (Inoue et al. 2002), and SAP97 (Wu et al. 2002). Recently, myosin VI was discovered between the actin core and the lateral membrane, alongside the actin paracrystal, suggesting that it might either translocate molecules along the paracrystal or shape and stabilize the tapered base of the stereocilia (Rzadzinska et al. 2004).

In the zebrafish, myosin VI is duplicated, with one form, *myo6a*, expressed ubiquitously during early development and at later stages, with highest expression in the brain, kidney, and gut (Seiler et al. 2004). The second form, *myo6b*, not only is expressed primarily in the sensory epithelium of the ear and lateral line during development (Fig. 7.2B), but mutations in this gene are responsible for the *satellite* phenotype. *satellite* mutants have irregular and disorganized

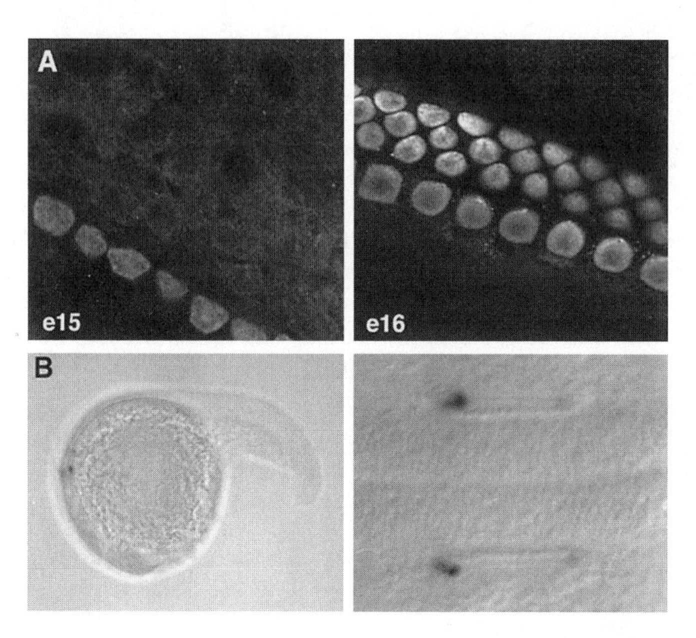

FIGURE 7.2. Myosin VI expression in (**A**) the E15 and E16 mouse sensory epithelium of the inner ear, using a myosin VI antibody (provided by Orit Ben-David, Tel Aviv University). (**B**) In zebrafish, myosin VIb is expressed in first hair cells of the otic placode at 24 hpf (adapted from Seiler et al. 2004).

stereocilia, eventually leading to fused stereocilia. Zebrafish are thus a suitable model for studying actin-based interactions of the plasma membrane, since *myo6b* is required for maintaining the integrity of the apical surface of hair cells.

2.3 MYO3A/DFNB30

Three mutations in the myosin IIIA (*MYO3A*) gene are associated with hearing loss in an Iraqi Israeli family (Walsh et al. 2002). Although the mode of inheritance was not clear upon examination of the family pedigree due to consanguinity, a genome scan revealed inheritance of two mutant alleles per hearing impaired individual. Once the critical region for the deafness locus *DFNB30* was defined to a 10-Mb region on human chromosome 10, candidates were examined for mutations segregating with the disorder. One nonsense and two splice mutations were found in affected members of the family, with either compound heterozygotes for the nonsense and one of two splice mutations or homozygotes for the nonsense mutation. The hearing impaired individuals are born with two defective copies of *MYO3A*, with a phenotype only appearing in the second or third decade of life. Since all mutations are predicted to cause a loss of function of *MYO3A*, it appears that myosin IIIA is not essential for development.

There is little information available regarding the role, or even expression, of myosin IIIA during development. In *Drosophila*, the myosin IIIA ortholog, NINAC, is one of several genes required for rapid deactivation of the photoreceptor in the eye following termination of light (Li et al. 1998). Found in the rhabdomere, NINAC interacts with actin filaments and the PDZ scaffolding protein INAD. Despite the role of NINAC in the eye, no visual abnormalities were detected in the family with *MYO3A* deafness (Walsh et al. 2002).

Expression of myosin IIIA by in situ hybridization has been examined and found as early as E16 (S. Vreugde and K. Avraham; unpublished observations), with higher expression at P0, and is found in both the inner and outer hair cells (Walsh et al. 2002). A mouse model is not yet available.

2.4 MYH9/DFNA17

The locus for NSHL DFNA17 was localized to the long arm of chromosome 22 containing 163 genes (Lalwani et al. 1999). As several other myosins were associated with hearing loss, the inner ear expression of one of the genes in the region encoding the nonmuscle-myosin heavy-chain A (*MYH9*), was examined to determine whether it was a relevant candidate. A member of the class II of myosins, *MYH9* is expressed in rat kidney, lung, and cochlea, revealed by reverse-transcriptase polymerase chain reaction (RT-PCR) (Lalwani et al. 2000). Further expression analysis by immunodetection demonstrated expression in the outer hair cells of the organ of Corti, the subcentral region of the spiral ligament, and Reissner's membrane. Subsequent mutation analysis of *MYH9* revealed a G to A transposition leading to a missense mutation, R705H. This arginine is well conserved, suggesting its functional importance in the resultant protein. Furthermore, this change occurs in a highly conserved linker region containing two free thiol groups that may play a role in conformational changes that occur in the myosin motor domain during force generation coupled to ATP hydrolysis.

In situ hybridization studies revealed that mouse *Myh9* is expressed within the epithelial layer of the otic vesicle at E10.5 and in the sensory cells of the developing cochlea at E16.5. Postnatally, Myh9 is expressed both within sensory hair cells and supporting cells, spiral ligament, and spiral limbus, but was not detected in the stria vascularis (Mhatre et al. 2004). There is no mouse model for the *MYH9* mutation.

2.5 MYO15A/DFNB3

Mutations in the myosin XVA (*MYO15A*) gene are associated with DFNB3 on chromosome 17p11.2, a recessively inherited form of sensorineural deafness (Wang et al. 1998). Originally found in a large extended family from Bengala, Bali, mutations in *MYO15A* have subsequently been identified in India and Pakistan (Liburd et al. 2001). Deaf members of the Balinese family are born with a profound hearing loss, suggesting that damage to the hair cells already occurs during embryonic development. Further insights are provided by the DFNB3 mouse model, shaker 2 (*sh2*) (Wang et al. 1998). Identification of the *sh2* gene facilitated the search for the DFNB3 locus. Transgenesis of a BAC containing the *Myo15a* gene "rescued" the deaf and circling phenotype of a *sh2* mutant, thereby leading to the identification of a mutation in this gene in the *sh2* mouse. As the mouse chromosome 11 region of the mouse is homologous to human chromosome 17p, *MYO15A* became an excellent candidate. Subsequent sequencing of the human gene in DFNB3 individuals revealed several mutations (Wang et al. 1998).

RNA analysis reveals expression in human fetal and adult brain, ovary, testis, kidney, and pituitary gland. Cochlea derived from 18 to 22-week-old human fetuses showed expression of *MYO15A* by RT-PCR of mRNA. In situ hybridization demonstrates mouse inner ear expression in the hair cells of the cochlea and the vestibular system, including the saccule, utricle, and crista ampullaris (Anderson et al. 2000). Immunodetection with an antibody raised against myosin XVA shows cochlear expression only in the hair cells, particularly in the stereocilia and tip link (Belyantseva et al. 2003). The earliest expression of myosin XVa was detected at E13.5, at the beginning stage of hair cell maturation (Fig. 7.3).

Mouse mutants with *Myo15a* mutations exhibit deafness and circling behavior

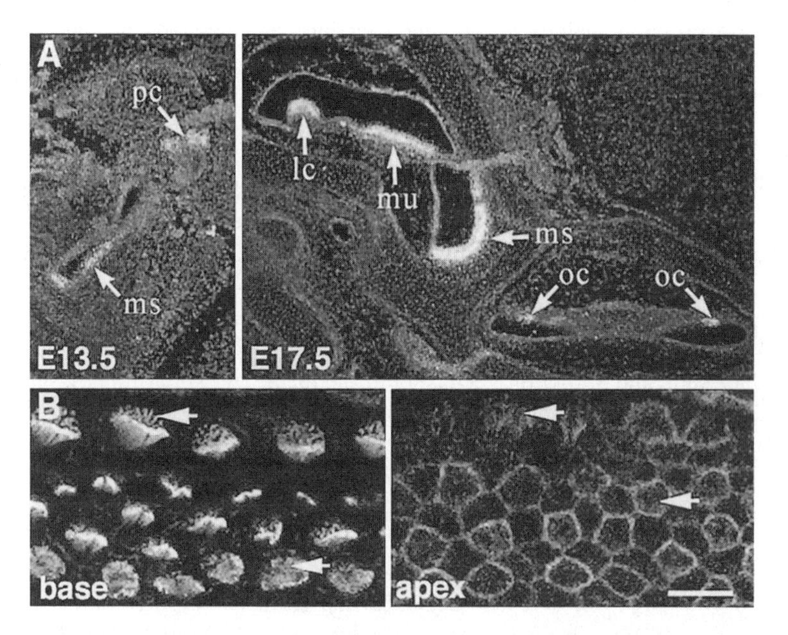

FIGURE 7.3. Myosin XVa expression seen by (**A**) in situ hybridization in the mouse inner ear at various developmental stages (adapted from Anderson et al. 2000). (**2**) Immunolabeling with a myosin XVa antibody reveals expression in an E18.5 mouse, with staining prominent in the tips of stereocilia in basal (top) but not found in apical (*bottom*) turn (adapted from Belyantseva et al. 2003). lc, Lateral cristae ampularis; ms, macula sacculi; mu, macula utriculi; oc, organ of Corti; pc, posterior cristae ampularis.

indicative of vestibular dysfunction (Wang et al. 1998). The stereocilia of shaker 2 (*sh2*) mice are shorter than normal, suggesting that their development is compromised due to *Myo15a* mutations. The arrangement of rows of outer and inner hair cells remains the same, although there are abnormal elongated structures with an actin core on both inner hair cells and vestibular hair cells. Thus myosin XVA appears to be essential for actin cytoskeleton organization.

3. Transcriptional Regulators

Transcription factors are key regulators of gene expression that control multiple developmental and physiological processes. Over two dozen transcription factors have been found to be essential for proper development of the mouse inner ear (MRC Institute of Hearing Research Mouse Mutants with Hearing or Balance Defects, http://www.sanger.ac.uk/deafmousemutants; Anagnostopoulos 2002) with roles ranging from the regulation of genes required to induce the development of the entire inner ear (e.g., *Hoxa1*), the semicircular canals and cochlear duct (e.g., *Otx1* and *Pax2*) or specific cell populations (e.g., *Math1*). To date, four transcriptional regulation genes have been associated with human NSHL (*EYA4*, *POU4F3*, *POU3F4*, and *TFCP2L3*), and several others with SHL (*EYA1*, *PAX2*, *FOXI1*, *MITF*, *PAX3*, and *SLUG*).

3.1 POU Transcription Factors

The POU-domain family of transcription factors was identified on the basis of amino acid sequence homology in the DNA binding domain of the transcription factors PIT1/GHF1, OCT1 and OCT2, and UNC86 (Herr et al. 1988). The POU domain consists of 147 to 156 amino acids and is comprised of two distinct DNA binding domains: a 69- to 78-amino-acid POU-specific domain located amino terminal to a 60-amino-acid POU homeodomain (Rosenfeld 1991). The two POU domains are separated by a variable linker, a flexible stretch of amino acids that increases the repertoire of the specific sequences to which these proteins can bind to and improves the kinetics of the binding (reviewed in Phillips and Luisi 2000).

The class III of POU domain genes were identified using PCR and degenerate oligonucleotides representing codons of the nine conserved amino acids in the original POU genes (He et al. 1989). All mammalian class III POU domain transcription factors are broadly expressed within the developing nervous system and assume more restricted expression patterns in the adult nervous system. Mouse knockout models for genes in this group reveal both the redundant and unique functions of these genes.

3.1.1 *POU3F4/DFN3*

The *POU3F4* (*BRN4*) gene has a clear role in development, manifested clinically, as children born with mutations in this gene suffer from congenital profound deafness. As described below, not only is the expression strong during early development of the embryo, but the *Pou3f4* mouse mutants have structural abnormalities in their auditory system during development. Mutations in the *POU3F4* transcription factor are associated with the *DFN3* locus on chromosome X (de Kok et al. 1995). The DFN3 phenotype is variable, and at the very least it is characterized by profound sensorineural hearing loss but is also often associated with conductive hearing loss and with stapes fixation. Furthermore, as *DFN3* maps to chromosome Xq21 in the region containing mental retardation and choroideremia, some patients with *POU3F4* mutations have additional symptoms due to deletions of larger portions of the chromosome. Both point mutations in the *POU3F4* gene, as well as larger deletions, duplications, and inversions of the Xq21 region are associated with the DFN3 phenotype. The deletions account for a little over half of *DFN3* mutations and many of them do not encompass the *POU3F4* coding region. A detailed molecular analysis of the region proximal to the *POU3F4* gene revealed small deletions 900 kb proximal to the gene that is associated with DFN3 deafness (de Kok et al. 1996). It was suggested that the DNF3 phenotype, without coding region mutations, could be caused by the loss of the *POU3F4* enhancer, repressor, or promoter.

Most of our knowledge regarding the role of POU3F4 during development comes from expression studies in the mouse, as well as from spontaneous and gene targeted mutagenesis models. Indeed, the *POU3F4* gene has a unique role in the development of the inner ear. During mouse development, *Pou3f4* is first expressed at E9.5 in the neural tube, and then in the otic capsule, the hindbrain, and the branchial arch mesenchyme (Phippard et al. 1998) (Fig. 7.4). Notably, the subcellular localization of this transcription factor shifts to the cytoplasm in areas of mesenchymal remodeling that will further develop to acellular structures, demonstrating a potential mechanism for crucial silencing of the gene during normal otic development.

Several *Pou3f4* mouse mutants are available and two of these, the sex-linked fidget mutation and a targeted null mutation, have both cochlear and temporal bone abnormalities (Phippard et al. 1999, 2000) (Fig. 7.4). These include a constricted superior semicircular canal, widening of the internal auditory meatus, thinning of various structures in the temporal bone; a misshaped stapes footplate; and a shortening of the cochlea, demonstrated as a reduction in the number of cochlear coils in most of the mutants. In addition, the mice have a generally hypoplastic cochlea, with widening of the scala tympani, and dysplasia of fibrocytes in the spiral limbus, that may lead to the hydrops observed in these animals. Notably, no structural abnormalities are observed in the organ of Corti. Since the Pou3f4 protein is expressed in mesenchymal tissue (Phippard et al. 1998), the widened structures may result from disruption of mesenchymal remodeling, the shortening of the cochlea from disruption of epithelial–mesenchymal interactions, and the mishaping of the stapes from disruption of mesenchymal–mesenchymal interactions. This phenotype resembles the temporal bone phenotype of the people suffering from *POU3F4* mutations.

Another mouse model created by a targeted deletion of the *Pou3f4* gene on

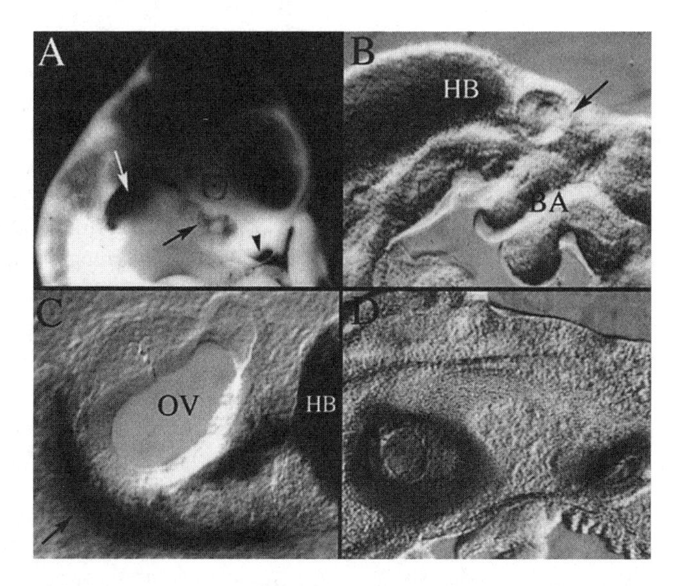

FIGURE 7.4. Gene-targeted mutagenesis, with the *lacZ* gene replacing *Pou3f4*, reveals its expression during development. (**A**) *Pou3f4* is expressed in most of the neuraxis and in some mesodermally derived tissues in the head, including the otic capsule. (**B**) *Pou3f4* is expressed in the hindbrain of a 9.5dpc embryo, but is not seen in the mesenchyme surrounding the otic vesicle. (**C**) In an E10.5 embryo, *Pou3f4* is expressed in the condensing mesenchyme of the otic vesicle. (**D**) Expression patterns of *lacZ* in a parasagittal section of a E14.5, *Pou3f4* is detected throughout the otic capsule (Phippard et al. 1999). BA, branchial arches; HB, hindbrain; OV, otic vesicle.

a different genetic background showed no gross temporal or inner ear defect, nor in the neuroepithelium or cochlear length, while the mice displayed profound deafness and a reduced endocochlear potential (Minowa 1999). The pathological effects appeared in the fibrocytes that line the stria vascularis that are hypothesized to be involved in potassium recycling back to the endolymph (Spicer and Schulte 1996).

3.1.2 *POU4F3/DFNA15*

A mutation in the *POU4F3* (*BRN3.1/BRN3C*) gene leads to progressive autosomal dominant hearing loss in an Israeli kindred (Vahava et al. 1998). While Pou4f3 has a clear role in development based on the recessive mouse mutant described below, in humans, there is only a late-onset effect. The *DFNA15* locus was discovered in a linkage project performed to determine the underlying cause of late-onset hearing loss in a family of Libyan descent, now living in Israel. Although close to the first autosomal dominant locus defined, *DFNA1*, on human chromosome 5q31, *DFNA15* defined a new locus. Examination of the mouse syntenic region revealed the presence of the *Pou4f3* gene, known to cause deafness when removed by gene-targeted mutagenesis in the mouse (Erkman et al. 1996; Xiang et al. 1996). Subsequent sequencing of the *POU4F3* gene in affected individuals revealed an eight-base-pair deletion in the second coding exon of the gene, which leads to a frameshift and premature stop codon. As a result, a bipartite nuclear localization signal is lost, and in a cell culture model of the mutation, leads to loss of expression in the nucleus (Weiss et al. 2003). The mutation does not appear to cause a dominant-negative effect, but rather may be unavailable to bind to its targets at a sufficient and necessary threshold over time.

In the mouse inner ear, Pou4f3 is a hair cell–specific protein (Fig. 7.5). In the adult mouse, all hair cell nuclei both in the auditory and vestibular systems express Pou4f3 (Erkman et al. 1996; Xiang et al. 1997). Pou4f3 protein can be detected as early as E12.5 in scattered cells of the developing otocyst (Xiang et al. 1998). Double staining of inner ear sections from bromodeoxy uridine (BrdU)-labeled embryos, with anti-BrdU and anti-Pou4f3 antibodies, revealed that Pou4f3 expression is confined to postmitotic cells (Xiang et al. 1998). Real-time RT-PCR of auditory sensory epithelia indicated that very low levels of *Pou4f3* mRNA can be detected as early as E12, and constantly increase thereafter, up to P3, the latest time point measured (Hertzano et al. 2004).

Mice with a targeted deletion of *Pou4f3* suffer from deafness and vestibular dysfunction, due to a loss of all hair cells by early postnatal days (Erkman et al. 1996; Xiang et al. 1997). Innervated hair cell–like cells with some sterociliary-like structures do form in the absence of *Pou4f3* and express hair cell markers such as myosin VI, myosin VIIa, parvalbumin, and calretinin (Xiang et al. 1998, 2003; Hertzano et al. 2004). However, the patterning of these cells is markedly disrupted and inner hair cell loss from the base of the cochleae can

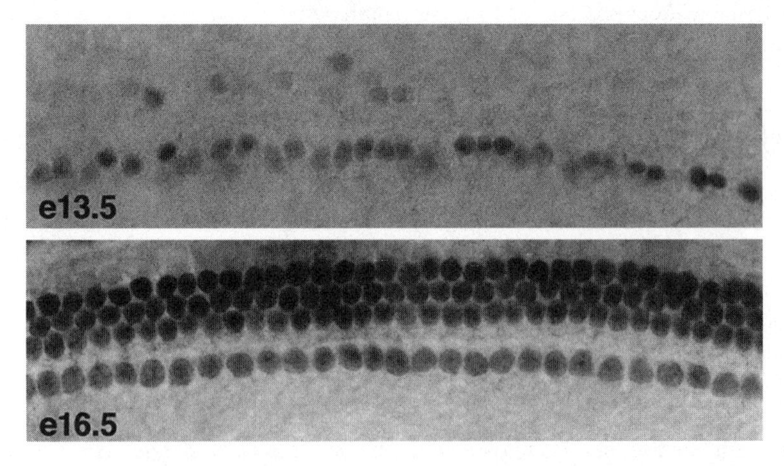

FIGURE 7.5. *Pou4f3* expression in E13.5 and E16.5 inner ears, demonstrated by immunostaining with a *Pou4f3* antibody (provided by Ronna Hertzano).

be detected as early as E16.5, accompanied by an increase in apoptotic cell death, as detected by a terminal deoxynucleotidyl transferase biotin-dUTP nick end labeling (TUNEL) assay, as early as E17.5 (Xiang et al. 1998; Hertzano et al. 2004). Reactive loss of sensory ganglion neurons and sensory epithelia supporting cells is also observed (Erkman et al. 1996; Xiang et al. 1997, 1998, 2003; Hertzano et al. 2004).

The gene encoding growth factor independence 1 (*Gfi1*), a known deafness gene in the mouse (Wallis et al. 2003), was recently identified as a target gene of Pou4f3 by using a transcription profiling approach of inner ears from wild type and *Pou4f3* mutant mice (Hertzano et al. 2004). *Gfi1* expression is markedly reduced in the ears of the *Pou4f3* mutant mice. A comparison of cochleae from *Pou4f3* and *Gfi1* mutants suggested that outer hair cell loss in the *Pou4f3*-deficient cochleae may result from the loss of *Gfi1* expression.

3.2 EYA4/DFNA10

The mammalian *Eya* gene family consists of four members that function as transcription coactivators, that share a highly conserved region called the eya-homologous region in the C-terminus and a less conserved transactivation domain at the N-terminus (Borsani et al. 1999). Mutations in the *Eyes absent 4* (*EYA4*) gene are associated with a dominant form of NSHL, DFNA10. Premature stop codons in the EYA4 protein cause late-onset progressive hearing loss in Belgian and American families (Wayne et al. 2001).

In situ hybridization results with a digoxigenin labeled RNA probe showed that at E9.5 *Eya4* is expressed in the otic vesicle, as well as in several other tissues including the nasal placode and a region above the developing forelimb bud (Borsani et al. 1999) (Fig. 7.6). Further characterization of the inner ear expression of *Eya4* by in situ hybridization with a radiolabeled RNA probe showed that in the rat inner ear *Eya4* is strongly expressed in the inner ear as early as E14.5, mainly in the upper half of the cochlear duct in cells that will

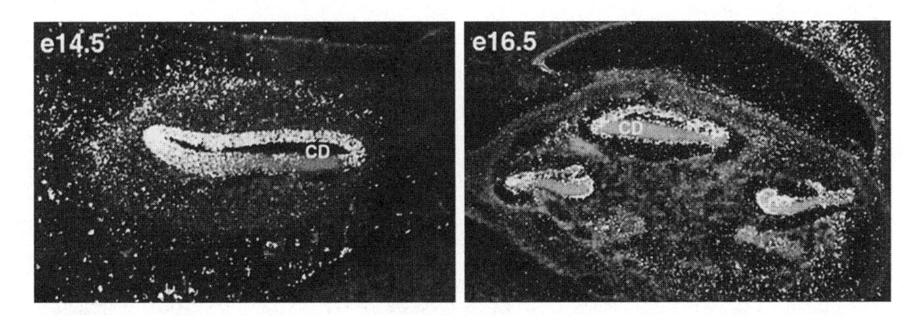

FIGURE 7.6. Eya4 expression in the developing rat cochlea at E14.5 and E16.5 in the upper half of the duct, cells that will form the stria vascularis and Reissner's membrane (adapted from Wayne et al. 2001). CD, Cochlear duct.

form the future stria vascularis and Reissner's membrane. A weak expression can also be detected in the mesenchyme surrounding the duct. Interestingly, toward E18.5, the expression of *Eya4* shifts toward the lower part of the cochlear duct, specifically the greater and lesser epithelial ridges and later to cells derived from the spiral limbus, the organ of Corti, and the spiral prominence. Expression is also detected in the cells of the ossifying bony capsule of the inner ear during the first 2 weeks after birth. In the developing vestibular system expression was reported to be localized mainly to the developing sensory epithelium (Wayne et al. 2001).

3.3 TFCP2L3/DFNA28

A frameshift mutation, leading to a premature translation stop codon in the transcription factor TFCP2L3 (transcription factor cellular promoter 2-like 3), causes autosomal dominant progressive NSHL in an extended American family (Peters et al. 2002). This five-generation family exhibits mild to moderate hearing loss, with the earliest reported onset at 7 years of age. A genome-wide scan defined a critical region of 1.4 cM on chromosome 8q22. While several candidate genes with known function were sequenced, no mutations were found. However, a previously uncharacterized gene, named *FLJ13782*, was found to contain a mutation that segregated with the hearing loss in the family. Thus a new member of the transcription factor cellular promoter family of genes was identified and named *TFCP2L3*.

The *TFCP2* family of transcription factors contain a novel DNA-binding domain and bind to many promoters, including HIV type 1 and simian virus 40 (Swendeman et al. 1994). *TFCP2L2*, another member of this family, is related to *Drosophila* grainyhead (*grh*), a gene involved in dorsal/ventral and terminal patterning of the newly fertilized embryo and later expressed in the central nervous system and in cuticle-producing tissues. Flies carrying mutations in *grh* have an embryonic lethal phenotype with flimsy cuticles, grainy and discontinuous head skeletons, and patchy tracheal tubes (reviewed in Wilanowski et al. 2002).

In situ hybridization results with a radiolabeled RNA probe show that in mouse inner ear, *Tfcp2l3* is expressed as early as E11.5 at the otocyst stage and later in development in the epithelia that surround the endolymph-containing compartments of the auditory and vestibular systems (Peters et al. 2002). While *TFCP2L3* is widely expressed in a variety of epithelial tissues, including prostate, thymus, kidney, mammary gland, pancreas, and digestive tract, the only pathological damage in the American family appears to be in the cochlear epithelial cells. Functional redundancy between members of the *TFCP2* transcription factors may protect other epithelial tissues in the family exhibiting hearing loss only. The progressive nature of the *TFCP2L3* mutation suggests that this transcription factor has a role in epithelial cell maintenance.

4. Gap Junction Proteins

One of the most dramatic discoveries in the field of hereditary hearing loss in recent years has been the large number of connexin mutations. Gap junction proteins encode the connexins, a component of connexons that allow molecules to pass from cell to cell. Two networks of gap junctions, the epithelial cell system and the connective tissue cell system, are functional in the cochlea (Kumar and Gilula 1996). Three connexins have been implicated in deafness: connexin 26 (*GJB2*), accountable for approximately 30% of childhood deafness (Denoyelle et al. 1997; Cohn et al. 1999); connexin 30 (*GJB6*) in NSHL (Lerer et al. 2001; del Castillo et al. 2002); and connexin 31 (*GJB3*) both in NSHL (Xia et al. 1998; Liu et al. 2000) and peripheral neuropathy and HL (Lopez-Bigas et al. 2001).

4.1 GJB2/DFNB1

The *DFNB1* locus, which is located on chromosome 13q11–12, was the first deafness recessive locus to be discovered. *DFNB1* has turned out to the most prevalent locus, and includes both the *GJB2* and *GJB6* genes. *GJB2* encodes connexin 26, and mutations in this gene account for 30% to 50% of congenital hearing loss (Denoyelle et al. 1997; Kelsell et al. 1997; Zelante et al. 1997; Kelley et al. 1998). More than 60 mutations have been discovered in the *GJB2* gene (Connexin-Deafness Homepage; http://www.crg.es/deafness/). One mutation, 35delG, accounts for the majority of mutant alleles and has been found in most parts of the world. The 167delT mutation is the second most prevalent *GJB2* mutation, and has been reported in both the Ashkenazi Jewish (Sobe et al. 1999) and Palestinian populations (Shahin et al. 2002).

Connexin 26 is expressed in the cochlea. At E14.5 in the mouse, connexin 26 is detected by in situ hybridization in the greater epithelial ridge, the region where Reissner's membrane and the tectorial membrane are attached (Buniello et al. 2004) (Fig. 7.7A). In the rat otocyst, connexin 26 was detected by immunofluorescence and appeared at E17 (Lautermann et al. 1999). By P3, connexin 26 was detected in the fibrocytes of the spiral limbus, the supporting cells of the neurosensory epithelium and between the stria vascularis and the spiral ligament. In a 22-week human embryo, connexin 26 was found in the cochlea.

A classical mouse knockout is available for *GJB2*, but its lethal phenotype has rendered it irrelevant for the human deafness phenotype. The mice show a lethal phenotype, with death at E11 due to a dysfunction of the placenta (Gabriel et al. 1998). A conditional knockout has played a much more relevant role in contributing to our understanding of the human pathology. Connexin 26 was inactivated using the Cre–*lox P* recombination system in order to remove this protein from the epithelial gap junction network that contains supporting cells and flanking epithelial cells (Cohen-Salmon et al. 2002). As a result, supporting cells of the inner hair cells succumbed to apoptotic cell death.

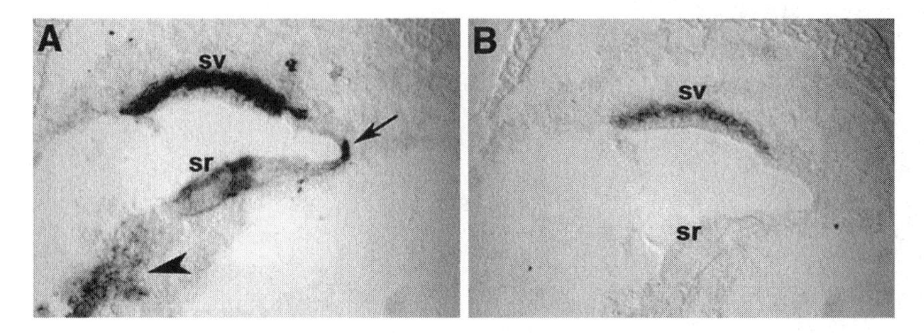

FIGURE 7.7. (**A**) Connexin 26 expression in the mouse otocyst at E14.5. (**B**) Connexin 30 expression in the mouse otocyst at E14.5 (adapted from Buniello et al. 2004). sr, Sensory region. sv, stria vascularis.

4.2 GJB6/DFNB1

Connexin 30 is encoded by the *GJB6* gene and is located on chromosome 13 in the *DFNB1* region. A 342-kb deletion that truncates the *GJB6* gene, del(*GJB6*–D13S1830), leads to autosomal recessive hearing loss (Lerer et al. 2001; del Castillo et al. 2002). However, this mutation is usually found in the heterozygous state in conjunction with a heterozygote *GJB2* mutation. A multicenter study revealed that this mutation is the second most prevalent mutation in Spain after the *GJB2* 35delG mutation, and significantly reduced the number of unexplained *GJB2* heterozygotes in Spain, France, and Israel (del Castillo et al. 2003). Furthermore, haplotype analysis suggests that there may be a common founder for individuals from countries in Western Europe.

In situ hybridization on E14.5 mice demonstrated that connexin 30 is expressed in the stria vascularis (Buniello et al. 2004) (Fig. 7.7B). In the rat otocyst at E17, connexin 30 was detected by immunofluorescence at low levels in the neurosensory epithelium and gradually increased to higher levels postnatally in the fibrocytes of the spiral limbus, the supporting cells of the neurosensory epithelium and between the stria vascularis and the spiral ligament. In a 22-week human embryo, connexin 30 was found in the cochlea (Lautermann et al. 1999).

The connexin 30 gene-targeted mutation in mice leads to severe hearing impairment, manifesting itself as a loss of the endocochlear potential (Teubner et al. 2003). Furthermore, the cochlear sensory epithelial cells degenerated by apoptosis.

5. Intercellular Adhesion Proteins

5.1 CDH23/DFNB12

Cadherin 23 (*Cdh23*) was first identified upon the discovery that mutations in this gene lead to stereocilia disorganization in the *waltzer* mouse mutant (Di Palma et al. 2001; Wilson et al. 2001). Both Usher syndrome type 1D and autosomal recessive DFNB12 are caused by mutations in *CDH23* (Bork et al. 2001). Cdh23 encodes a large single-pass transmembrane protein, with 20 cadherin repeat (EC) domains.

Expression of Cdh23 is apparent in the P0 and P5 hair bundle and in Reissner's membrane (Siemens et al. 2004). By P42, expression is still in the cochlear hair cells, but is confined to the stereociliary tips. Together with the fact that cadherin 23 forms a complex with myosin 1c, a member of the mechanotransduction apparatus, suggests that this protein is involved in regulating the activity of mechanically gated ion channels in hair cells. In the zebrafish, cadherin 23 affects the tip links in the sensory hair cells, found in the sputnik mutant (Sollner et al. 2004).

5.2 PCDH15/DFNB23

The protocadherin 15 gene (*PCDH15*) encodes a protein that is a member of the cadherin superfamily of calcium-dependent cell adhesion molecules and is thought to be involved in neural development, neural circuit formation, and formation of the synapse (Suzuki 2000). The gene is localized to chromosome 10q11.2–q21. Mutations in *PCDH15* were first identified in Usher syndrome type 1F patients (Ahmed et al. 2001; Alagramam et al. 2001b). These patients have congenital profound sensorineural hearing loss, vestibular areflexia, and retinitis pigmentosa that begins around puberty. The mutations include several premature stop codons and a splice mutation. *PCDH15* missense mutations were subsequently found to lead to autosomal recessive DFNB23 as well (Ahmed et al. 2003b). As in other cases, a mouse mutant was instrumental in the human gene identification. *Pcdh15* was first identified when it was found to be the causative gene for the deafness and circling in the *Ames waltzer* (*av*) mouse (Alagramam et al. 2001a). After *Pcdh15* mutations were found in the *av* mutant, the human homolog, *PCDH15,* became a candidate for USH1F, as a result of the homology between human chromosome 10q and mouse chromosome 10.

Pcdh15 mutant mice have inner ear defects in the cochlea and the saccule (Alagramam et al. 1999; Alagramam et al. 2001a). The stereocilia of both inner and outer hair cells of P10 mice are disorganized. Protocadherin 15 may play a role in regulating planar polarity in the sensory epithelium of the inner ear.

Protocadherin 15 is expressed in the neurosensory epithelium in both the eye and ear, consistent with its association with damage to both organ systems (Alagramam et al. 2001b; Ahmed et al. 2003b). Protocadherin 15 was found in human and monkey photoreceptors of the retina, revealed using an antibody specific for this protein. Immunolocalization in the inner ear detected protocadherin 15 in both the mouse organ of Corti and vestibular hair cells. Specifically, expression was found along the length of the stereocilia, in the cuticular plate, and diffused in the cytoplasm of both inner and outer hair cells. Protocadherin 15, examined from E16 in the organ of Corti, was seen in the stereocilia as soon as they become apparent on the apical surface of the hair cells.

6 Miscellaneous

6.1 WFS1/DFNA6/A14/A38

Mutations in wolframin (*WFS1*) contribute to Wolfram syndrome defined by juvenile diabetes mellitus, optic atrophy, and hearing loss, as well as DFNA6/A14/A38 progressive high-frequency NSHL (Bespalova et al. 2001; Young et al. 2001). Several different types of mutations are associated with WS, including stop, frameshift, deletion, and missense mutations (Khanim et al. 2001). Five different heterozygote missense mutations are associated with NSHL localized to human chromosome 4. Interestingly, in the DFNA38 family, one individual is homozygote for the missense mutation and in addition to hearing loss, has clinical features of WS.

WFS1 encodes an integral endoglycosidase H-sensitive membrane glycoprotein with eight to ten predicted transmembrane domains that is speculated to be involved in membrane trafficking, protein processing and/or calcium homeostasis (Takeda et al. 2001). The subcellular localization of WFS1 was examined in cultured cells and in rat brain. WFS1 is expressed in the endoplasmic reticulum and in neurons in the hippocampus CA1, amygdaloid areas, olfactory tubercle, and superficial layer of the allocortex. Immunohistochemistry and in situ hybridization analysis in the postnatal inner ear reveals expression in many cell types, including vestibular hair cells, and in the cochlea; in the inner and outer hair cells; spiral ganglion; external and inner sulcus cells; marginal cells of the stria; Hensen, Claudius, Deiter's and interdental cells; and at lower levels in Reissner's membrane and pillar cells (Cryns et al. 2003). The strong expression at postnatal day 1 suggests that wolframin has a role in inner ear development, although its persistence (albeit reduced) until and through maturation suggests that it also has a maintenance role. WFS1 also colocalizes with the ER marker calreticulin. These data are in accordance with the cellular expression described above, suggesting that WFS1 may play a role in inner ear ion homeostasis as maintained by the canalicular reticulum.

6.2 TMC1/DFNB7/B11/A36

Mutations in the gene encoding the transmembrane cochlear 1 (TMC1) protein are associated both with recessive and dominant inherited hearing loss, DFNB7/B11 and DFNA36 (Kurima et al. 2002). Fortuitously, mutations in the mouse orthologue, *Tmc1*, are associated with deafness in two mouse mutants, deafness (*dn*) and Beethoven (*Bth*) (Vreugde et al. 2002). Neither of these mice circle, indicating that *Tmc1* mutations are associated only with hearing loss and hair cells of the cochlea.

Tmc1 does not appear to have an early developmental role, based on the mutant mouse phenotypes and expression analysis. Real-time quantitative RT-PCR analysis demonstrated low levels of expression in temporal bone RNA from E14-P0, with a rise at P5 that levels off at P10 and is reduced again at P20 (Kurima et al. 2002). In situ hybridization revealed *Tmc1* expression in both cochlea and vestibule; specifically, in inner and outer hair cell of the cochlea and neurosensory epithelia of the vestibular end organs.

6.3 STRC/DFNB16

Mutations in the gene *STRC* encoding the hair bundle protein stereocilin are associated with DFNB16 deafness (Verpy et al. 2001). Stereocilin was identified during a candidate gene approach to identify genes expressed in the inner ear. Subtracted mouse inner ear cDNA libraries were constructed from whole cochlea or sensory regions of the vestibule. A clone with homology to human genomic clones from chromosome 15q15 was identified. Fortuitously, the *DFNB16* locus had already been mapped to this region (Campbell et al. 1997). Two frameshift mutations and a large deletion were found in families linked to the *DFNB16* locus.

Expression analysis demonstrated that stereocilin is located in the six sensory areas of the inner ear during the stages tested, from P6 to P20 (Verpy et al. 2001). Based on the expression analysis and amino acid sequence analysis, stereocilin is predicted to be an integral protein of the stereocilia or a cell surface protein associated with the hair bundle. As the onset of stereocilin expression is relatively late, it is unlikely to play a role in the early development of the inner ear, and rather have a function associated with mature hair bundles.

7. Conclusion

Identification of human deafness genes and elucidation of the function of the proteins these genes encode has revealed a plethora of details that only begin to form our understanding of the intricacies of the development of the inner ear. While the relevant clinical information is derived from the discovery of the mutations, manifested in genetic counseling for affected families and early detection for habilitation (Greinwald and Hartnick 2002), our understanding of inner ear function comes primarily from expression in the mouse and analysis of mutant mice. Eventually, elucidation of the developmental pathways and mechanisms of auditory function will lead to successful intervention and therapeutics for alleviating hearing loss, including replacement of lost sensory hair cells (Minoda et al. 2004).

Acknowledgments. Research support in the Avraham laboratory is provided by the Israel Science Foundation, NIH grant R01 DC005641, the Israel Ministry of Science, the German-Israeli Foundation for Scientific Research and Development (G.I.F), and a gift from B. and A. Hirschfield. We thank I. Belyantseva, E.B. Crenshaw, M. Xiang, C. Seiler, O. Ben-David, R.J. Smith, and V. Marigo for contributing figures.

References

Ahituv N, Avraham KB (2002) Mouse models for human deafness: current tools for new fashions. Trends Mol Med 8:447–451.

Ahmed ZM, Riazuddin S, Bernstein SL, Ahmed Z, Khan S, Griffith AJ, Morell RJ, Friedman TB, Wilcox ER (2001) Mutations of the protocadherin gene *PCDH15* cause Usher syndrome type 1F. Am J Hum Genet 69:25–34.

Ahmed ZM, Morell RJ, Riazuddin S, Gropman A, Shaukat S, Ahmad MM, Mohiddin SA, Fananapazir L, Caruso RC, Husnain T, Khan SN, Riazuddin S, Griffith AJ, Friedman TB, Wilcox ER (2003a) Mutations of *MYO6* are associated with recessive deafness, DFNB37. Am J Hum Genet 72:1315–1322.

Ahmed ZM, Riazuddin S, Ahmad J, Bernstein SL, Guo Y, Sabar MF, Sieving P, Griffith AJ, Friedman TB, Belyantseva IA, Wilcox ER (2003b) PCDH15 is expressed in the neurosensory epithelium of the eye and ear and mutant alleles are responsible for both USH1F and DFNB23. Hum Mol Genet 12:3215–3223.

Alagramam KN, Kwon HY, Cacheiro NL, Stubbs L, Wright CG, Erway LC, Woychik RP (1999) A new mouse insertional mutation that causes sensorineural deafness and vestibular defects. Genetics 152:1691–1699.

Alagramam KN, Murcia CL, Kwon HY, Pawlowski KS, Wright CG, Woychik RP (2001a) The mouse Ames waltzer hearing-loss mutant is caused by mutation of *Pcdh15*, a novel protocadherin gene. Nat Genet 27:99–102.

Alagramam KN, Yuan H, Kuehn MH, Murcia CL, Wayne S, Srisailpathy CR, Lowry RB, Knaus R, Van Laer L, Bernier FP, Schwartz S, Lee C, Morton CC, Mullins RF, Ramesh A, Van Camp G, Hageman GS, Woychik RP, Smith RJ, Hagemen GS (2001b) Mutations in the novel protocadherin *PCDH15* cause Usher syndrome type 1F. Hum Mol Genet 10:1709–1718.

Anagnostopoulos AV (2002) A compendium of mouse knockouts with inner ear defects. Trends Genet 18:499.

Anderson DW, Probst FJ, Belyantseva IA, Fridell RA, Beyer L, Martin DM, Wu D, Kachar B, Friedman TB, Raphael Y, Camper SA (2000) The motor and tail regions of myosin XV are critical for normal structure and function of auditory and vestibular hair cells. Hum Mol Genet 9:1729–1738.

Aschenbrenner L, Lee T, Hasson T (2003) Myo6 facilitates the translocation of endocytic vesicles from cell peripheries. Mol Biol Cell 14:2728–2743.

Avraham KB, Hasson T, Steel KP, Kingsley DM, Russell LB, Mooseker MS, Copeland NG, Jenkins NA (1995) The mouse *Snell's waltzer* deafness gene encodes an unconventional myosin required for structural integrity of inner ear hair cells. Nat Genet 11:369–375.

Belyantseva IA, Boger ET, Friedman TB (2003) Myosin XVa localizes to the tips of inner ear sensory cell stereocilia and is essential for staircase formation of the hair bundle. Proc Natl Acad Sci USA 100:13958–13963.

Bespalova IN, Van Camp G, Bom SJ, Brown DJ, Cryns K, DeWan AT, Erson AE, Flothmann K, Kunst HP, Kurnool P, Sivakumaran TA, Cremers CW, Leal SM, Burmeister M, Lesperance MM (2001) Mutations in the Wolfram syndrome 1 gene (*WFS1*) are a common cause of low frequency sensorineural hearing loss. Hum Mol Genet 10:2501–2508.

Biemesderfer D, Mentone SA, Mooseker M, Hasson T (2002) Expression of myosin VI within the early endocytic pathway in adult and developing proximal tubules. Am J Physiol Renal Physiol 282:F785–794.

Bitner-Glindzicz M, Lindley KJ, Rutland P, Blaydon D, Smith VV, Milla PJ, Hussain K, Furth-Lavi J, Cosgrove KE, Shepherd RM, Barnes PD, O'Brien RE, Farndon PA, Sowden J, Liu XZ, Scanlan MJ, Malcolm S, Dunne MJ, Aynsley-Green A, Glaser B (2000) A recessive contiguous gene deletion causing infantile hyperinsulinism, enteropathy and deafness identifies the Usher type 1C gene. Nat Genet 26:56–60.

Boeda B, Weil D, Petit C (2001) A specific promoter of the sensory cells of the inner ear defined by transgenesis. Hum Mol Genet 10:1581–1589.

Boeda B, El-Amraoui A, Bahloul A, Goodyear R, Daviet L, Blanchard S, Perfettini I, Fath KR, Shorte S, Reiners J, Houdusse A, Legrain P, Wolfrum U, Richardson G, Petit C (2002) Myosin VIIa, harmonin and cadherin 23, three Usher I gene products that cooperate to shape the sensory hair cell bundle. EMBO J 21:6689–6699.

Bork JM, Peters LM, Riazuddin S, Bernstein SL, Ahmed ZM, Ness SL, Polomeno R, Ramesh A, Schloss M, Srisailpathy CR, Wayne S, Bellman S, Desmukh D, Ahmed Z, Khan SN, Kaloustian VM, Li XC, Lalwani A, Bitner-Glindzicz M, Nance WE, Liu XZ, Wistow G, Smith RJ, Griffith AJ, Wilcox ER, Friedman TB, Morell RJ (2001) Usher syndrome 1D and nonsyndromic autosomal recessive deafness DFNB12 are caused by allelic mutations of the novel cadherin-like gene *CDH23*. Am J Hum Genet 68:26–37.

Borsani G, DeGrandi A, Ballabio A, Bulfone A, Bernard L, Banfi S, Gattuso C, Mariani M, Dixon M, Donnai D, Metcalfe K, Winter R, Robertson M, Axton R, Brown A, van Heyningen V, Hanson I (1999) EYA4, a novel vertebrate gene related to *Drosophila eyes absent*. Hum Mol Genet 8:11–23.

Buniello A, Montanaro D, Volinia S, Gasparini P, Marigo V (2004) An expression atlas of connexin genes in the mouse. Genomics 83:812–820.

Bunn RC, Jensen MA, Reed BC (1999) Protein interactions with the glucose transporter binding protein GLUT1CBP that provide a link between GLUT1 and the cytoskeleton. Mol Biol Cell 10:819–832.

Buss F, Kendrick-Jones J, Lionne C, Knight AE, Cote GP, Luzio J (1998) The localization of myosin VI at the golgi complex and leading edge of fibroblasts and its phosphorylation and recruitment into membrane ruffles of A431 cells after growth factor stimulation. J Cell Biol 143:1535–1545.

Buss F, Arden SD, Lindsay M, Luzio JP, Kendrick-Jones J (2001) Myosin VI isoform localized to clathrin-coated vesicles with a role in clathrin-mediated endocytosis. EMBO J 20:3676–3684.

Campbell DA, McHale DP, Brown KA, Moynihan LM, Houseman M, Karbani G, Parry G, Janjua AH, Newton V, al-Gazali L, Markham AF, Lench NJ, Mueller RF (1997) A new locus for non-syndromal, autosomal recessive, sensorineural hearing loss (DFNB16) maps to human chromosome 15q21–q22. J Med Genet 34:1015–1017.

Cohen-Salmon M, Ott T, Michel V, Hardelin JP, Perfettini I, Eybalin M, Wu T, Marcus DC, Wangemann P, Willecke K, Petit C (2002) Targeted ablation of connexin26 in the inner ear epithelial gap junction network causes hearing impairment and cell death. Curr Biol 12:1106–1111.

Cohn ES, Kelley PM, Fowler TW, Gorga MP, Lefkowitz DM, Kuehn HJ, Schaefer GB, Gobar LS, Hahn FJ, Harris DJ, Kimberling WJ (1999) Clinical studies of families with hearing loss attributable to mutations in the connexin 26 gene (GJB2/DFNB1). Pediatrics 103:546–550.

Cryns K, Thys S, Van Laer L, Oka Y, Pfister M, Van Nassauw L, Smith RJ, Timmermans JP, Van Camp G (2003) The *WFS1* gene, responsible for low frequency sensorineural hearing loss and Wolfram syndrome, is expressed in a variety of inner ear cells. Histochem Cell Biol 119:247–256.

de Kok YJ, van der Maarel SM, Bitner-Glindzicz M, Huber I, Monaco AP, Malcolm S, Pembrey ME, Ropers HH, Cremers FP (1995) Association between X-linked mixed deafness and mutations in the POU domain gene *POU3F4*. Science 267:685–688.

de Kok YJ, Vossenaar ER, Cremers CW, Dahl N, Laporte J, Hu LJ, Lacombe D, Fischel-Ghodsian N, Friedman RA, Parnes LS, Thorpe P, Bitner-Glindzicz M, Pander HJ, Heilbronner H, Graveline J, den Dunnen JT, Brunner HG, Ropers HH, Cremers FP (1996) Identification of a hot spot for microdeletions in patients with X-linked deafness type 3 (DFN3) 900 kb proximal to the DFN3 gene *POU3F4*. Hum Mol Genet 5: 1229–1235.

del Castillo I, Villamar M, Moreno-Pelayo MA, del Castillo FJ, Alvarez A, Telleria D, Menendez I, Moreno F (2002) A deletion involving the connexin 30 gene in nonsyndromic hearing impairment. N Engl J Med 346:243–249.

del Castillo I, Moreno-Pelayo MA, Del Castillo FJ, Brownstein Z, Marlin S, Adina Q, Cockburn DJ, Pandya A, Siemering KR, Chamberlin GP, Ballana E, Wuyts W, Maciel-Guerra AT, Alvarez A, Villamar M, Shohat M, Abeliovich D, Dahl HH, Estivill X, Gasparini P, Hutchin T, Nance WE, Sartorato EL, Smith RJ, Van Camp G, Avraham KB, Petit C, Moreno F (2003) Prevalence and evolutionary origins of the del(GJB6-D13S1830) mutation in the DFNB1 locus in hearing-impaired subjects: a multicenter study. Am J Hum Genet 73:1452–1458.

Denoyelle F, Weil D, Maw MA, Wilcox SA, Lench NJ, Allen-Powell DR, Osborn AH, Dahl HH, Middleton A, Houseman MJ, Dode C, Marlin S, Boulila-ElGaied A, Grati M, Ayadi H, BenArab S, Bitoun P, Lina-Granade G, Godet J, Mustapha M, Loiselet J, El-Zir E, Aubois A, Joannard A, Levilliers J, Garabedian E-N, Mueller RF, McKinlay Gardner RJ, Petit C (1997) Prelingual deafness: high prevalence of a 30delG mutation in the connexin 26 gene. Hum Mol Genet 6:2173–2177.

Di Palma F, Holme RH, Bryda EC, Belyantseva IA, Pellegrino R, Kachar B, Steel KP, Noben-Trauth K (2001) Mutations in *Cdh23*, encoding a new type of cadherin, cause stereocilia disorganization in waltzer, the mouse model for Usher syndrome type 1D. Nat Genet 27:103–107.

El-Amraoui A, Schonn JS, Kussel-Andermann P, Blanchard S, Desnos C, Henry JP, Wolfrum U, Darchen F, Petit C (2002) MyRIP, a novel Rab effector, enables myosin VIIa recruitment to retinal melanosomes. EMBO Rep 3:463–470.

Erkman L, McEvilly RJ, Luo L, Ryan AK, Hooshmand F, O'Connell SM, Keithley EM, Rapaport DH, Ryan AF, Rosenfeld MG (1996) Role of transcription factors Brn-3.1 and Brn-3.2 in auditory and visual system development. Nature 381:603–606.

Ernest S, Rauch GJ, Haffter P, Geisler R, Petit C, Nicolson T (2000) *Mariner* is defective in *myosin VIIA*: a zebrafish model for human hereditary deafness. Hum Mol Genet 9: 2189–2196.

Friedman TB, Griffith AJ (2003) Human nonsyndromic sensorineural deafness. Annu Rev Genomics Hum Genet 4:341–402.

Gabriel H, Jung D, Butzler C, Temme A, Traub O, Winterhager E, Willecke K (1998) Transplacental uptake of glucose is decreased in embryonic lethal connexin26-deficient mice. J Cell Biol 140:1453–1461.

Gibson F, Walsh J, Mburu P, Varela A, Brown KA, Antonio M, Beisel KW, Steel KP, Brown SD (1995) A type VII myosin encoded by the mouse deafness gene *shaker-1*. Nature 374:62–64.

Greinwald JH, Jr., Hartnick CJ (2002) The evaluation of children with sensorineural hearing loss. Arch Otolaryngol Head Neck Surg 128:84–87.

Hasson T, Mooseker MS (1994) Porcine myosin-VI: characterization of a new mammalian unconventional myosin. J Cell Biol 127:425–440.

Hasson T, Gillespie PG, Garcia JA, MacDonald RB, Zhao Y, Yee AG, Mooseker MS, Corey DP (1997) Unconventional myosins in inner-ear sensory epithelia. J Cell Biol 137:1287–1307.

He X, Treacy MN, Simmons DM, Ingraham HA, Swanson LW, Rosenfeld MG (1989) Expression of a large family of POU-domain regulatory genes in mammalian brain development. Nature 340:35–41.

Herr W, Sturm RA, Clerc RG, Corcoran LM, Baltimore D, Sharp PA, Ingraham HA, Rosenfeld MG, Finney M, Ruvkun G, Horvitz HR (1988) The POU domain: a large conserved region in the mammalian pit-1, oct-1, oct-2, and *Caenorhabditis elegans unc-86* gene products. Genes Dev 2:1513–1516.

Hertzano R, Montcouquiol M, Rashi-Elkeles S, Elkon R, Yucel R, Frankel WN, Rechavi G, Moroy T, Friedman TB, Kelley MW, Avraham KB (2004) Transcription profiling of inner ears from $Pou4f3_{ddl/ddl}$ identifies Gfi1 as a target of the Pou4f3 deafness gene. Hum Mol Genet 13:2143–2153.

Inoue A, Sato O, Homma K, Ikebe M (2002) DOC-2/DAB2 is the binding partner of myosin VI. Biochem Biophys Res Commun 292:300–307.

Kellerman KA, Miller KG (1992) An unconventional myosin heavy chain gene from *Drosophila melanogaster*. J Cell Biol 119:823–834.

Kelley PM, Harris DJ, Comer BC, Askew JW, Fowler T, Smith SD, Kimberling WJ (1998) Novel mutations in the connexin 26 gene (GJB2) that cause autosomal recessive (DFNB1) hearing loss. Am J Hum Genet 62:792–799.

Kelsell DP, Dunlop J, Stevens HP, Lench NJ, Liang JN, Parry G, Mueller RF, Leigh IM (1997) Connexin 26 mutations in hereditary non-syndromic sensorineural deafness. Nature 387:80–83.

Khanim F, Kirk J, Latif F, Barrett TG (2001) *WFS1*/wolframin mutations, Wolfram syndrome, and associated diseases. Hum Mutat 17:357–367.

Kros CJ, Marcotti W, van Netten SM, Self TJ, Libby RT, Brown SD, Richardson GP, Steel KP (2002) Reduced climbing and increased slipping adaptation in cochlear hair cells of mice with *Myo7a* mutations. Nat Neurosci 5:41–47.

Kumar NM, Gilula NB (1996) The gap junction communication channel. Cell 84:381–388.

Kurima K, Peters LM, Yang Y, Riazuddin S, Ahmed ZM, Naz S, Arnaud D, Drury S, Mo J, Makishima T, Ghosh M, Menon PS, Deshmukh D, Oddoux C, Ostrer H, Khan S, Deininger PL, Hampton LL, Sullivan SL, Battey JF, Jr., Keats BJ, Wilcox ER,

Friedman TB, Griffith AJ (2002) Dominant and recessive deafness caused by mutations of a novel gene, *TMC1*, required for cochlear hair-cell function. Nat Genet 30:277–284.

Kussel-Andermann P, El-Amraoui A, Safieddine S, Hardelin JP, Nouaille S, Camonis J, Petit C (2000a) Unconventional myosin VIIA is a novel A-kinase-anchoring protein. J Biol Chem 275:29654–29659.

Kussel-Andermann P, El-Amraoui A, Safieddine S, Nouaille S, Perfettini I, Lecuit M, Cossart P, Wolfrum U, Petit C (2000b) Vezatin, a novel transmembrane protein, bridges myosin VIIA to the cadherin–catenins complex. EMBO J 19:6020–6029.

Lalwani AK, Luxford WM, Mhatre AN, Attaie A, Wilcox ER, Castelein CM (1999) A new locus for nonsyndromic hereditary hearing impairment, DFNA17, maps to chromosome 22 and represents a gene for cochleosaccular degeneration. Am J Hum Genet 64:318–323.

Lalwani AK, Goldstein JA, Kelley MJ, Luxford W, Castelein CM, Mhatre AN (2000) Human nonsyndromic hereditary deafness DFNA17 is due to a mutation in nonmuscle myosin *MYH9*. Am J Hum Genet 67:1121–1128.

Lautermann J, Frank HG, Jahnke K, Traub O, Winterhager E (1999) Developmental expression patterns of connexin26 and-30 in the rat cochlea. Dev Genet 25:306–311.

Lerer I, Sagi M, Ben-Neriah Z, Wang T, Levi H, Abeliovich D (2001) A deletion mutation in *GJB6* cooperating with a *GJB2* mutation in *trans* in non-syndromic deafness: a novel founder mutation in Ashkenazi Jews. Hum Mutat 18:460.

Li HS, Porter JA, Montell C (1998) Requirement for the NINAC kinase/myosin for stable termination of the visual cascade. J Neurosci 18:9601–9606.

Libby RT, Steel KP (2001) Electroretinographic anomalies in mice with mutations in *Myo7a*, the gene involved in human Usher syndrome type 1B. Invest Ophthalmol Vis Sci 42:770–778.

Liburd N, Ghosh M, Riazuddin S, Naz S, Khan S, Ahmed Z, Liang Y, Menon PS, Smith T, Smith AC, Chen KS, Lupski JR, Wilcox ER, Potocki L, Friedman TB (2001) Novel mutations of *MYO15A* associated with profound deafness in consanguineous families and moderately severe hearing loss in a patient with Smith–Magenis syndrome. Hum Genet 109:535–541.

Liu XZ, Walsh J, Mburu P, Kendrick-Jones J, Cope MJ, Steel KP, Brown SD (1997a) Mutations in the myosin VIIA gene cause non-syndromic recessive deafness. Nat Genet 16:188–190.

Liu XZ, Walsh J, Tamagawa Y, Kitamura K, Nishizawa M, Steel KP, Brown SD (1997b) Autosomal dominant non-syndromic deafness caused by a mutation in the myosin VIIA gene. Nat Genet 17:268–269.

Liu XZ, Xia XJ, Xu LR, Pandya A, Liang CY, Blanton SH, Brown SD, Steel KP, Nance WE (2000) Mutations in connexin31 underlie recessive as well as dominant non-syndromic hearing loss. Hum Mol Genet 9:63–67.

Lopez-Bigas N, Rabionet R, Arbones ML, Estivill X (2001) R32W variant in Connexin 31: mutation or polymorphism for deafness and skin disease? Eur J Hum Genet 9:70.

Melchionda S, Ahituv N, Bisceglia L, Sobe T, Glaser F, Rabionet R, Arbones ML, Notarangelo A, Di Iorio E, Carella M, Zelante L, Estivill X, Avraham KB, Gasparini P (2001) *MYO6*, the human homologue of the gene responsible for deafness in *Snell's waltzer* mice, is mutated in autosomal dominant nonsyndromic hearing loss. Am J Hum Genet 69:635–640.

Menger DJ, Tange RA (2003) The aetiology of otosclerosis: a review of the literature. Clin Otolaryngol 28:112–120.

Mhatre AN, Li J, Kim Y, Coling DE, Lalwani AK (2004) Cloning and developmental expression of nonmuscle myosin IIA (*Myh9*) in the mammalian inner ear. J Neurosci Res 76:296–305.

Miller KG (2004) Converting a motor to an anchor. Cell 116:635–636.

Minoda R, Izumikawa M, Kawamoto K, Raphael Y (2004) Strategies for replacing lost cochlear hair cells. NeuroReport 15:1089–1092.

Minowa O, Ikeda K, Sugitani Y, Oshima T, Nakai S, Katori Y, Suzuki M, Furukawa M, Kawase T, Zheng Y, Ogura M, Asada Y, Watanabe K, Yamanaka H, Gotoh S, Nishi-Takeshima M, Sugimoto T, Kikuchi T, Takasaka T, Noda T (1999) Altered cochlear fibrocytes in a mouse model of DFN3 nonsyndromic deafness. Science 285:1408–1411.

Mohiddin SA, Ahmed ZM, Griffith AJ, Tripodi D, Friedman TB, Fananapazir L, Morell RJ (2004) Novel association of hypertrophic cardiomyopathy, sensorineural deafness, and a mutation in unconventional myosin VI (*MYO6*). J Med Genet 41:309–314.

Montcouquiol M, Kelley MW (2003) Planar and vertical signals control cellular differentiation and patterning in the mammalian cochlea. J Neurosci 23:9469–9478.

Mooseker MS, Cheney RE (1995) Unconventional myosins. Annu Rev Cell Dev Biol 11:633–675.

Peters LM, Anderson DW, Griffith AJ, Grundfast KM, San Agustin TB, Madeo AC, Friedman TB, Morell RJ (2002) Mutation of a transcription factor, TFCP2L3, causes progressive autosomal dominant hearing loss, DFNA28. Hum Mol Genet 11:2877–2885.

Petit C (2001) Usher syndrome: from genetics to pathogenesis. Annu Rev Genomics Hum Genet 2:271–297.

Phillips K, Luisi B (2000) The virtuoso of versatility: POU proteins that flex to fit. J Mol Biol 302:1023–1039.

Phippard D, Heydemann A, Lechner M, Lu L, Lee D, Kyin T, Crenshaw EB, 3rd (1998) Changes in the subcellular localization of the *Brn4* gene product precede mesenchymal remodeling of the otic capsule. Hear Res 120:77–85.

Phippard D, Lu L, Lee D, Saunders JC, Crenshaw EB, 3rd (1999) Targeted mutagenesis of the POU-domain gene *Brn4/Pou3f4* causes developmental defects in the inner ear. J Neurosci 19:5980–5989.

Phippard D, Boyd Y, Reed V, Fisher G, Masson WK, Evans EP, Saunders JC, Crenshaw EB, 3rd, Lu L, Lee D, Heydemann A, Lechner M, Kyin T (2000) The *sex-linked fidget* mutation abolishes *Brn4/Pou3f4* gene expression in the embryonic inner ear. Hum Mol Genet 9:79–85.

Riazuddin S, Castelein CM, Ahmed ZM, Lalwani AK, Mastroianni MA, Naz S, Smith TN, Liburd NA, Friedman TB, Griffith AJ, Wilcox ER (2000) Dominant modifier DFNM1 suppresses recessive deafness DFNB26. Nat Genet 26:431–434.

Rosenfeld MG (1991) POU-domain transcription factors: pou-er-ful developmental regulators. Genes Dev 5:897–907.

Rzadzinska AK, Schneider ME, Davies C, Riordan GP, Kachar B (2004) An actin molecular treadmill and myosins maintain stereocilia functional architecture and self-renewal. J Cell Biol 164:887–897.

Sahly I, El-Amraoui A, Abitbol M, Petit C, Dufier JL (1997) Expression of myosin VIIA during mouse embryogenesis. Anat Embryol (Berl) 196:159–170.

Seiler C, Ben-David O, Sidi S, Hendrich O, Rusch A, Burnside B, Avraham KB, Nicolson T (2004) Myosin VI is required for structural integrity of the apical surface of sensory hair cells in zebrafish. Dev Biol 272:328–338.

Self T, Sobe T, Copeland NG, Jenkins NA, Avraham KB, Steel KP (1999) Role of myosin VI in the differentiation of cochlear hair cells. Dev Biol 214:331–341.

Sellars JR (1999) Myosins. Oxford: Oxford University Press.

Shahin H, Walsh T, Sobe T, Lynch E, King MC, Avraham KB, Kanaan M (2002) Genetics of congenital deafness in the Palestinian population: multiple *connexin 26* alleles with shared origins in the Middle East. Hum Genet 110:284–289.

Siemens J, Lillo C, Dumont RA, Reynolds A, Williams DS, Gillespie PG, Muller U (2004) Cadherin 23 is a component of the tip link in hair-cell stereocilia. Nature 428: 950–955.

Sobe T, Erlich P, Berry A, Korostichevsky A, Vreugde S, Shohat M, Avraham KB, Bonné-Tamir B (1999) High frequency of the deafness-associated 167delT mutation in the connexin 26 (*GJB2*) gene in Israeli Ashkenazim. Am J Med Genet 86:499–500.

Sollner C, Rauch GJ, Siemens J, Geisler R, Schuster SC, Muller U, Nicolson T (2004) Mutations in *cadherin 23* affect tip links in zebrafish sensory hair cells. Nature 428: 955–959.

Spicer S, Schulte B (1996) The fine structure of spiral liagment cells relates to ion return to the stria and varies with place-frequency. Hear Res 100:80–100.

Street VA, Kallman JC, Kiemele KL (2004) Modifier controls severity of a novel dominant low frequency myosin VIIA (MYO7A) auditory mutation. J Med Genet 41:e62.

Suzuki ST (2000) Recent progress in protocadherin research. Exp Cell Res 261:13–18.

Swendeman SL, Spielholz C, Jenkins NA, Gilbert DJ, Copeland NG, Sheffery M (1994) Characterization of the genomic structure, chromosomal location, promoter, and development expression of the alpha-globin transcription factor CP2. J Biol Chem 269: 11663–11671.

Takeda K, Inoue H, Tanizawa Y, Matsuzaki Y, Oba J, Watanabe Y, Shinoda K, Oka Y (2001) *WFS1* (Wolfram syndrome 1) gene product: predominant subcellular localization to endoplasmic reticulum in cultured cells and neuronal expression in rat brain. Hum Mol Genet 10:477–484.

Tamagawa Y, Ishikawa K, Ishida T, Kitamura K, Makino S, Tsuru T, Ichimura K (2002) Phenotype of DFNA11: a nonsyndromic hearing loss caused by a myosin VIIA mutation. Laryngoscope 112:292–297.

Teubner B, Michel V, Pesch J, Lautermann J, Cohen-Salmon M, Sohl G, Jahnke K, Winterhager E, Herberhold C, Hardelin JP, Petit C, Willecke K (2003) Connexin30 (*Gjb6*)-deficiency causes severe hearing impairment and lack of endocochlear potential. Hum Mol Genet 12:13–21.

Vahava O, Morell R, Lynch ED, Weiss S, Kagan ME, Ahituv N, Morrow JE, Lee MK, Skvorak AB, Morton CC, Blumenfeld A, Frydman M, Friedman TB, King M-C, Avraham KB (1998) Mutation in transcription factor *POU4F3* associated with inherited progressive hearing loss in humans. Science 279:1950–1954.

Velichkova M, Guttman J, Warren C, Eng L, Kline K, Vogl AW, Hasson T (2002) A human homologue of *Drosophila* kelch associates with myosin-VIIa in specialized adhesion junctions. Cell Motil Cytoskeleton 51:147–164.

Verpy E, Leibovici M, Zwaenepoel I, Liu XZ, Gal A, Salem N, Mansour A, Blanchard S, Kobayashi I, Keats BJ, Slim R, Petit C (2000) A defect in harmonin, a PDZ domain-containing protein expressed in the inner ear sensory hair cells, underlies Usher syndrome type 1C. Nat Genet 26:51–55.

Verpy E, Masmoudi S, Zwaenepoel I, Leibovici M, Hutchin TP, Del Castillo I, Nouaille S, Blanchard S, Laine S, Popot JL, Moreno F, Mueller RF, Petit C (2001) Mutations in a new gene encoding a protein of the hair bundle cause non-syndromic deafness at the DFNB16 locus. Nat Genet 29:345–349.

Vreugde S, Erven A, Kros CJ, Marcotti W, Fuchs H, Kurima K, Wilcox ER, Friedman TB, Griffith AJ, Balling R, Hrabe De Angelis M, Avraham KB, Steel KP (2002) Beethoven, a mouse model for dominant, progressive hearing loss DFNA36. Nat Genet 30:257–258.

Wallis D, Hamblen M, Zhou Y, Venken KJ, Schumacher A, Grimes HL, Zoghbi HY, Orkin SH, Bellen HJ (2003) The zinc finger transcription factor *Gfi1*, implicated in lymphomagenesis, is required for inner ear hair cell differentiation and survival. Development 130:221–232.

Walsh T, Walsh V, Vreugde S, Hertzano R, Shahin H, Haika S, Lee MK, Kanaan M, King MC, Avraham KB (2002) From flies' eyes to our ears: mutations in a human class III myosin cause progressive nonsyndromic hearing loss DFNB30. Proc Natl Acad Sci USA 99:7518–7523.

Wang A, Liang Y, Fridell RA, Probst FJ, Wilcox ER, Touchman JW, Morton CC, Morell RJ, Noben-Trauth K, Camper SA, Friedman TB (1998) Association of unconventional myosin *MYO15* mutations with human nonsyndromic deafness *DFNB3*. Science 280: 1447–1451.

Warner CL, Stewart A, Luzio JP, Steel KP, Libby RT, Kendrick-Jones J, Buss F (2003) Loss of myosin VI reduces secretion and the size of the Golgi in fibroblasts from Snell's waltzer mice. EMBO J 22:569–579.

Wayne S, Robertson NG, DeClau F, Chen N, Verhoeven K, Prasad S, Tranebjarg L, Morton CC, Ryan AF, Van Camp G, Smith RJ (2001) Mutations in the transcriptional activator *EYA4* cause late-onset deafness at the DFNA10 locus. Hum Mol Genet 10: 195–200.

Weil D, Blanchard S, Kaplan J, Guilford P, Gibson F, Walsh J, Mburu P, Varela A, Levilliers J, Weston MD, Kelley PM, Kimberling WJ, Wagenaar M, Levi-Acobas F, Larget-Piet D, Munnich A, Steel KP, Brown SDM, Petit C (1995) Defective myosin VIIA gene responsible for Usher syndrome type 1B. Nature 374:60–61.

Weil D, Levy G, Sahly I, Levi-Acobas F, Blanchard S, El-Amraoui A, Crozet F, Philippe H, Abitbol M, Petit C (1996) Human myosin VIIA responsible for the Usher 1B syndrome: a predicted membrane-associated motor protein expressed in developing sensory epithelia. Proc Natl Acad Sci USA 93:3232–3237.

Weiss S, Gottfried I, Mayrose I, Khare SL, Xiang M, Dawson SJ, Avraham KB (2003) The DFNA15 deafness mutation affects POU4F3 protein stability, localization, and transcriptional activity. Mol Cell Biol 23:7957–7964.

Wells AL, Lin AW, Chen LQ, Safer D, Cain SM, Hasson T, Carragher BO, Milligan RA, Sweeney HL (1999) Myosin VI is an actin-based motor that moves backwards. Nature 401:505–508.

Wilanowski T, Tuckfield A, Cerruti L, O'Connell S, Saint R, Parekh V, Tao J, Cunningham JM, Jane SM (2002) A highly conserved novel family of mammalian developmental transcription factors related to *Drosophila grainyhead*. Mech Dev 114:37–50.

Wilson SM, Householder DB, Coppola V, Tessarollo L, Fritzsch B, Lee EC, Goss D, Carlson GA, Copeland NG, Jenkins NA (2001) Mutations in *Cdh23* cause nonsyndromic hearing loss in *waltzer* mice. Genomics 74:228–233.

Wu H, Nash JE, Zamorano P, Garner CC (2002) Interaction of SAP97 with minus-end directed actin motor myosin VI: implications for AMPA receptor trafficking. J Biol Chem 277:30928–30934.

Xia JH, Liu CY, Tang BS, Pan Q, Huang L, Dai HP, Zhang BR, Xie W, Hu DX, Zheng

D, Shi XL, Wang DA, Xia K, Yu KP, Liao XD, Feng Y, Yang YF, Xiao JY, Xie DH, Huang JZ (1998) Mutations in the gene encoding gap junction protein beta-3 associated with autosomal dominant hearing impairment. Nat Genet 20:370–373.

Xiang M, Gan L, Zhou L, Klein WH, Nathans J (1996) Targeted deletion of the mouse POU domain gene Brn-3a causes selective loss of neurons in the brainstem and trigeminal ganglion, uncoordinated limb movement, and impaired suckling. Proc Natl Acad Sci USA 93:11950–11955.

Xiang M, Gan L, Li D, Chen ZY, Zhou L, O'Malley BW, Jr., Klein W, Nathans J (1997) Essential role of POU-domain factor Brn-3c in auditory and vestibular hair cell development. Proc Natl Acad Sci USA 94:9445–9450.

Xiang M, Gao W-Q, Hasson T, Shin JJ (1998) Requirement for Brn-3c in maturation and survival, but not in fate determination of inner ear hair cells. Development 125: 3935–3946.

Xiang M, Maklad A, Pirvola U, Fritzsch B (2003) Brn3c null mutant mice show long-term, incomplete retention of some afferent inner ear innervation. BMC Neurosci 4:2.

Yerkes RM (1907) The Dancing Mouse. New York: Macmillan.

Young TL, Ives E, Lynch E, Person R, Snook S, MacLaren L, Cater T, Griffin A, Fernandez B, Lee MK, King MC, Cator T (2001) Non-syndromic progressive hearing loss *DFNA38* is caused by heterozygous missense mutation in the Wolfram syndrome gene *WFS1*. Hum Mol Genet 10:2509–2514.

Zelante L, Gasparini P, Estivill X, Melchionda S, D'Agruma L, Govea N, Mila M, Monica MD, Lutfi J, Shohat M, Mansfield E, Delgrosso K, Rappaport E, Surrey S, Fortina P (1997) Connexin26 mutations associated with the most common from of non-syndromic neurosensory autosomal recessive deafness (DFNB1) in Mediterraneans. Hum Mol Genet 6:1605–1609.

Index

SPRINGER HANDBOOK OF AUDITORY RESEARCH *(continued from page ii)*

Volume 22: Evolution of the Vertebrate Auditory System
Edited by Geoffrey A. Manley, Arthur N. Popper and Richard R. Fay

Volume 23: Plasticity of the Auditory System
Edited by Thomas N. Parks, Edwin W Rubel, Arthur N. Popper, and Richard R. Fay

Volume 24: Pitch: Neural Coding and Perception
Edited by Christopher J. Plack, Andrew J. Oxenham, Richard R. Fay, and Arthur N. Popper

Volume 25: Sound Source Localization
Edited by Arthur N. Popper and Richard R. Fay

Volume 26: Development of the Inner Ear
Edited by Matthew W. Kelley, Doris K. Wu, Arthur N. Popper and Richard R. Fay

For more information about the series, please visit www.springer-ny.com/shar.